ATOMIC
BOMBSHELLS

ATOMIC BOMBSHELLS

HOW PLASTICS SHAPED POSTWAR BODIES

ISABELLE HELD

DUKE UNIVERSITY PRESS ◉ DURHAM AND LONDON ◉ 2026

© 2026 DUKE UNIVERSITY PRESS
All rights reserved
Project Editor: Lisa Lawley
Designed by Matthew Tauch
Typeset in Garamond Premier Pro and Morganite
by Westchester Publishing Services

Library of Congress Cataloging-in-Publication Data
Names: Held, Isabelle, [date] author.
Title: Atomic bombshells : how plastics shaped postwar bodies /
    Isabelle Held. Description: Durham : Duke University Press,
    2026. | Includes bibliographical references and index.
Identifiers: LCCN 2025022003 (print)
LCCN 2025022004 (ebook)
ISBN 9781478033103 (paperback)
ISBN 9781478029656 (hardcover)
ISBN 9781478061854 (ebook)
Subjects: LCSH: Plastics industry and trade—United States—
    History—20th century. | Silicone industry—United
    States—History—20th century. | Surgery, Plastic—
    United States—History—20th century. | Beauty culture—
    United States—History—20th century. | Gender
    expression—United States—History—20th century.
Classification: LCC HD9661.U6 H45 2026 (print) |
    LCC HD9661.U6 (ebook)
LC record available at https://lccn.loc.gov/2025022003
LC ebook record available at https://lccn.loc.gov/2025022004

Cover art: Don English, *Lee Merlin: "Miss Atomic Bomb,"* 1957.
Las Vegas News Bureau. LVCVA Archive, Las Vegas.

For my dream wife, Melissa.

I could not have written this book without you.

CONTENTS

SILICONE / FLUID

ILLUSTRATIONS

ABBREVIATIONS

AMA	American Medical Association
CGW	Corning Glass Works
DCCAMR	Dow Corning Center for Aid to Medical Research
DPAD	DuPont Advertising Department
FDA	Food and Drug Administration
GE	General Electric
QLF	Queens Liberation Front
R & D	Research and development
SCAP	Supreme Commander for the Allied Powers
STS	Science and technology studies

INTRODUCTION

One day, toward the end of World War II, a young white American man sank into the soft seat of a captured German bomber and marveled at its plastic foam padding. The upholstery's plush synthetic material was unlike anything he had experienced before. Sitting in its cushioned embrace felt particularly good to him after the hours he had spent observing plastic surgery for the Royal Canadian Air Force in Toronto. A sergeant pointed to a cutaway in the seat, explaining that this was a new type of lightweight synthetic foam developed by German chemists—a replacement for foams made from rubber.

A decade later, in 1950s Hollywood, a blonde striptease artist, frustrated with her flat chest, sought the services of a famous cosmetic surgeon known as the Beauty Surgeon. She requested that he permanently enhance her body to emulate that of a famed brunette bombshell: "I want you should give me a chest like Jane Russell's."[1] The surgeon regretfully replied that he could do nothing for her at that time; while surgeries existed to reduce breast size, methods to increase the bust, such as fat transplants, had proven unsuccessful to date.

When searching for a solution to her request, the Beauty Surgeon recalled a moment, earlier in his career, when he experienced the novel pleasure of the bomber's wartime upholstery and its uniquely spongy properties. He tracked down a sample of the bomber seat and, after a series of experiments, determined it was ideal for use in breast implants. This is the origin story that Robert Alan Franklyn repeatedly told in countless publications throughout the 1960s to frame his research for developing "Surgifoam," the synthetic material he designed for his "Breastplasty" operation to augment the bustline cosmetically.

Expedited during wartime and now freed from military applications, plastics were becoming increasingly available on the US consumer market in the postwar era.[2] The whitestream press celebrated them as miracles of Big Science.[3] These plastics were widely considered chemically inert—and therefore understood as benign. A postwar plastics boom also resulted in more medical applications. Franklyn noted, "Along with surgeons all over the world, I had rejoiced in the age of plastics."[4] The blonde who entered his office would "never have to wear any kind of padded brassiere again."[5] Instead, the Beauty Surgeon would perfect her body with plastic foam inserts sheathed in nylon—another synthetic material—implanted within the body. Theoretically, this surgery offered a permanent solution for achieving the fashionable, fuller bust that many women desired, eliminating the need for an external shaping device. It thereby destabilized seemingly fixed limits of the body. This technoscientific tale of self-actualization used military-industrial technology to position the postwar American female body as plastic in new and different ways.

From bullet bras to bazookas, Franklyn's Breastplasty origin story fits into a broader cultural context of referencing military technology in relation to the sexually desirable female body. The fashionable white, cisgender, heteronormative feminine ideal promoted by Hollywood and whitestream media was increasingly sheathed in synthetics and commonly referred to as a "bombshell." Hollywood and entertainment culture exported the glamorous, curvaceous bombshell to audiences worldwide. But there was never just one bombshell.

Atomic Bombshells challenges the usual narrative and trajectory of how war technologies enter domestic use. Starting in the late 1930s, it traces the early development of a set of synthetic materials to show how these technologies were originally created and imagined by decision-makers within a white, male-dominated industry. Continuing through the Cold War, I analyze the context of how these surplus materials were commercialized and marketed as domestic innovations for capitalist consumption that largely targeted a narrow definition of woman. The typical technology narrative within material science is complicated by incorporating a wider range of makers and users who used plastics to (re)shape gender.

This book follows plastics on their journey into women's bodies. It explores the effects of military-industrial science, as manifested in various synthetic products—nylon, silicone, and plastic foams—on embodied and expressive configurations of gender, sexuality, and race. By tracing materials across multinational networks, I show that the links between women's bodies and wartime technologies were not simply symbolic but materially and corporeally manifest via plastics research and development.

Through working closely with plastics as an artifact and focusing on their materiality, my analysis moves from *outside* the body to *inside* the body. I follow these materials across networks of production and consumption to show both their hegemonic and counterhegemonic uses. In mapping makers and users, I highlight the multitude of actors that influenced the many different uses and meanings of plastics' application in reimagining and reshaping femininities. Ultimately, this book argues that the histories of femininity and the fashioned female body cannot be understood as purely binary, as they are always entangled with intersectional queer and trans histories. Concepts of gender and the body are never fixed, but rather shaped by the cultural, historical, sociopolitical, and material conditions in which they are created.

The primary focus of this book is the United States from the late 1930s, with the launch of nylon—widely celebrated as the world's first fully synthetic fiber—to the late 1970s, when shifts in policy relating to the dangerous health consequences of implantable plastics, such as polyurethane foam and silicones, were enacted. Franklyn's anecdote, located at the center of this book's chronology, provides a useful starting point for retracing the biopolitical history of plastics as a technology of the gendered body. It raises many interconnected questions about the provenance of plastics, practices of self-fashioning the feminine body, and American postwar constructions of racialized gender.

While Franklyn is a known figure in the history of US cosmetic surgery, the cultural-historical significance of the military-industrial materials he used—and their intimate relationship to the shaping of feminine body ideals—has remained largely unexplored. Perhaps this critical absence of attention to the material reflects the omnipresence of plastics today. They have infiltrated every aspect of contemporary life—and the body—making it challenging to imagine a world without them. But for this precise reason, it is essential to examine their history more closely. These materials and practices continue to circulate and affect bodies across a wide range of communities and social groups today. This makes it all the more urgent to critically revisit their histories.[6] In order to guide the search for more sustainable and equitable alternatives, it is essential to improve our understanding of the materials' behaviors and provenance, as well as make the power structures in which they are created more visible.

The history of materials is not neutral. Histories of science, technology, and medicine do not exist in a vacuum; they are shaped by actors involved and the sociopolitical conditions of their time. This book contrasts with material science histories that are gender- and race-neutral, devoid of social or cultural context beyond a technoscientific one. It joins feminist STS scholars who have challenged such approaches to history and highlighted the power relations within which science and technology are created.[7]

This book is structured around materials. Centering on the transnational story of three types of materials used for military and industrial applications during World War II—nylon, foam plastics, and silicone—I trace their development, actors, networks, and their applications in shaping feminine bodies. Nylon, unveiled by the explosives manufacturer DuPont in 1939, was promoted as the world's first fully synthetic fiber and introduced to the domestic women's intimate apparel market. Plastic foams made from wartime rubber substitutes were molded into 3D objects and used to augment bodies through padded bra inserts known as "falsies," as well as padded foundationwear, prosthetics, and implants. Silicone, an engine lubricant developed to aid the US war effort, was later used in cosmetics, aesthetic body contouring, and breast augmentation surgery.

By giving greater attention to the materials themselves, I am able to trace how their uses and meanings change as they circulate through a wider network of users and include histories that have been less documented. In this book, I untangle a complex network of people, including chemical company representatives, cosmetic and plastic surgeons, individuals working without medical licenses, chemists, beauty salon workers, film directors, actresses, sex workers, go-go dancers, journalists, nightclub owners, dermatologists, fashion and industrial designers, and Hollywood agents and producers—all part of the wider network that brought plastics to (and often into) postwar bodies. Only by identifying, mapping out, and critically analyzing the embedded structural inequities of these materials—from their development to their access—can one begin to deconstruct them, write new histories, and reimagine and reclaim futures. This process of deconstruction can also make visible the ways in which a wider range of users and makers conformed to, challenged, complicated, and subverted the power structures that influenced plastics creation. Through this process, *Atomic Bombshells* also highlights how complex power relations affected the applications and cultural meanings of the materials themselves.

Mapping Plastics

The thread of this book's story is nylon. In 1938, DuPont announced the creation of nylon, a new synthetic material that they promoted as the world's first *fully* synthetic fiber. Unlike earlier "artificial silks," such as rayon, a semisynthetic fiber made from plant-based cellulose, nylon was made from coal derivatives and unlocked a new range of possibilities, eventually leading to the commercial launch of other revolutionary synthetic materials, including polyester and Lycra.[8] Nylon was first made available as a synthetic replacement for pig bristles and fishing lines. However, it was the more sensual launch of this material as women's stockings that captured the attention of the American public.

Nylon's shifting identity moved from the lab to glamorous, intimate fashion items, to wartime parachutes, and back again. Resistant to mold, rot, and pests, nylon embodied DuPont's slogan, "Better Things for Better Living . . . Through Chemistry," and contributed to an imagined synthetic future, independent of foreign natural fibers such as silks and rubbers. By the late 1930s, plastics like nylon were no longer understood as mere imitations of nature but as superior materials, scientifically engineered by chemists for specific purposes. The history of plastics and other synthetics is rooted in such hierarchical binaries and tension between culture and nature, as well as the "natural" and "unnatural." In this dualist discourse of science conquering nature, malleable plastics were frequently gendered as feminine, while white men were associated with power and control. Chemical companies like DuPont advertised plastics as materials that man designed to his own specifications. This discourse's gendered, racialized, and colonial rhetoric reveals the heteropatriarchal power structures and ideological hierarchies within which plastics were originally created.

Nylon's history establishes connections between military-industrial materials R & D and women's bodies through its promotion and domestication. I explore how the historical moment of nylon's public launch, along with the discourse surrounding it, established a paradigm of racialized, gendered bodies within a web of science, industry, fashion, and the military. While nylon's history and relation to the military-industrial complex has been clearly charted in ongoing scholarship, the history of plastic foams and silicones, much like these materials' properties, is less easily defined and discussed.[9] The high level of recognition that nylon has received is partly due to the familiarity of its brand name and association with DuPont.

Polyurethane foam and silicone, on the other hand, are wider categories: Unlike nylon, they do not bear the name of a singular synthetic patent licensed by a particular chemical company and are thus more challenging to chart. But it is important to do so, as they also have similar ties to military and industry and continue to affect self-fashioning practices today.

At the dawn of World War II, the demand for natural rubbers and textiles outgrew supply amid disintegrating colonial networks, prompting an international push for cheaper synthetic replacements. When the United States entered the war, key military decision-makers annexed American industrial plastics R & D, demanding near-exclusive use of the latest developments, such as nylon, for parachutes and mosquito nets. US plastics production rapidly intensified during the war, almost quadrupling from 213 million pounds in 1939 to 818 million pounds in 1945.[10] Government funding during World War II supported the rise of what came to be known as the military-industrial complex, part of a wider historical legacy of the US government as a major sponsor and agent of technological change.[11] Interconnections between technology, warfare, government, and industrial R & D changed profoundly during World War II and the postwar years.[12] World War II was the first time US scientists and engineers were almost fully mobilized for a common effort.[13] In 1941, President Franklin Roosevelt established the Office of Scientific Research and Development (OSRD), a new organization that employed engineers and scientists to support the crucial role R & D would play in the war effort.

US military demand for weapons and equipment in World War II resulted in a proliferation of new materials, products, and technologies. This included silicone fluid, which was developed in secrecy as an engine lubricant by Dow Corning under urgent navy orders. While polyurethane foam was originally developed in Germany during wartime, its rapid postwar transfer to the United States via intelligence reports was driven by political military-industrial power structures, a history of US-Germany chemical industry competition, and the potential for civilian applications. All these materials were initially independently developed in industrial labs for commercial production. Only after their creation did the military increasingly take note and commission them for war supply, expediting their development while simultaneously restricting their commercial uses.

After World War II, the US military no longer needed large quantities of nylon, plastic foams, and silicones. With wartime restrictions lifted, chemical companies could finally make these products commercially available and were keen to establish new markets for peacetime conversion, particularly targeting the domestic market and female consumers. Chemical companies frequently promoted their materials as having withstood the test of war, reasoning that if the materials were safe and suitable for military use, they would also meet the demands of the American housewife.[14] Consumers found that some of the plastics they previously encountered before World War II had been improved and further developed during wartime. Design historian Cynthia Lee Henthorn refers to America's military-industrial R & D cycle "as a strain of commercial fallout," where better living and defense became synonymous.[15]

This was the Age of More and Better Plastic Things for Better Living Through Big Science. Synthetic materials, once developed for particular industrial and military uses, now proliferated on the American market during the booming postwar economy. The plastics manufacturing industry grew exponentially. By 1960, more than six billion pounds of plastics were produced annually.[16] Advertising executives, journalists, and designers, among others, celebrated plastics as modern miracles of science that grew from military research. The abundance of synthetics on the consumer market in the Atomic Age delivered the plastics utopia that chemical companies had promised the American public throughout World War II. Nylons were finally available in large quantities. Fashionable padded silhouettes, such as the conical "Sweater Girl" look (figure I.1), which became popular in the late 1930s, were made more accessible by synthetics that had been refined during wartime and were now available on the domestic market. Silicone lotions protected hardworking hands, pan glaze improved baking, and injections reportedly enhanced feminine gender presentation. A postwar surplus of synthetic materials flooded the consumer market, becoming more intimately entangled with everyday life and the body.

Plastic materials were not the only scientific miracles celebrated during this period. Cosmetic and plastic surgery were also hailed as miracles of science.[17] Cosmetic surgery, still in its nascent stages, was the commercial by-product of plastic surgery, a discipline that grew in response to changes in warfare during World War I. Before the 1930s, materials for medical applications were largely limited to those found naturally, such as ivory, glass, silk, wood, and metals. Wartime advances in polymer science unlocked a greater range of previously unavailable materials. These synthetic discoveries differed greatly from their predecessors, offering physical properties more closely resembling biological tissue and greater ease of sterilization. Therefore, they found popular applications as implantable materials. The

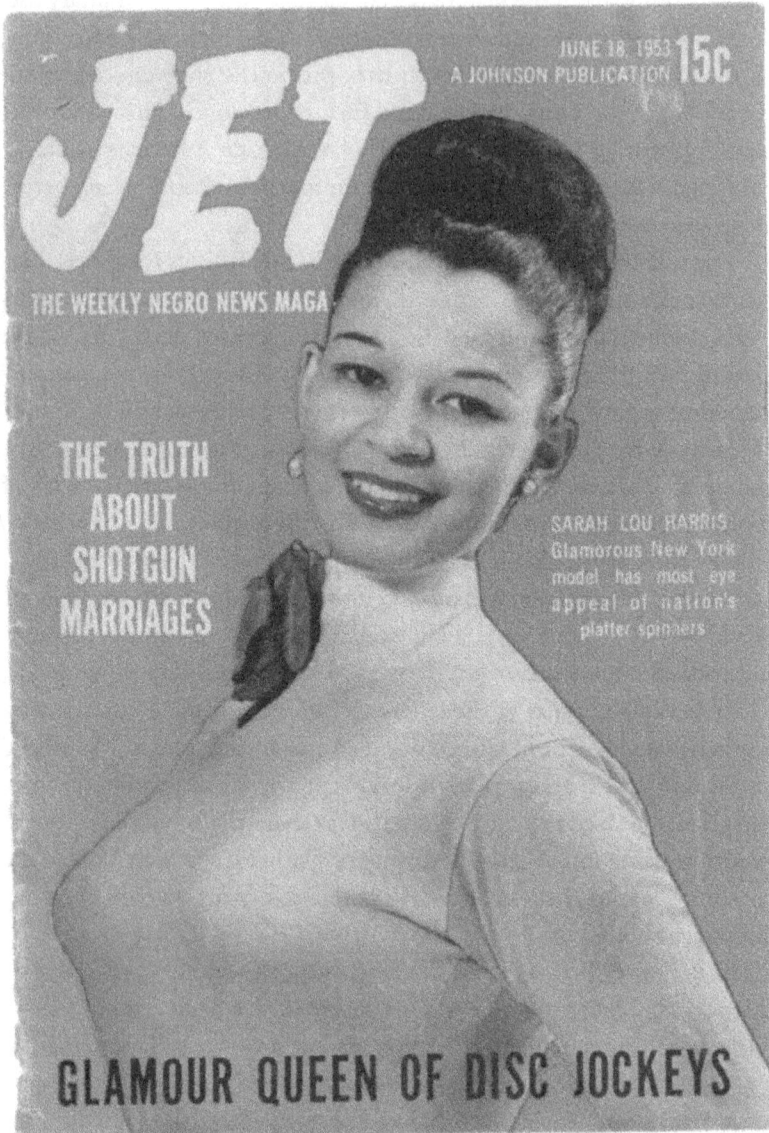

I.1 A Sweater Girl look modeled by Sarah Lou Harris on the cover of *Jet*, June 18, 1953. New York Public Library. https://digitalcollections.nypl.org/items /d3a0fac0-81dc-0135-cde1-5dceb882b926.

use of plastics grew as the development of antibiotics reduced mortality rates, thereby increasing the demand for prostheses. Surgeries benefited from synthetic materials that became available as plastics developments and the disciplines of plastic and cosmetic surgery flourished.

Postwar US bodies were increasingly transformed as part of a larger shift toward medical consumerism during the productive economy of the 1950s.[18] However, the relationship between medicine and consumerism was not new. In the early twentieth century, historian David Serlin argues, "medical products came to be understood as amenities of a prosperous economy and a modern social self."[19] Plastic surgery is an example of this. In the 1920s, it was used to surgically reconstruct World War I veterans, but high-income individuals also accessed it for aesthetic purposes. Identity became increasingly rooted in self-presentation throughout the twentieth century.[20] Wartime developments in cosmetic surgery meant that the face and body could be shaped acquisitively, promising social mobility and self-actualization, as well as reflecting "competing aims between customization and standardization."[21] In the postwar United States, newspapers, magazines, and journals celebrated medical miracles as offshoots of military R & D.[22] This discourse of life-changing medical advancements through postwar science and plastics affected cosmetic surgery's reception in popular culture. Medical science came to embody a postwar utopian concept of better living and progress without conflict.[23]

The interplay between military-industrial materials R & D and postwar bodies continued in foundationwear and spacesuit design. Synthetic foam rubbers were likely the first military transfer technology used in foundationwear.[24] Companies like Maidenform and Frederick's of Hollywood advertised products padded with new and improved wartime "poly foams" to shape the body into a curvaceous silhouette. In the postwar United States, racialized and gendered corporeal technologies of foundationwear were also applied to the design of spacesuits and the development of high-performance aviation materials. The industrial division of Playtex, a company that specialized in foundationwear, was commissioned to design the spacesuit for the Apollo mission.[25] The spacesuit is an example of how technologies once widely available to the public were transformed into classified R & D materials for enhancing corporeal performance.

The book's central narrative ends around 1976. By that point, it was becoming clear that implantable materials, initially considered inert and therefore benign, could have significant and even fatal health consequences. In the thirty-seven years following DuPont's introduction of nylon through its "Test Tube Girl," plastics increasingly shaped bodies from within. This eventually resulted in laws regulating the industry of implanted materials and permanent body contouring, directly affecting millions of Americans. The 1976 Medical Device Regulation Act marked a pivotal shift in US implant legislation.[26] It intended to provide reasonable assurance, risk-based classification, and formal safeguarding criteria for medical device safety and effectiveness. Regulatory pathways and procedures, as well as postmarket requirements, were introduced for new medical devices. For the first time, the FDA was given the authority to ban devices, thereby legally regulating what could and could not be implanted in the body. This marked the end of a largely unregulated market of actors in chemical companies, medicine, and cosmetic surgery experimenting with plastic materials in the body. However, many materials from this period were grandfathered in under the 1976 Medical Device Regulation Act and are still used in body-contouring practices today.

Queering Bombshells

In this book, I queer existing narratives of postwar women's bodies. To do so, it is important to first establish the conventional history of the bombshell, as it has been told to date. The cultural concept of the bombshell is intertwined with narratives comparing women with weaponry. During World War II, US pilots often named their bombers after their female sweethearts and painted the sides with pin-ups and topless women. Aggressively pointed conical bra designs were named "missile" or "bullet." The highly revealing new two-piece style of women's swimwear was called the "bikini," after the nuclear testing site. Yet the most enduring link is perhaps the "bombshell," which often denoted a slender yet curvaceous woman. The term entered common usage to describe a glamorous, attractive, and often sexually alluring woman with the release of the Hollywood movie *Blonde Bombshell* (1933), starring Jean Harlow, a bottle-blonde, white actress.[27] The word was later applied to countless other white actresses, such as 1950s blonde icons Jayne Mansfield and Marilyn Monroe. As military-industrial technologies changed, the meaning and scale of the bombshell shifted. For instance, in *Gilda* (1946), Rita Hayworth epitomized the femme fatale bombshell. A nuclear bomb tested at Bikini Atoll was named after the film and had Hayworth's photograph pinned to it.[28] Scholars have explored how the pin-up functioned as a military morale-booster and, on a larger scale, as an icon of modern American female sexuality.[29] This book extends this thinking by moving beyond the purely representational links and exploring how wartime advancements in materials research and development affected the shaping of feminine bodies, both symbolically and materially.

The image of the bombshell acquired new meaning with the arrival of the Atomic Age and the Cold War. A wide range of publications featured imagery that likened seminude female models to nuclear weaponry. One of the most famous images associated with the atomic bombshell is that of Lee Merlin, who had been crowned "Miss Atomic Bomb" in 1957 (figure I.2). Posing in the Nevada desert, she appears to be nude except for a fluffy white cotton mushroom cloud that erupts upward from between her legs, almost entirely concealing a light nude-colored bikini bottom. This image was created in the context of Las Vegas atomic viewing gatherings, when Nevada's nuclear bomb testing spawned a spectator culture of "dawn bomb parties," "atomic cocktails," and the "Miss Atomic Bomb" beauty pageant, starring white women dressed as mushroom clouds.[30] Merlin was the last model to be featured as "Miss Atomic Bomb," a marketing campaign conceptualized by Vegas nightclub owners that featured their dancers to give atomic testing a sexy image.[31] Other, lesser-known examples include Edy Rich, a light-skinned nylon stocking- and bikini-clad dancer from Miami, who in 1956 in San Francisco was billed as "TNT Girl" and referred to in the press as an "H-Bomb Beauty," whose potent sexuality was off the seismic charts.[32] "Miss Bomarc," named after the US missile, was a white, blonde bombshell from Utah who captured the attention of journalists when she entered a 1958 hairstyling competition.[33]

In the Atomic Age, the normative white cisheteropatriarchal domestic sphere was associated with new cultural and political connotations. During the Cold War, America's nuclear dominance was cast not only as a threat but also as progress driven by science. Cultural anxieties around gender roles also marked the period. National ideals of female identity were transformed during wartime and once again in the postwar United States. Wartime work and the reduction in the number of men available for work within the United States gave women, particularly white, middle-class, able-bodied, cisgender women, increased opportunities. However, women's growing sexual and economic emancipation threatened established racialized, cisheteropatriarchal norms and social structures. Female

I.2 Don English, *Lee Merlin: "Miss Atomic Bomb"* (1957), Las Vegas News Bureau. LVCVA Archive, Las Vegas.

sexual potency was seen as explosive when it was outside the home, with no strong male authority to control it.[34] The atomic bomb, and thus the bombshell, was characterized by institutions such as the American Social Hygiene Association as a potential carrier of sexually transmitted infections and linked to Communism and careerism, serving as a threat to the social order embodied by the white, heteronormative nuclear family.[35] The bombshell posed an increasing threat to postwar American gender stability.

In this book, I explore how the bombshell trope was both a threat to patriarchal norms and a product of it. The vivacious, disarming sexual allure of the postwar bombshell arguably aided acceptance of Big Science products while simultaneously illustrating the complexities of the Atomic Age. In the Cold War context, the conflation of the bombshell with US consumerism and democracy in the face of Communism became political. Design historians argue there was a "doubling" effect in the Big Science context of Cold War culture, where a "threat" was also framed as "progress" or even desire.[36] Similarly, the bombshell embodied the tension between spectacular hyperfemininity and overt sexuality, often constructed by and contained within restrictive foundationwear and American gender norms, but one that also had the contradictory potential power to disrupt them.

Postwar American cultural narratives equated female sexuality with the bomb: Just as nuclear power could be harnessed for the greater good, so, too, could the bombshell.[37] Marriage and domesticity could put her sexual power to "good use," benefiting US Cold War society by appeasing husbands and raising children according to traditional gender roles. A sexualized, commodified female became an ideological, biopolitical, and weaponized symbol of US freedom, one that could tackle the threat of sexually "perverse" Communism as the "other" to wholesome American patriarchal, capitalist values.[38] The threatening potential of the bombshell was thus harnessed and controlled through cultural channels, including propaganda, entertainment, and domestication.

The domestication of female sexuality is shown in the blonde striptease artist of Franklyn's story. After reinventing herself as "Chesty," she excelled in her career. Franklyn recalled her transformation: "Chesty's profession was not the most dignified in the world but that didn't bother me. What mattered was that just a month later she dropped by to tell me she was going to be married. Now that she didn't have to hide the fact that Nature had neglected her, she wanted to marry and settle down. And today, there isn't a more ardent PTA mother in San Fernando Valley than the ex-Miss 'Chesty' of Main Street, Los Angeles."[39] Scientific advancements enabled "Chesty" to transform multiple times: Her physical transformation supported economic growth and social mobility, facilitating a shift from sex work to the socially acceptable realm of heteronormative feminine domesticity and motherhood.

Histories of the bombshell tend to focus almost exclusively on an unmarked white, cisgender, heterosexual, and often blonde bombshell.[40] In this book, I seek to actively queer this historiography and highlight a range of bombshells with intersecting identities. From the 1930s onward, journalists in whitestream and African American publications celebrated a number of Black female performers as "bronze bombshells," including Vivian Henderson, Joyce Bryant, Vickie Henderson, Leslie Uggams, Tina Turner, Lena Horne, and Eartha Kitt.[41] Further examples of bombshells include the Japanese-born Jewish performer Marie Misakura, known as the "Oriental bombshell," and the Portuguese-born Carmen Miranda, who was billed as the "Brazilian bombshell."[42] In September 1945, *Life* magazine dubbed Linda Christian, a Mexican actress wearing a two-piece swimsuit, the "anatomic bomb."[43] The language of the bombshell also circulated in underground print media, including *Bombshells Burlesque* and various trans feminine publications. In an issue of *Female Mimics*, a white model named Vicky (figure I.3) was described as "power packed, explosive . . . these are just a few of the words that have made Vicky known as *the Bombshell*."[44]

An internationally famous blonde glamour girl of the Cold War era is Christine Jorgensen, widely recognized as the first transsexual person to gain significant press coverage.[45] In December 1952, Jorgensen made worldwide headlines when a *New York Daily News* front-page story titled "Ex-GI Becomes Blonde Beauty" reported that she had received gender-affirming surgery in Denmark. The American media initially celebrated Jorgensen's story as a postwar medical miracle, "another testament to the magnitude of modern science," and its power of transformation.[46] Journalists described Jorgensen as a "GI turned glamour girl," stressing the respectability of her military background while simultaneously demonstrating their acceptance of her as an American bombshell beauty by applying the same language they reserved for Hollywood stars like Marilyn Monroe.[47] Here, as Susan Stryker notes, a "macho archetype such as 'the soldier' could be transformed into a stereotypically feminine 'blonde bombshell.'"[48]

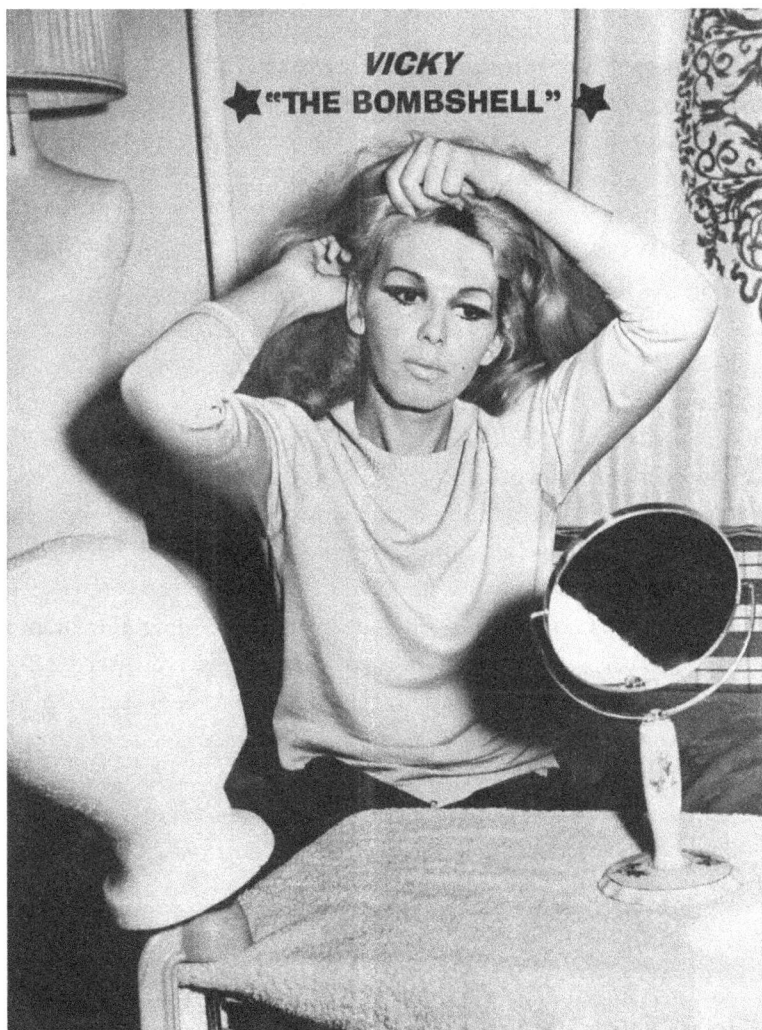

1.3 "Vicky 'the Bombshell,'" *Female Mimics* 1, no. 11 (1968): 52.

Trans visibility increased in the postwar period, but its representation in US whitestream media often subscribed to established racialized gender norms.[49] While the stories of a select number of trans feminine people who transitioned within professional medical structures were highlighted, most trans individuals had very different experiences and pursued DIY forms of transition.[50] Jorgensen played a central role in the American media's construction of what historian Emily Skidmore critically describes as the "good transsexual," a racialized image that embodied "the norms of white womanhood, most notably domesticity, respectability, and heterosexuality."[51] Through this lens, Jorgensen could be considered a bombshell put to "good use." This vision of malleability and capacity for transformation was frequently limited in its application in whitestream media to individuals who fit within established rigid, racialized gender norms that continued to dominate in the 1950s.

Coverage of Jorgensen and other trans feminine individuals arguably also contributed to destabilizing dominant cultural understandings of sex and gender as fixed. Ultimately, as Stryker notes, in this era "questions of what made a man a man or a woman a woman, and what their respective roles in life should be, were very much up for debate."[52] As figure I.4 shows, the postwar abundance of gendered consumer goods made from the latest plastics—including padded foundationwear, prosthetic "falsies," and implants—were part of a wider politicized debate and deeply rooted anxieties about the shifting nature of feminine gender presentation and sexuality. Historian Jules Gill-Peterson argues that the United States' postwar geopolitical, economic, and cultural hegemony, paired with its position as ideologically opposed to Communism, "relied upon a vision of a new scientific malleability to the American body in the Atomic Age, one for which the idea of transsexuality was well suited."[53]

Increased access to new materials in the postwar US also enabled users operating outside military-industrial structures to develop their own design innovations using plastics. For example, trans feminine publications circulated advice on making foam falsies and foundations using the latest plastics R & D and techniques for dyeing these objects to match one's skin tone. Ultimately, increased access enabled a wider range of users and makers to engage with plastics in various body-shaping practices that upheld, conformed to, challenged, subverted, or resisted dominant postwar American racialized gender ideals and unmarked norms of womanhood.

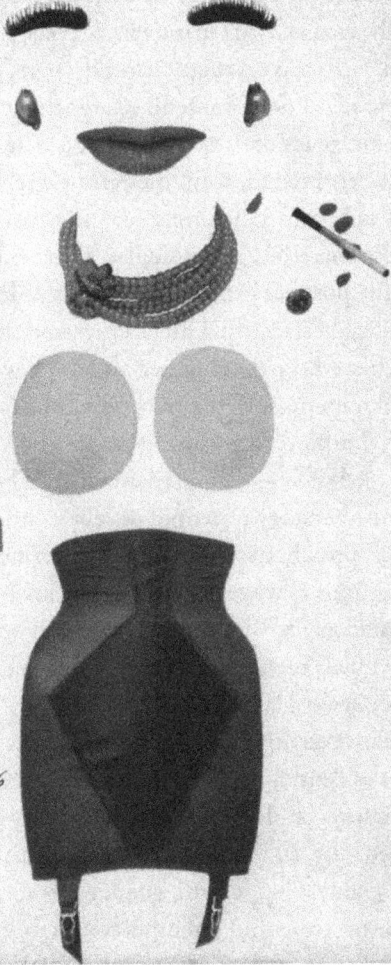

1.4 "What Are American Women Made Of?," *Pageant*, May 1953.

The Plastic Biopolitics of Bombshells

The concept of the desirable, healthy, "natural," and ideal body is always so-cially, politically, and culturally constructed. Feminist STS scholars, critical theorists, and historians have shown the biopolitical dimensions of defin-ing the human body in universal terms, particularly when bodies are made ideal.[54] There is no single, natural, "pure," or "original" body; all bodies are modified.[55] I do not intend to argue that bodies only become plastic with the emergence of soft plastics that can be used in a variety of body-shaping practices. Indeed, from diet and exercise to adornment, people engage daily in bodily transformation and have done so throughout history using various materials, technologies, and techniques.

The postwar explosion of plastics, developed within Big Science, de-stabilized established hierarchies and boundaries of natural/artificial. I articulate how these materials, along with other scientific and medical developments, offered new possibilities for permanently reshaping the body. I explore how these possibilities—and the ways a wide range of ac-tors used them—disrupted established borders of the body, opening it up to the liberatory potential of self-authorship.[56] In the Cold War United States, prosthetics, implants, and cosmetic surgery became increasingly recognized as what Alka Menon describes as "part of a deliberately fash-ioned identity."[57] In theory, these and other plastic technologies enabled individuals to exercise agency over the body and acquisitively determine their outward appearance.

However, it was not so straightforward. I explore how, when technol-ogies of gender made from the latest synthetic materials emerged, access to these technologies and the possibility of self-authorship was not meted out equally. In the postwar United States, as gender norms shifted and the "realness" of sex and gender was increasingly questioned, biopoliti-cal restrictions and attempts to restabilize sex and gender were simulta-neously put firmly in place by medical, legal, and political institutions where decision-makers were usually not the ones using the technologies.[58] *Atomic Bombshells* discusses how gatekeepers frequently imagined and allowed only the gendered body to be malleable within their rigid estab-lished structures and binary ideals. Ultimately, I argue that intersecting forms of oppression informed material access and use, privileging the le-gitimacy and safety of certain bodies and gender expressions over others.

To do this work, I draw on feminist, queer, and trans studies. Femininity is frequently positioned as inherently malleable, artificial, fake, and frivolous, whereas masculinity is understood as fixed, natural, real, and serious. These binary positions, as I explore throughout this book, are also often employed in the gendered and racialized promotion of plastics and their application in feminine body-shaping practices. In addition to the gender binary, other related dualisms—such as the nature-culture and mind-body divide—are persistent in Western traditions and structures, where they are used to legitimize forms of oppression.[59]

While femininity is disparaged irrespective of whom it is performed by, society imposes a gender hierarchy where trans expressions of femininity are deemed as more artificial and with heightened suspicion.[60] The oppression this is based on is directly related to the biopolitical plasticity of definitions of sex and gender across other identity categories.[61] For example, as Stryker has critiqued, "'Woman' typically has been mobilized in ways that advance the specific class, racial, national, religious, and ideological agendas of some feminists at the expense of other women."[62] But as Black, Indigenous, and feminists of color have established, there is no universal definition of woman, womanhood, or femininity.[63] This thinking is foundational to my argument.

Feminist scholars' use of metaphor also informs this analysis. In response to exclusionary essentialist definitions of "natural" womanhood within some strands of feminist thinking in the 1970s and 1980s, Donna Haraway famously proposed the concept of the cyborg as a hybrid entity—neither entirely technological nor fully organic—to challenge and undo binary concepts and oppressive systems. While the cyborg is often considered to be the most famous metaphor in feminist STS studies, it also has a complicated and violent history rooted in the military-industrial complex and pathologization.[64] Recent feminist STS scholarship has further critiqued the limitations of this figure while also acknowledging the important groundwork it laid out for future discussions and reimaginings.[65] Haraway reminds us, "The main trouble with cyborgs, of course, is that they are the illegitimate offspring of militarism and patriarchal capitalism."[66] However, Haraway's formulation argues that these particular technocultural origins paradoxically hold the radical potential to be subverted and can work to destabilize the very power relations they were created within. This line of thinking is particularly relevant to my exploration of technologies of femme embodiment with military-industrial origins.

In this book, I use the term "bombshell" as a metaphor and an identity that moves beyond its military origins. Feminist STS scholars argue that "it is useful to consider *how* metaphors (are) matter."[67] The bombshell is both discursive *and* real. Individuals referred to themselves as bombshells and continue to do so. I leverage the bombshell as a way of more closely exploring how the discursive informed the material, and vice versa. I aim to show how plastics as a form of military-industrial technology shaped feminine bodies discursively *and* physically. I argue that the impact of World War II on technologies of embodiment is reflected in how the bombshell ideal shifts as technologies and gender norms change over time in the postwar United States. Thus, the bombshell can offer a helpful feminist model for rethinking technocultural bodies and the power relations within which they are created. By showing the hegemonic history of materials used in feminine self-fashioning, the book makes all the more clear *how* histories of science, technology, and medicine are not neutral and *why* counterhegemonic histories are so important.

From Bullet Bras to Bombshells

The research journey behind this project began with an object that signified a major design change: the bullet bra, a colloquial term used to describe a type of conical bra, often padded with foam and finished with whirlpool stitching to create an enhanced, structured point to the bust. As a high-femme-presenting fashion and design historian who has worn a range of padded foundationwear, I had long been fascinated by this pointy silhouette, which first emerged in the 1930s and appeared to reach peak popularity by the mid-1960s. Since then, there have been smaller-scale revivals including within vintage subcultural contexts and high-end fashion. For example, in the late 1980s, designer Jean Paul Gaultier created his now-infamous structured pieces of underwear as outerwear, popularized by Madonna during her Blond Ambition tour, which were recently manufactured and sold again. Yet, the conical bustline ideal seemed to never return to the same widespread popularity of earlier decades. As this book shows, the conical silhouette that persisted for decades in bra design and Sweater Girl looks ranged in extremes, from more discreet stitching to thick foam padding. But why hasn't this conical style come back in quite the same way? How did social, political, cultural, and technological changes affect the construction of this ideal? And how did these factors shape material bodies?

This book is driven by materials and a desire to connect seemingly disparate but closely interconnected disciplines. I use a wide range of interdisciplinary archival materials to explore a diverse set of actors, including doctors, designers, drag queens, and decision-makers in the US military. To map this complex network, I use a variety of approaches, drawing on histories of fashion, design, technology, science, medicine, and cultural studies. Building on established methods within these disciplines, along with engagement with the feminist, queer, and trans critical theory outlined earlier, I employ extensive archival research with various sources— including foundationwear company papers, surgeons' papers, plastics manufacturers' corporate records, magazines, and plastic objects—to generate the arguments presented in the book. I also work expressly with a number of archives, from established collections within museums and other heritage institutions, as well as community-driven collaborative resources such as the Digital Transgender Archive.

A note on language: I draw on an intersectional approach to deconstruct unmarked definitions of womanhood that are prevalent in dominant visual and material culture, as well as archival sources. I am a white, queer, high-femme-presenting nonbinary lesbian with an invisible disability from a mixed-class background. As a feminist historian, I seek to push back against received exclusionary histories, "unmarked norms" of woman, and repeated problematic myths within scholarship. However, it is important to note that many of the sources I engage with do not acknowledge gender outside the binary, gender fluidity, or intersectionality. Most of these primary and secondary sources use the term "woman" in a singular way, while I aim to use it intersectionally.

Attention to language is important to this book's methods. The widespread usage of the word *transgender* has only been adopted in recent decades to describe "a range of gender-variant identities and communities within the United States in the early 1990s."[68] Just as its core definition is rooted in movement away from an assigned, unchosen gender position, its meanings are continuously being shaped and reimagined. While Stryker defines it as "the widest imaginable range of gender-variant practices and identities," she also notes that the abbreviated term "trans" is more expansive, as "contemporary connotations of transgender are often limited."[69] Hence, the abbreviated term is used in this book. Moreover, Gill-Peterson states that "trans" is also used to "mark a *political* distinction from medical or pathological meanings that have accrued to the term 'transgender' in recent years, many of which have been borrowed from the earlier term 'transsexual.'"[70]

Terminology plays a key role in trans studies and in telling stories about the political history of gender variance that are not limited to one experience.[71] Terms have changed over time and will continue to do so. Wherever available, I aim to include the original language and pronouns that a person used to describe themselves: for example, terms such as "transvestite," "transsexual," "drag queen," "cross-dresser," and "female impersonator." While "female impersonator" was used in the past, it is no longer commonly used in trans studies today.[72] However, it is critical to include the original terminology and categories for historical context.[73] People who referred to themselves as female impersonators covered a wide range of intersecting identities, including self-identifying as gay, lesbian, bisexual, straight, transsexual, transvestite, drag queen, and cross-dresser. Some female impersonators additionally described themselves as "professional" and "amateur." It is key to note that the way people described themselves in the historical sources used throughout this book is only a snapshot of a particular moment, and how they identify may have changed over time.

While I endeavor to make the histories of individuals who have been left out of historical narratives more visible, much of the historical content in *Atomic Bombshells* focuses on often-unmarked white actors. There are four main reasons why this occurs. First, it is important to establish the origin stories of plastic materials and the ways white men in Big Science shaped racialized gender ideals. Second, due to the historical nature of my research, people in decision-making roles within the military-industrial complex tended to be white men. Third, due to inequitable societal power structures in the postwar United States, those who were able to afford these new materials and procedures when they were commercially launched within established structures tended to be middle-class white cisgender women. Fourth, archives and collecting institutions have largely privileged the documenting of these actors. I establish the whitestream, heteronormative network of postwar plastics in the United States to show how these structures were disrupted and queered by individuals who have been left out of historical narratives.

Atomic Bombshells is a road map to the actors, plastic materials, and politics that shaped femme bodies and gender expression in the postwar United States. The book is arranged by material and chronology in three parts—nylon, plastic foams, and silicone. Tracing the material's relation to R & D and women's bodies, each part starts with the material's origins and progressively explores key actors, gender, the body, and agency. *Atomic Bombshells* sequentially follows plastics' journey into the body, beginning with their use on the skin's surface with the launch of nylon and progressing to plastic foams, which start on the surface and eventually move inside the body. The final part focuses on silicone, which was initially injected into the body before being used in breast implants and external prostheses.

The first part, on nylon, shows how this material's history connects racialized gendered bodies, chemical companies, and military-industrial materials. By doing so, it weaves a key conceptual thread that runs throughout the book: the relationship between women's bodies and emerging plastics. In the first chapter, I establish the explosive history of nylon, launched by DuPont as the world's first entirely synthetic fiber. In the 1930s, DuPont executives were eager to reinvent the company's public image, shifting from that of the explosive "powder people" associated with war and destruction to providers of "Better Things for Better Living . . . Through Chemistry." Some individuals initially feared nylon, concerned that this new material, "conjured from coal and air," might be hazardous, potentially exploding or melting upon contact with the skin. I trace and analyze the reception of nylon's materiality, from its earliest displays to the politics surrounding its wartime and postwar applications. Nylon was used as a substitute for silk imports from Japan and came to embody promises of an American postwar plastic utopia. I show how nylon's associations with women's bodies are key to understanding changing attitudes toward synthetic materials, their political dimensions, and their proximity to the body.

In 1939, DuPont unveiled nylon stockings to the public at two US world's fairs. DuPont's "Wonder World of Science" exhibitions featured a white female figure known as the Test Tube Girl, who embodied the latest in scientific developments, including nylon. In the second chapter, I contextualize the Test Tube Girl and related imagery within wider cultural discourse around racialized gender within fashion, streamline design, and eugenics at the time. I critically examine a range of examples of visual and material culture to demonstrate how whiteness was presented as the unmarked norm. Racialized gender was coded into experimental synthetic objects, such as white skin-toned "wear-test" underwear. I also present a counterhistory, highlighting how individuals pushed back against these exclusionary practices in the ensuing decades and crafted nylon hosiery for a wider range of consumers, including Black, queer, and trans feminine people.

The second part of the book is dedicated to foam plastics and padding. It explores polyurethane foam, which offers a distinct materiality compared to nylon and silicones: lightweight, fluffy, pliable, spongy, compressible, and filled with interconnected air pockets. It "breathes," becomes lifelike, and is anthropomorphized, expanding into US markets after World War II. In chapter 3, I trace the history of polyurethane foams, starting with a series of World War II US military intelligence reports that recommended the postwar transfer of plastic foams from Germany to the "soft power" politics of US postwar domesticity. I explore how, in the postwar United States, polyurethane foam's materiality and soft appeal to the body aided its movement from the lab to domestic spaces. Actors central to the promotion of plastic foams, including chemical company advertising staff, journalists, and industrial designers, harnessed these sensual properties for foam's postwar domestic promotion and applications, often presenting a carefully constructed ideal of white womanhood and heteronormativity.

Chapter 4 covers polyurethane foam's corporeal application in foundationwear, falsies, and cosmetic surgery. Polyurethane foam's soft materiality and seemingly endless array of molded forms made it easily malleable, allowing it to be shaped into a point that defined the conical bustline ideal. The chapter begins with a discussion of external shaping devices designed to achieve this look, such as the bullet bra and padded girdles. It discusses different types of bombshells and the role of foam plastics research and development in shaping fashions, including falsies, butt pads, and padded girdles. I demonstrate how these body-contouring designs appealed to a wide range of consumers and were featured in the wardrobes of stage and screen celebrities, those who survived breast cancer, contestants in drag balls, models, and sex workers. Polyurethane foam's soft pliable properties also attracted cosmetic surgeons, who crafted implants that resembled this fashionable silhouette. A range of archival materials is used to illustrate the close resemblance between these implants and the foundationwear and prostheses worn externally on the skin's surface.

Chapter 5 explores how women's bodies in the Atomic Age were culturally compared to weaponry, domestic interiors, and automobiles. Moving beyond such symbolic cultural comparisons, I examine how plastic foams produced by chemical companies supplying the automobile and furniture industries were being cosmetically implanted in women's bodies in the postwar United States. These examples demonstrate how an amalgamation of technological changes and shifting attitudes to the body as a site of consumer improvement were made possible by plastics, resulting in permanent modifications to women's bodies using foam padding sourced from industrial and military suppliers.

The final part turns to silicone, which offered a new kind of inert and "othered" fluidity for shaping the body. In contrast to foam plastics, early silicone applications aimed at shaping the body were injected directly, moved beyond borders, and could not be contained within easily classifiable objects. Fluid in both materiality and classification, it slipped through cracks in established medical structures and regulations, went underground, and quickly reached sex workers as well as queer and trans feminine communities.

Chapter 6 focuses on silicone's complex military-industrial origins and its immediate postwar entry into the commercial market. I examine silicone's provenance, R & D, and changing meanings and applications through a variety of primary archival sources, ranging from chemistry engineering textbooks to Dow Corning's promotional materials and women's magazines. The chapter traces how new postwar applications of silicone, as promoted in these texts, moved increasingly closer to the body, in both representation and reality. Promoted as inert and an American miracle of peacetime conversion, silicone inevitably began making its way more overtly into the body.

Building on the previous chapter's exploration of silicone *on* the skin's surface, chapter 7 offers a deeper exploration of silicone *beneath* the surface. *Atomic Bombshells* utilizes new archival evidence to complicate the problematic, unreferenced, and frequently repeated Japanese origin myth of silicone bust-augmentation injections. Silicone was racialized and othered by various actors—including licensed American medical practitioners and journalists—who cast it as something distasteful, associated with Japanese sex workers. I further contextualize this process of othering and Orientalism within the wider discourse of postwar American politics and popular culture by analyzing representations of Japanese women, including Akiko Kojima—crowned Miss Universe in 1959—and the "Hiroshima Maidens."

Chapter 8 explores how silicone shaped West Coast US strip culture and the "topless craze." Silicone's visceral materiality, purported inertness, and the immediacy of its fluid injection into the body forever changed the landscape of body-shaping practices and cosmetic surgery. Implantable silicone materials proved challenging for the FDA to categorize: Was the silicone material a liquid (drug) or an object (prosthetic)? As an injectable fluid, it could operate outside traditional medical structures, leading to an underground market in silicone injections more readily accessible to trans feminine and queer communities, as well as sex workers. In this chapter, I illustrate how silicone, fluid and other in its material identity, also became a means for simultaneously complicating and upholding gender norms. Using trans feminine publications, I show how queer and trans feminine communities shared information about various methods of shaping the body with silicones, as well as its potential harms. The chapter builds up to the 1976 US Medical Device Regulation Act, which granted the FDA federal jurisdiction over the regulation of medical devices and had a groundbreaking effect on the largely unregulated implantable plastics market.

The legacy of plastics developed, initially by chemical companies for industrial and military applications, endures in practices of self-fashioning femme bodies. Today, plastics, including nylon, polyurethane foams, and silicones, continue to shape and affect the body. In the epilogue, I turn to this legacy by exploring examples that model the contemporary queering of these materials and the lasting impact of the bombshell.

NYLON / THREAD

1
SPINNING NYLON /
AN EXPLOSIVE HISTORY
FROM LABS TO LEGS

In 1935, an industrial laboratory in Wilmington, Delaware, was entangled in an expansive, glistening web. Five years after Julian Hill, a chemist on Wallace Carothers's team at DuPont's Experimental Station, pulled the first taffy-like fiber-forming polymer from a beaker, the same team used this technique to draw out nylon 6-6.[1] Upon hearing this news, excited chemists began testing out the fiber to see how much farther they could draw it out, spinning what looked like a giant spider's web.[2] Carothers and his team literally spun new strands of research, turning their years of work into the latest development and cocooning the lab in what came to be known as the world's first fully synthetic fiber—a revolutionary thread that would later lead to further developments in plastics, ranging from polyester to Lycra.

DuPont executives quickly identified nylon's commercial potential and fast-tracked its development from the test tube to a commercial plant in five years. Nylon, first available as synthetic replacements for pig bristles and fishing lines, is explored for its role in the 1939 launch of stockings, particularly in relation to racialized gender. I chart nylon's changing uses and meanings, from an industrial material spun out from the labs of DuPont, the world's largest explosives manufacturer at the time, to a political material associated with American independence, women's bodies, and the US military. I closely examine the materiality of nylon and its intimate relationship with the body, analyzing how touch and wearers' experiences were key to the acceptance of this new material.

In the mid-1930s, DuPont executives were eager to reinvent the company's public image from the "powder people"—a nickname tied to its history with explosives and wartime destruction—to promoters of the slogan "Better Things for Better Living . . . Through Chemistry." They promoted this new slogan by investing in public engagement activities, such as radio programming and exhibits that showcased the latest advancements from their laboratories and promised to improve the lives of Americans. Amid escalating global conflict, the company's synthetic alternatives to imported natural materials were an important area of development that lent itself well to public relations. The Second Sino-Japanese War in 1937 led to the US movement to boycott Japanese goods. Actors in public relations, advertising, and the media framed nylon as a synthetic alternative to silk in a bid for US independence from raw silk imports from Japan, which accounted for 90 percent of America's total raw silk imports, valued at $100 million in 1938.[3]

This chapter uses archival material to show how DuPont's Advertising Department used the company's "Wonder World of Chemistry" displays at the 1939 world's fairs in New York and San Francisco as conceptual labs for testing ideas around nylon with the public. Nylon, frequently the most requested item on display, quickly emerged as an excellent material for showcasing a more sensual and glamorous side of DuPont's R & D. In response, the company expanded the original small, mechanized display of nylon stockings by featuring a live white female model. Visitors surrounded her daily, asking what it felt like to wear them, and some women asked if they could feel the stocking on her leg. Others were initially fearful of nylon, worried that this new material, "conjured from coal and air," could potentially explode or melt on the skin. But by the end of DuPont's exhibit run, the new material was generally accepted and feared less, as evidenced by the nylon riots and "N-Day" sales, as discussed later in this chapter. When the US military needed nylon to make parachutes, tents, and mosquito nets, nylon stockings (known as "nylons") were no longer available on the domestic market, and the material's meaning transformed once more to a promise of a postwar era of plastic abundance. Nylon helped to successfully promote DuPont's new "softened" image as a provider of essential, safe, and even sensuous civilian items. This repositioning is key to understanding shifting attitudes in the US toward synthetic materials and their proximity to the body.

Important to this exploration are DuPont's company archives at the Hagley Museum and Library, the DuPont Nylon Collection and Trade Literature Collection at the Smithsonian National Museum of American History, and the Josef Labovsky and Wallace Carothers Collections at the Science History Institute. (Labovsky was Carothers's assistant.) This chapter largely draws on papers from DuPont's Advertising Department at the Hagley Museum, which kept extensive records on nylon's presentation at the San Francisco Golden Gate International Exposition and New York World's Fair in 1939. I have chosen to focus on these sources because they give invaluable insight into some of the public's earliest reactions to nylon and its materiality, particularly from female visitors. These archival sources show how, at the time of nylon's launch, some visitors were concerned about its association with an explosives company—a factor that had been largely forgotten during its postwar return to the market. This point is important to the book's charting of changing attitudes toward synthetics. Furthermore, these sources provide powerful documentation of the importance of touch in the history of synthetic materials. In this chapter, I reconnect nylon's technocultural history more closely with the racialized, gendered body and provide new analysis of archival sources to demonstrate how touch and embodiment were integral factors in nylon's success. DuPont's Advertising Department closely monitored female visitor reception and adjusted its display accordingly, hoping to gain greater interest and support, thereby complicating arguments about agency and women's bodies. These sources reveal the public and industry's first interactions with this material, as well as the narrative DuPont chose to weave about it. The history of the DuPont company is rooted in connections between military and industry. Understanding this history and its power relations is important to this book's exploration of the provenance of plastics and their relationship to shaping discursive and material femininities.

"Better Things for Better Living . . . Through Chemistry"

E. I. du Pont de Nemours and Company, commonly known as the DuPont company, was established in 1802 in Wilmington, Delaware, by the French-American chemist and industrialist Éleuthère Irénée du Pont, a Huguenot Protestant who fled to the United States after the French

Revolution. A former gunpowder apprentice for French government powder works, du Pont strategically chose the banks of the Brandywine Creek in Wilmington to set up a gunpowder mill, using imported French specialist machinery. The site provided power, trees to make charcoal for powder, and proximity to the Delaware River, on which other key ingredients like sulfur and saltpeter could be shipped. E. I. du Pont de Nemours and Company's Wilmington gunpowder factory became the largest explosives manufacturer in the world.

Seeking to expand its business beyond explosives via materials science research and development, the company set up the DuPont Experimental Station in 1903. Located across the Brandywine Creek from the original powder factory site, this industrial laboratory was dedicated to nonexplosive products, including dyes and lacquers. To better promote its latest offerings, the company set up the DuPont Advertising Department in 1911 and the Publicity Bureau in 1916. These departments worked toward communicating a vision of DuPont as "an explosive manufacturer whose products were intended to enhance rather than extinguish life."[4] Campaigns included "Farming with Dynamite," in which the DPAD suggested that explosives could support fruit tree cultivation, ditches could be excavated quickly, and trees could be "ploughed" rather than felled.[5]

More successful in promoting a diversification of nonexplosive products were DuPont's investments in materials science research. In December 1926, Charles M. A. Stine, chemist and director of DuPont's Chemical Department, proposed to the DuPont Executive Committee a program dedicated to "pure science or fundamental research," with the goal of making scientific discoveries.[6] As part of DuPont's commitment to nonexplosive research and materials science, the committee approved Stine's detailed funding application in March 1927 for his project and for the construction of a new lab, which some colleagues skeptically dubbed "Purity Hall." Shortly thereafter Stine, who had an interest in polymerization, hired the chemist and Harvard instructor Wallace Carothers to oversee organic chemistry and polymer research. One of Carothers's earliest projects at DuPont was directing the commercial development of neoprene, a type of synthetic rubber.[7]

Many developments were happening at Purity Hall and the Experimental Station, but by the mid-1930s, DuPont was finding it difficult to manage its public image. At the close of World War I, DuPont, which had supplied the US military in the first war involving chemicals, reported a steep rise in profits. This later haunted the company in 1934, when the US Senate

Munitions Investigating Committee, led by Gerald P. Nye, accused it of wartime profiteering and of continuing its influence in the international munitions industry. Contemporary journalists referred to DuPont and other accused companies as "merchants of death" that had overcharged the US military for powder and explosives in World War I.[8]

Consequently, DuPont employed the services of Bruce Barton, founding member of Batten, Barton, Durstine & Osborn, known as "the most famous advertising man in America at the time."[9] Barton's new approach to DuPont's marketing focused on public perception and recognition. He found that even though 98 percent of DuPont's business came from "peacetime" products, the public still thought of DuPont as "the powder people."[10] In 1935, Barton and his agency launched DuPont's memorable slogan "Better Things for Better Living . . . Through Chemistry," stressing the company's domestic role and chemistry's contribution to society through an increasing range of life-improving consumer goods. Barton secured a budget of $650,000 to reinvent the company's image, which included sponsoring *Cavalcade of America*, a popular radio series.[11] The programs, which began with the new DuPont slogan, were dedicated to a narrative of patriotic humanitarian progress through technological advances, with tales of life-enhancing wonders invented by DuPont's chemists mixed in. Public engagement and exhibits were a central part of bringing Barton's message of "Better Things for Better Living . . . Through Chemistry" to a wider audience via the corporate showmanship for which his agency was famed.[12]

DuPont's "Wonder World of Chemistry" exhibit premiered at the 1936 Texas Centennial Exhibition, showcasing to 1.5 million attendees how agricultural raw materials were converted into chemical products.[13] DuPont's rhetoric communicated that its chemists were "improving upon nature" and that the company, in turn, was providing consumer goods to improve people's lives. This marked a turning point in DuPont's approach to communication, which would increasingly engage with the public through corporate displays and advertising. Manufacturing a highly technical product was no longer enough for DuPont; Barton advised the company that this burgeoning range of materials needed to be "explained" and contextualized to the consumer to recuperate the company's image. With Barton's help, the DPAD created a vision of heroic innovators improving civilian life—for example, solving everyday problems such as insect infestations, medical equipment shortages, and the need for waterproof clothing. These explanations needed to be innovative, engaging, and increasingly spectacular to make the company's materials R & D more relatable, accessible, and desirable. DuPont's Texas presentation demonstrated that chemical manufacturers were becoming the direct link between the laboratory and the consumer.

In 1935, the same year the "Better Things for Better Living . . . Through Chemistry" slogan was launched, another key development occurred in DuPont's history. Nylon 6-6—also secretly named "Rayon 66," "Polymer 66," or just "sixty-six" during its experimental phase and later known as "Fiber 66" (1937) and "nylon" (1938)—was first drawn at the DuPont Experimental Station by Carothers's team. After lengthy deliberations—and after quickly abandoning suggestions such as DUPAROOH (an acronym for "DuPont Pulls a Rabbit Out of a Hat") and "norun" (discarded when nylon hosiery was found to ladder)—DuPont's naming committee settled on "nylon." In October 1938, after a decade of R & D, "nylon," the generic name chosen for materials of the synthetic polyamide type, was formally announced to the public at a New York World's Fair preview press event. By autumn 1938, nylon bristles were being sold in toothbrushes across the United States, and in January 1939, nylon fishing lines were introduced. Shortly afterward, construction began on a large-scale plant for the commercial manufacture of nylon polymer and yarn in Seaford, Delaware. When nylon's large-scale production began, DuPont had already invested $27 million into its research, development, engineering, and production.[14] In February 1939, nylon stockings were revealed to the public at the 1939 Golden Gate International Exposition in San Francisco.[15] The DPAD leveraged nylon's soft, sensual qualities—distinct from its early uses in toothbrush bristles and fishing lines—to emphasize its intimate associations with women's bodies.

New Labs for Nylon in the Wonder World of Chemistry

In 1939, on the eve of World War II, DuPont presented a reworked version of its 1936 Texas "Wonder World of Chemistry" exhibit at two world's fairs in the United States: the Golden Gate International Exposition in San Francisco and the New York World's Fair. Nylon stockings were first introduced to the public at DuPont's "Wonder World of Chemistry" display at the San Francisco exposition, which attracted over two million visitors during its run.[16] Later that year, the 1939 New York "Wonder World of Chemistry" exhibit drew nearly ten million visitors in total.[17] These visitor demographics are not broken down by identity categories such as race, ethnicity, or gender. Nor do DuPont's detailed reports on exhibit attendance, discussed later in this chapter, categorize adult visitors by identity categories, except by assumed gender. However, it is important to note that most visitors were white.[18]

Material in DuPont's archives shows that visitor numbers exceeded predictions, which the company press office presented as "demonstrating an increasing public interest in new chemical developments."[19] DuPont's decision to present an accessible research lab to the public, staffed by actors posing as chemists, aimed first to increase knowledge of its products and second to help to dispel its "merchants of death" image. In keeping with the company's efforts to dissociate itself from explosives and the spoils of war, the display on munitions was partially obscured, with only one small column dedicated to it.[20]

The exhibit lab also functioned on a third, hidden level: as a lab for experimenting with and testing ideas on the public about materials emerging from it. As papers at the Hagley Museum and Library reveal, the DPAD invested much time, effort, and resources into monitoring visitors to their 1939 display. The exhibit floor staff, also known as lecturers, wrote daily reports, and a selection of their observations were sent weekly to DuPont's headquarters. The DPAD analyzed these reports and made changes to improve the display accordingly, aiming to appeal to a wider audience. What is so interesting about this material is that the DPAD specifically responded to women's views on nylon and wove them into its own PR and narrative.

As the weeks passed, DuPont's exhibit staff and advertising office soon discovered that nylon was consistently the most visited part of the display.[21] In an early weekly report from New York, an exhibition supervisor reflected, "When the exhibit first opened, it was expected that we would get a fair number of questions on munitions. These have not developed. The great battle cry is: 'how soon can we buy nylon hose?'"[22] Original blueprints and sketches by designer Walter Dorwin Teague, press shots of the display, and lecturers' reports all reveal that neither DuPont staff nor Teague had anticipated this interest.

When the company's Wonder World of Chemistry exhibits opened in 1939, nylon was originally relegated to a small display, one of dozens dedicated to DuPont's chemical products and processes. The size of the display contrasted with the overwhelming enthusiasm from journalists and the public following DuPont's 1938 announcement of nylon.[23] After DuPont's New York exhibit opened, a lecturer seeking to improve nylon's visibility suggested, "It would be a good idea to have [nylon] products lettered in large letters over the first section such as neoprene and other products in the second section are pointed out."[24] The original exhibit design featured nylon alongside other materials such as Lucite—a clear plastic that the DPAD believed would grab the public's attention, thanks to its novel, shiny, and transparent materiality. Lucite was presented in applications such as dentistry and as heels for shoes; however, it failed to capture the attention and imagination of the audience in the same way that nylon did. No one at DuPont could have predicted the spectacular rise in interest in nylon generated by journalists or the timely opportunity that the Japanese silk boycott presented for reinventing the company's image.

Nylon and Nationalism: The 1937 Silk Boycott

The world's fairs, new public relations activities, and positive coverage from journalists helped to improve and soften DuPont's public image, particularly in relation to its latest developments in synthetic materials. DuPont was further boosted by nylon and the solution it offered to the Japanese silk boycott that began in 1937. The United States and Japan were locked in economic and political conflict, and, as scholars have described, women's hosiery played an unexpectedly important part in this.[25]

Silk stockings were a social convention for women's everyday dress in the United States during this time. But 90 percent of America's raw silk was sourced from Japan, with imports totaling $100 million in 1938, 75 percent of which were used for hosiery production.[26] A study at the time noted that "American women are definitely sold on full-fashioned silk hosiery," while only "small amounts" went into men's and women's seamless hosiery, as well as men's full-fashioned silk hose, by comparison.[27] Though developments in synthetic fibers such as semisynthetic rayon had made a significant reduction in silk imports for woven fabrics across all apparel, "rayon has made little headway in women's hosiery," with just 6 percent used in "hosiery of all kinds."[28] In 1938, 1.55 million pairs of stockings were sold in the United States every day, totaling an annual expenditure of $475 million. Before nylon's 1939 launch, Japanese silk dominated the market.

World War I had already shown the United States the importance of increasing its national independence from imported chemical products and raw materials. This included imports from Germany, which remained a key technological hub, excelling in organic chemistry, dye production, pharmaceuticals, and countless other products.[29] Other imports included iodine and rubber, as well as nitrates, which were essential for producing both explosives and fibers; fertilizers were imported from Chile. Politicians considered obtaining increased material independence a matter of security and survival for the United States and its industry. DuPont clearly recognized this need and demonstrated its allegiance to the US government, commenting later, "Until this country could build up its own resources, its needs could be met only at the whim of a foreign power . . . such things were irritating in peace, for they forced the country to pay a sort of ransom to attain the living standard it desired. And in war this dependence on foreign technology could well prove fatal."[30] World War I left a legacy of anxiety in the United States about achieving and maintaining technological prowess without dependency on imports. The importance of material independence was emphasized.

News reports welcomed the advent of nylon as a timely invention that could secure US independence from imported Japanese silk for stockings, highlighting its impact on the chemical industry.[31] Historian Lawrence Glickman notes that the US movement to boycott Japanese goods began in August 1937.[32] In July 1937, the Second Sino-Japanese War broke out, following Japan's decades-long imperialist strategy of military and political expansion.[33] In an October 1937 US public opinion poll, 1 percent of respondents voiced support for Japan, 59 percent supported China, and the remaining 40 percent had no opinion.[34] This public view intensified after Japan began its undeclared war against China in July 1937 and Chinese representatives lobbied the United States. In contrast to the less successful boycott of Nazi Germany, which did not single out a particular product, the antisilk campaign connected military imperialism with a specific import.[35]

A wide range of US groups and individuals—including the American Federation of Labor, the Congress of Industrial Organizations, YMCAS, students, Chinese Americans, African Americans, veterans, consumer groups, manufacturers' associations, celebrities, progressive women's groups, and liberal and leftist organizations—campaigned for the boycott of Japanese silk.[36] Simultaneously, the American Federation of Hosiery Workers, an industrial union, protested the boycott, arguing that the principles of ethical consumption and fashion trends required the continued purchase of silk to prevent the loss of US jobs. Despite the Great Depression, the consumption of women's silk hosiery remained fairly stable. It slowly "advanced from 30 million dozen pairs in 1929 to 31.6 million in the first eleven months of 1935, setting an all-time record in the latter year."[37] However, at the time of the silk boycott, many hosiery workers were already facing unemployment, having suffered from precarious employment in recent years due to strikes and the general impact of the ongoing economic crisis. Although the silk boycott was far-reaching and aided by press coverage of women's stockinged legs and their actions protesting without stockings, consumers seemed to have heeded the calls of the hosiery workers. Furthermore, consumers complained that hosiery made from natural materials found in the United States, such as lisle from cotton and rayon from cellulose, lacked elasticity, was less soft, and was not as silky to touch or as comfortable on the skin. In 1936, a researcher commented, "American hosiery manufacturers, especially the full-fashioned group, cannot readily substitute other fibers for silk [hosiery]."[38] Ultimately, before nylon's arrival, the silk boycott wasn't quite as enticing to stocking-wearers.

A journalist writing for the *New York World Telegram* viewed the discovery of nylon as "having the utmost social and economic significance . . . it won't be so difficult to popularize a boycott of Japanese silk when women can obtain stockings from DuPont's mechanical silkworm that are not only equally attractive but wear longer."[39] Nylon's introduction to the general public offered a "homegrown," sensuous, synthetic alternative to imported silk, while simultaneously securing employment for the country's hosiery workers. Furthermore, nylon's promise of being a "manmade," new-and-improved version of silk meant that a boycott wouldn't be necessary; instead, nylon would eventually replace silk stockings.

Lecturers' reports from the 1939 fairs show that DuPont did not explicitly present its nylon stockings as a means of boycotting Japanese silk. However, visitors frequently politicized nylons as a form of exerting economic sanction against Japan and, perhaps influenced by news reports, attached this meaning to the nylons themselves. A lecturer observed, "Most women who have asked me about [nylon] 'can't wait' for these stockings to be put on the market and consider it their patriotic duty to buy them in preference to Japanese silk."[40] Many female visitors attributed to nylon a powerful agency—the potential to affect Japan's economy and military capabilities.

In contrast to nylon's soft materiality, the political discourse about it in relation to Japan often connoted violence. One lecturer noted, "[Every day] I get ... at least half a dozen people who say I'd like to see us *knock* Japan out of business"; according to another, "Many [visitors to the exhibit] have been expressing considerable joy at the idea of *hurting* Japan through replacements of her silk markets with nylon."[41] References to nylon inflicting pain on Japan are frequent in lecturers' notes, sometimes adopting the militarized language of ammunition and liberation: "Wistful thinking, potentially realized, seizes a great many women upon hearing of nylon. And if that person is politically minded, they usually see in nylon a powerful weapon for freeing the Far East and Western United States from the Japanese military caste."[42]

In this context, nylon became an ideological, economic, and highly politicized weapon in the United States' World War II arsenal. Notably, DuPont's live demonstration of Japanese beetle extermination in its pesticides section was a well-attended feature of the display. This can be understood as another means of performing and consuming the infliction of pain on Japan.[43] Journalists frequently targeted Japanese beetles, labeling them "saboteurs" and, after Pearl Harbor, joining entomologists in renaming them "the Jap beetle."[44]

On the eve of Europe entering World War II, nylon became increasingly desirable, and its consumption became an empowering act of retaliation against Japan. A lecturer noted of a female retailer, "In nylon she saw freedom from Japanese silk," as it could "liberate" her from the dilemma of consuming an imported material from an enemy country.[45] Lecturers' reports reveal that nylon's timely unveiling appealed to the public's emotions, and it was hoped this would translate into sales. Two female buyers from Macy's commented, "The patriotic feelings of the American people will be an important factor in the sale of nylon."[46] Purchasing nylons became a fashionable political act, and visitors recognized it as such. One lecturer observed, "It is especially interesting that many of the persons to whom I talked about nylon immediately think of the Japanese silk business and express the hope that nylon will replace foreign silk. One lady displayed a cotton-clad leg, [and] said she would not wear silk stockings and groaned when I told her that we expected nylon hose to be on the market in about six months. Several other persons have asked me where they can buy nylon now."[47]

There was an urgency in wanting to purchase nylon stockings for political and aesthetic reasons, as well as comfort. However, full-scale commercial production was still in its infancy. Some female consumers saw their boycotting of silk and adopting of nonsilky cotton as a burdensome, if patriotic, act of selflessness and grew impatient as they waited for DuPont's synthetic alternative to hit the market. The lecturer continued, "Personally, I never make any reference to Japan when discussing nylon but stress the beauty and strength of the fiber or fabric and mention its uses besides stockings. However, this patriotism is apparently second only to nylon's wearing qualities."[48] Nylons provided a comfortable and viable alternative to silk stockings. They also effectively legitimized the consumption of a cutting-edge technology during times of economic uncertainty as a form of activism.

In response to increased global conflict, DuPont's Public Relations Department adjusted the rhetoric of the New York exhibit. Over time, lecturers noted a decline in questions about nylons, with occasional inquiries instead focusing on DuPont's capacity as an arms manufacturer. Under the DPAD's watchful eye, new material identities, such as nylon, could still be shaped to accommodate the public's desires based on exhibition feedback. In May 1940, it announced a reworking of the exhibit that depicted "progress of the chemical industry toward making America self-sufficient."[49] Over a hundred representatives from newspapers, magazines, and trade publications attended a glamorous preview on the eve of its opening. DuPont was keen to present itself as a key contributor to self-sufficiency, economic independence, and American technological progress. An exhibition press release emphasized the political elements of DuPont's products: "Displays indicated an economic security and industrial independence sharply contrasting with the precarious reliance on foreign sources of supply in 1914–1918."[50] Crowd-pleaser nylon and its soft materiality provided an ideal vehicle for achieving American self-sufficiency.

Dangerous Dress and the Power of Touch

Fashion historian Susannah Handley claims that, in 1939, "DuPont's image was totally metamorphosed in the public's perception after the introduction of nylon—no longer behind munitions, instead it was a company behind lovely legs."[51] Yet, feedback from lecturers' reports shows that some female consumers were suspicious of nylon's potential adverse effects and health consequences, given that it was developed by a munitions company. For example, inquiries and concerns around nylon allergies were not uncommon.[52] A female professional buyer noted "three cases of allergy which a doctor had explained to her did not come from the yarn, but from the dyes."[53] Another woman asked if exposure to nylon could poison the bloodstream, echoing the skepticism others had about the material's chemical aspects.[54] Indeed, as dress historian Alison Matthews David has demonstrated, chemical dyes—such as nineteenth-century arsenic greens and aniline mauves—have a long history of toxicity and harm to the body.[55] Like nylon in the United States, aniline mauve had been celebrated in Europe as a synthetic alternative to expensive, imported natural dyestuffs and pigments. Careful consideration of safety was paramount to nylon's success.

Chemical dyes used for nylon stockings could irritate the skin, but so, too, could nylon. In their work on nylon, DuPont's research and development and DPAD staff needed to factor in the body. As part of its research and development, DuPont tested nylon yarns, hosiery, underwear, and other apparel before launching it on the commercial market. Indeed, as noted in a DuPont promotional booklet on health and safety, "Any chemical composition or chemically-made fiber used in intimate contact with delicate skin becomes a specialized and challenging problem."[56] Reflecting on the importance of testing new materials like nylon, the company continued, "Human skin breathes and perspires, chafes and itches. It is subject to allergies and peculiar sensitivities." The surface of the body itself is a living, breathing element, reacting to whatever it encounters. Reflecting on nylon's 1939 launch, DuPont claimed, "Before a strand of nylon could touch a single square millimeter of so fragile an area . . . it was put through a complete program of testing." Initially, patches were applied to laboratory animals; once these results were "prove[n] harmless," nylon was tested on "a selected group of volunteers, to whom small pieces were applied to arms and legs to see if irritation would result."[57]

Once hundreds of tests had been successfully completed without causing blemished skin, nylon clothing could be wear-tested. The next round of tests focused on sizing and dyes, since earlier cases of dermatitis were linked to "finishes that proved to be irritating." DuPont was wary of nylon's various finishes, both dermatologically and aesthetically, understanding the importance of perfecting these to better compete with rayon, which had high-shine finishes and lacked elasticity, leaving American consumers dissatisfied. The company also understood the importance of nylon's intimate relationship to the self-fashioned body: "In effect, nothing is left to chance to make sure that nylon is something which, with safety as well as pleasure[,] the skin loves to touch."[58]

The safety of nylon was an ongoing concern for some women attending DuPont's 1939 world's fair displays. Despite DuPont's efforts to downplay its role as an explosives manufacturer, the Nye committee's charges clearly remained on some visitors' minds. Another female visitor, apparently uneasy about the material's origins as an invention by a munitions manufacturer, was worried that nylons could explode.[59] She connected nylon's materiality to DuPont's role as a munitions manufacturer and questioned its safety, given that it was produced by the same company. Lecturers responded by gendering and ridiculing her inquiry. However, her question about DuPont's involvement with munitions were echoed by other visitors, who may have been wary of this latest effort by a chemical company to spin "artificial silk," considering the lethal failures of earlier attempts.

In my close examination of these reports, I want to slow our reading: It is important to attend more closely to these dismissed voices, as they reveal an initial resistance to and skepticism of new synthetic materials, which is often glossed over in the historiography of nylon. This is particularly helpful for understanding the research and development of plastics within complex power structures and charting people's changing attitudes toward synthetics. Furthermore, it is important to contextualize these concerns within a longer history of experimentation and failure in artificial materials research and development.

It is crucial to stress that nylon wasn't the first "artificial silk." The quest for silk replacements and the production of semisynthetic fibers has a longer history, which scholars have traced to eighteenth-century chemists.[60] However, nylon was more popular than its predecessors as a material to replicate silk stockings, including Chardonnet silk, a type of cellulose-based silk from France that began commercial production in 1891.[61] Complaints about Chardonnet silk included that "it was glossier than silk but had an undesirable metallic lustre that made it look cheap"; it wasn't as soft, light, or elastic as silk; it was challenging to dye; it was "sensitive to moisture and couldn't be washed"; and it didn't keep you warm. Moreover, the components were "too expensive and remained highly flammable."[62] The real turning point for artificial silks came in the early twentieth century, when the British textiles company Courtaulds purchased patents for the "viscose" chemical process of producing "artificial silk yarn" from wood pulp, later known internationally as rayon. Handley notes that "in a global sense [Courtaulds' patent acquisition] marked the real launch of the man-made fiber industry."[63] Indeed, it quickly proved superior to other methods, including Chardonnet's, and remedied earlier problems, such as dye adherence, thereby securing Courtaulds' position as "the pioneer of the world's most successful rayon business."[64] By buying out viscose patent-holding firms in the United States, Canada, France, and Germany, Courtaulds sought to further secure its leading status.[65] However, DuPont, a powerful rayon-making enterprise, remained Courtaulds' main competitor in the United States.[66]

Rayon eventually dominated the market, replacing silk in many textile applications. By 1936, 85 percent of dresses purchased in the United States were made from rayon.[67] However, rayon was deemed too shiny, not stretchy enough, and inadequately sheer for the American stocking market, which continued to be dominated by Japanese silk hosiery.[68] Furthermore, wearers of rayon stockings complained that they bagged at the knees and ankles.[69] Nylon's soft materiality and light elasticity would remedy many of these complaints. Unlike rayon, a semisynthetic cellulose fiber sourced from living plants, nylon was a "true synthetic"—a product of the petrochemical industry, rather than from a natural source.[70] The coal tar derivative ingredients required to produce nylon were similar to those needed in the manufacture of dyes and explosives, giving DuPont, an established explosives manufacturer, a unique competitive edge.

However, DuPont was less keen on promoting this advantage, and the DPAD worked hard to distract from its explosives history. In July 1939, lecturers noted, "There have been few questions about DuPont and munitions. This week, however, a question about how so many different products would spring from the manufacture of powder was asked."[71] DuPont wanted to display the variety of products it offered, extolling its motto "Better Things for Better Living . . . Through Chemistry." However, in the aftermath of the Nye hearings, the company did not go into detail about how these new and exciting synthetic materials—though they set DuPont apart from competitors—could arguably be linked to its role as a munitions manufacturer.

For many members of the public, mystery and confusion surrounded the production of nylon, a new type of material that appeared to be magically conjured by an explosives company from natural elements. A female visitor was suspicious that nylon stockings were derived from coal, air, and "acids taken from corpses," claiming she knew the person who harvested the macabre ingredient.[72] The experience of wearing nylon remained a mystery for most until production increased. Meanwhile, articles circulated about nylons melting in bus exhaust, a concern also reflected in some comments to the lecturers.[73]

Women's concerns and experiences, largely recorded by male lecturers, often reveal a misogynist bias and belittling of their embodied experiences. When a female buyer shared that "several women found [nylons] to be cold and were afraid they would be too cold for winter wear," a lecturer dismissed her experiences: "We decided that was due to imagination and could easily be explained away. Women always have to have something to complain about anyway, and the minute they find out something is made from coal, air, and water they think there must be a draught somewhere."[74] In keeping with dominant gender roles of 1930s America, male lecturers appear to have considered women visitors unfit to understand the complexity and "superiority" of science, dismissing their corporeal interactions with the material. Instead, the mind and "logic" were privileged over the body, matter, and embodied experience.

Nonetheless, touch played a central role in the public's acceptance of nylon. Visitors to DuPont's displays at the 1939 world's fairs were likely to have heard of nylon and its potential in hosiery, as this had been announced in October 1938. However, the public had not yet been able to touch or see nylon in person, insofar as they had learned of it through the press or DuPont's sponsored *Cavalcade of America* radio show. Journalists circulated stories of nylons being "run-proof" and "strong as steel, yet fine as a spider's web," adding to the hype of an invincible, supernatural material.[75] But DuPont did not seem to have fully anticipated nylon's popularity at the exhibits or women's requests for a closer look at the material. In June 1939, a few months after opening in New York, nylon's formal representation at the display remained limited. Women seeking out nylons were often told to return to the information desk, as DuPont's receptionist was wearing nylons. Hoping to improve this "inadequate" experience, lecturers stressed the importance of supplying free nylon hose to female exhibit staff, who wore these products on the exhibit floor: "Showing the nylon hose is an important part of their work as many came to this building specifically to see nylon."[76]

DuPont had previously relied on mannequins to model fashions—sometimes in miniature form, as seen at the New York Science Display of 1937 (figure 1.1)—and, as early photographs reveal, initially at the 1939 Golden Gate International Exposition.[77] In keeping with 1930s corporate

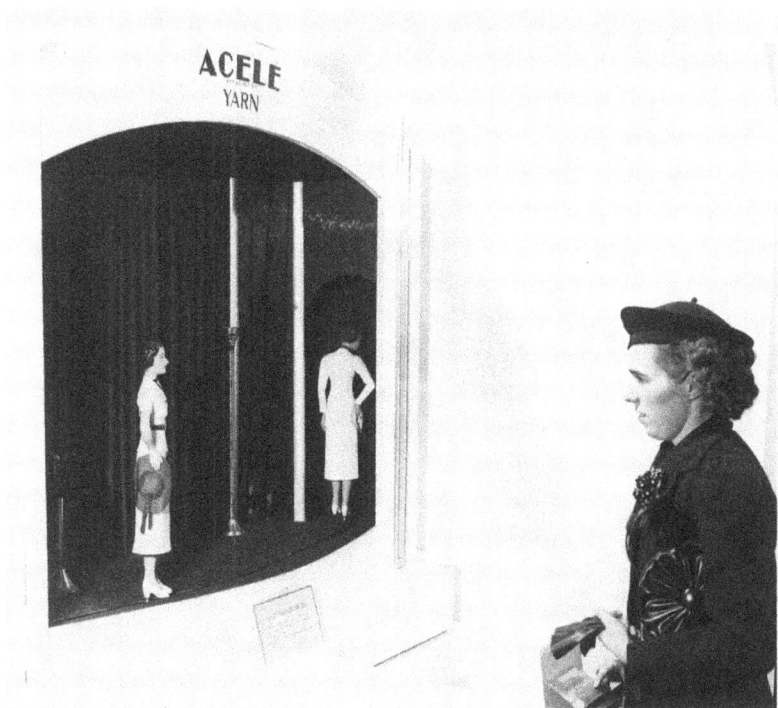

1.1 Mannequins at the DuPont exhibition, New York Museum of Science and Industry, 1937. Hagley Museum and Library, Wilmington, DE.

showmanship, these mechanical displays were activated with a push button. This device limited physical interaction to a preset mechanical response. However, the "army of behavioral psychologists, sociologists, and advertising experts" employed to study visitor engagement at science displays considered it a highly effective tool for audience engagement, particularly designed to appeal to an ideal visitor described as a "young white male."[78]

Instead of designing displays that allowed visitors to touch nylon stockings, DuPont had originally installed a display featuring a mechanical pair of hands endlessly tugging at a pair of nylons back and forth, presumably to demonstrate their longevity and ability to hold their shape (figure 1.2). Echoing earlier displays, the focus here was on a corporate spectacle of "scientific" mechanical testing of materials, rather than on interactive touch from the public. By this disembodied logic, the machine replica of a human hand could determine the reliability of the material

1.2 DuPont's mechanical hands-on display (*see background*) at the 1939 San Francisco Golden Gate International Exposition. Hagley Museum and Library, Wilmington, DE.

better than the human hand itself could. In this hierarchy, the company's equipment and scientific expertise are prioritized over the agency of consumers to decide for themselves. The promotional photograph shows two white women tugging at a stocking in front of the display, perhaps in response to visitors' complaints of not being able to handle nylon stockings with their own hands. By creating a photograph that used live models to demonstrate nylon, beyond the Lucite-enclosed display behind them, the DPAD softened its image while also making nylon more tangible and real to the viewer.

A few months after opening, DuPont responded to the public's countless requests for nylon stockings at their Wonder World of Tomorrow displays. At the New York World's Fair, Katherine Mitton, a white woman originally employed by DuPont as an exhibition receptionist, was soon promoted to nylon hose demonstrator and model.[79] This suggests that the DPAD incorporated feedback from female visitors who wanted to see nylons on the body and reacted by spontaneously changing Mitton's role from receptionist to a model in a live nylon demonstration. This strategic deployment of the white female body also played a major role in assuaging concerns about this invention and its proximity to the body.

Members of the public were eager to obtain their own pair of stockings to experience what they were like. Reports also show that they enjoyed speaking with DuPont representatives, like Miss Mitton, who were already wearing them. Mitton noted in her reports, "The chief questions on the stockings are still 'when will they be on the market?' and 'how much will they be?'"[80] Some women reportedly asked if they could touch her leg to feel what nylon was like on the skin. Others were keen to quiz her on her experiences of wearing nylons—What do they feel like? Are they less prone to runs? Do they keep you warm?[81] No longer something purely scientific and intangible conjured from coal, air, and tar, nylon could now be touched. Unlike DuPont's earlier mechanical mannequins, live white female models now wore and embodied nylon's silky-soft texture, with exhibit staff serving as brand ambassadors.

Lecturers consistently noted a high interest in nylon's materiality. Visitors had *heard* about nylon in the run-up to its presentation at the 1939 fairs. Some had even *seen* it depicted in imagery promoting the sensuality of its touch, such as in DuPont's film *A New World Through Chemistry* (1939).[82] Visitors now wanted to see and feel nylon for themselves. Touch was central to DuPont's wear-tests and consumer marketing studies that tried to better understand how to quantify and harness its power for profit. As shown in figure 1.2, the DPAD also understood the underlying sensuality of touch—the affective way it can tap into desire and the promise of pleasure—a key element of advertising.

Dissatisfaction at being unable to experience the material up close ran high. One lecturer noted, "The request that we turn down most often is 'can I feel that nylon stocking?' A close second is 'where can I see the stocking at close view?'"[83] Another voiced a frequent complaint: "Most people are disappointed in that they can't handle a nylon stocking."[84] Indeed, it must have been an anticlimax to visit the DuPont display—seeing nylons stretched endlessly by mechanical hands behind glass or worn by Mitton—yet be unable to touch the material or interact with it directly. How was one supposed to know whether they were "real" or not? Or how they compared with earlier offerings of "artificial silk"? Or what they felt like to touch—how they smelled, reacted to temperature and skin, their stretch and softness, or how they moved when handled? Touching nylons would enable both general and trade visitors to assess their quality. As business historian Regina Lee Blaszczyk notes, veteran textile salespeople in the 1930s "frequently referred to the 'hand' or feel of the fabric."[85] In this period, tacit corporeal knowledge of textiles and scrutiny of fabric bolts extended beyond fashion districts and textile mills, with many women making their own clothes at home.

Materiality and touch played a vital role in nylon's promotion and success, enabling visitors to evaluate this new material through haptic and instinctual knowledge, distinguishing between success or failure. Lecturer Herbert S. Chason recognized the potential publicity dangers of not allowing visitors to touch the stockings: "The obvious reaction to the frustration of not being able to see the stocking closer is the raising of doubt in the visitors' mind of the success of the stocking."[86] Another lecturer repeatedly stressed that "the word-of-mouth advertising we could secure by allowing people to feel the stocking would be incalculable."[87] If, as the company and press reported, nylons would be as "strong as steel" and "sheer as cobwebs," this could surely only be tested by touching and interacting with the material itself.[88] Chason noted that such doubt could easily be avoided by providing a sample of nylon cloth and a nylon stocking in a translucent Lucite box. Unlike nylon stockings, nylon cloth was less prone to being damaged by touch, so he suggested "the cloth would be removed so that people could feel it and the stocking could be kept in the box to prevent it from being pulled by over-anxious women."[89]

Both touch and "hysteria" over nylon are gendered in the reports. Lecturers and DPAD staff evidently feared that women were overzealous and emotional in their handling of nylons, which had very limited commercial production capacity at the time. Cultural historian of touch Constance Classen describes how touch is central to the way we experience the world around us, yet it has been largely neglected from scholarship.[90] Classen traces the omission of touch in academia to racist nineteenth-century tropes that claimed vision was the superior sense, while touch was racialized as an inferior modality for understanding the world. The privileging of sight and othering of touch is evident in DuPont's displays, which were carefully designed to welcome touch from certain visitors but to discourage it from others. Still, the body and touch played a vital role in the success of nylon; it was ultimately the embodied experiences of nylon's wearers that would determine whether it was truly a viable alternative to silk for stockings and therefore a commercial success.

In addition to failing to anticipate nylon's multifaceted and timely appeal, it is likely that DPAD staff were initially unable to supply sample stockings for handling or nylon samples for visitors to take with them, because, when DuPont introduced nylons at the 1939 fairs, the company had only a limited amount available—its chemists and engineers were still working out production logistics. In 1939, DuPont was displaying sample batch runs of stockings. A lecturer noted, "Questions about nylon continue to outdraw all others," and enthusiastically exclaimed, "There is more interest in nylon than in any other product before being placed on the market." However, "present demand ought to require many months of plant operation before satisfied."[91] It is likely that the heightened interest in nylon at the 1939 fairs encouraged DuPont to concentrate more efforts on the material's R & D and production methods. As the 1939 exhibits continued, nylon's production moved from small-scale lab sample batches to larger-scale commercial production in 1940.

When DuPont first unveiled nylon stockings to the public at the 1939 fairs, the yarn used in these demonstration stockings came from experimental production runs. DuPont's exhibit evolved alongside the unfolding development of nylon's R & D and production. Eventually, nylon was presented from its "birth in the laboratory to its appearance in the form of dainty hosiery."[92] By April 1940, "machinery showing the steps in the manufacture of nylon stockings [was soon to be] assembled, actually knitting the hose on the exhibit floor."[93] On May 15, 1940, nylon was first made available to the general public—dubbed "N-Day"—when hosiery manufacturers offered four million pairs for simultaneous sale across the country. The supply of the most popular sizes was exhausted within four days.[94]

Nylon's popularity in stores reflects a shift in attitude toward the first fully synthetic fiber, compared to earlier concerns about its potential dangers; nylon became a commercial success. In early September 1940, a lecturer noticed a change from previous weeks; nylon feedback was now largely "based on actual experience with the stockings. It appears that most women have been able to secure and try out at least one pair of nylon hose."[95] Inquiries increasingly related to maintenance, such as "How should they be washed?" and concerns that nylon would be rationed: "Is the government going to take over the production of nylon?"[96] There were also complaints, including how hard it was to get them, that "they run very fast once started," "colors fade in time, usually after a month or two," sizing issues, that "they bag on thin ankles," that they "are too short," that the "heel [is] too low," and that "they are hard to repair."[97] Often, complaints corroborated one another and pointed to a flaw in the stockings' structure; for example, many women noted "a weakness of the topping corner of the heel." Those attempting to repair this new material were likely disappointed that what the press presented as utopian science could not prevent runs and snags from forming.

Positive nylon feedback included that "they last longer than silk stockings," "they hold their shape and go back when washed," "they wash easily and dry quickly," "they look more sheer," and "many women like the way they feel." Lecturers also noted a fairly recent phenomenon in which women wore nylons every day for periods of time ranging from two to three months, to "test" them, effectively carrying out their own independent wear-tests to determine the quality of this new product. Reports confirmed nylon's success: "The only complaints received from these women were that either the color had faded, or that they were 'sick' of looking at them."[98]

"When the Nylons Bloom Again": Wartime Use of Nylon

Following the US entry into World War II in 1942, DuPont's entire nylon output was allocated by the War Production Board for vital needs from February 1942 to August 1945. Additionally, unknown to the American public at that time, many of DuPont's top nylon engineers worked on the Manhattan Project.[99] Consumers were encouraged to donate used silk and nylon stockings for "Uncle Sam" to use for gun powder bags.[100] Here, nylon's primary application site shifted once more, this time from women's bodies to the US Armed Forces. This essentially returned nylon to its original military and industrial context, as a highly technological and scientific material developed by a company known for manufacturing explosives and supplying the US Army. Nylon's military and naval applications included tires for bombers, towropes for gliders, parachutes, Arctic tents, and jungle equipment. DuPont saw this as a marketing opportunity, enthusing, "[Nylon is] an all-American peacetime development—and a most timely one—now playing a vital role in winning a global war."[101] Nylon was presented as far from the home front, deployed in combat, just as US troops were absent.

In the wartime United States, nylon became a symbol of hope and aspiration for a technoscientific, synthetic utopian future free from war and rationing. Jazz musician and entertainer Fats Waller and screenwriter George Marion Jr. memorialized these sentiments in the aptly titled musical number "When the Nylons Bloom Again." Marion's nostalgic lyrics placed hope in a future of synthetic abundance—a harvest of desirable plastics that would never run dry. In 1943, when their musical comedy *Early to Bed* premiered, nylon stockings were increasingly hard to come by, regardless of socioeconomic status. The song's lyrics lamented, "Now poor or rich, we're enduring instead, woolens which itch, rayons that spread." Nylons were not "ersatz" replacements for natural materials like silk. Instead, natural materials and semisynthetic fibers were described as inferior—undesirable, uncomfortable, and unerotic wartime sacrifices that had to be endured.

At a time when shifting gender roles accelerated changes in women's dress, such as wearing pants and overalls for work, the song also articulated a desire to return to more feminine styles and gender norms: "I'll be happy when the nylons bloom again / Cotton is monotonous to men / . . . Keep on smiling 'til the nylons bloom again / And the WACs come back to join their men." Nylon's novelty and its gendered, lightweight, transparent materiality were contrastingly positioned as thrilling to touch: "Only way to keep affection fresh; get some mesh for your flesh." For many Americans, such dreams of silky nylons and a return to the excitement of "Better Things . . . for Better Living" offered a welcome escape from war.

In the early 1940s, purchasing nylons was possible only for those who could afford it and were prepared to shop outside formal economies, where they sold for over ten dollars a pair—around $200 in today's currency.[102] Some women who could not afford stockings used a trompe l'oeil approach, painting seams on their legs and coating them in "liquid silk" makeup to create a stocking-like illusion. Memphis-based Keystone Cosmetics was one company that offered leg makeup at this time. It was founded in 1923 by Morris Shapiro and Joseph Menke, two white Jewish chemists who developed and sold many products aimed at an African American clientele, including "Hi-Brown Liquid Face Powder and Leg Make-Up" (figure 1.3). Elaine Cyrus wrote in her column "Facts About Fashion" for the African American weekly newspaper *Pittsburgh Courier* that "leg make-up has been trying to take the place of stockings. Baggy, rayon stockings have done much to skyrocket the liquid to its present acceptance." However, she speculated that this trend would be short-lived: "When post-war stockings are available leg make-up will most likely lose its appeal . . . some will start fighting their way back to store counters for sheer durable stockings." Cyrus hoped this day wasn't too far off, as she felt that "most legs do not look well without stockings. They need to

1.3 La Jac Hi-Brown Liquid Face Powder and Leg Make-Up, 1923–49.
Science History Institute, Philadelphia.

be under cover."[103] Indeed, nylon stockings were considered valuable, and some of the remaining pairs from peacetime production were even sold in bond rallies. Movie stars like Betty Grable, known for her legs, auctioned off her worn stockings, garnering bids as high as $40,000.[104] Meanwhile, nylon-stockinged pin-ups reportedly boosted morale among US troops.[105]

On the home front, promises of new resistant material were aimed at lifting the spirits of female consumers. A DuPont booklet from 1942 focused on the soft yet indestructible nature of nylon, which provided an antidote to the destruction of war and ravages of nature, promising that this was within reach: "Post-war nylon: this spun nylon sweater: soft and warm as fine wool, has the advantage of not shrinking when laundered. . . .

Introduced on a small scale before the war, sweaters, socks, undergarments, fleece coats and other woven and knitted wool-like garments will return after Victory. Nylon is resistant to attack by moths in that moth larvae do not feed on nylon itself."[106] In this wartime rhetoric, nylon's novel synthetic materiality would return to improve postwar domestic life.

Foundationwear companies also promoted the promise of future innovative materials, which had to be temporarily sacrificed as part of the war effort.[107] Tomorrow's peacetime promise of a perfect female body wrapped in an abundance of chemically conjured synthetics was often stressed for its transformative and restorative properties. One advertisement reassured women on the home front that "miraculous fabrics [are] doing a wartime job today, but earmarked to glamorize you when the war is won. Then as always, you'll be a vision of loveliness in your foundations by American Lady."[108] Stretchy synthetic fibers were promoted to an assumed heterosexual female consumer eagerly awaiting the homecoming of male troops as having the advantage of stopping the clock or restoring "youth." Such promises of a return to "normalcy" and domestic bliss were frequent; provocative images of war, including soldiers in action, were often juxtaposed with images of white domesticity (see, for example, figure 1.4). Textron promised consumers a technocultural future of self-fashioning and corporeal control: "Soon you may wear the Textron gossamer that rides the skies in parachutes today . . . soon you may mold your slimness in the delicate-strong fibers that are twisted into tow ropes for gliders!"[109] Here, postwar racialized, gendered bodies are shaped in the same "miracle fabrics" that served to secure the US military victory. Although these are representations, the proximity of military technologies to the unclothed white female body—and their role in shaping it—is striking. Such advertising promoted synthetics as having the power to protect an ideal white American femininity, transforming today's stresses of war into tomorrow's promises of peace and prosperity. Like the biopolitical shift in gender roles during wartime, synthetics could be transformed from the domestic to the military and back again.

After World War II, nylon promotion became associated with a material measure of peacetime prosperity, in contrast with earlier efforts that used a gendered appeal to promote wartime sacrifice. In August 1945, just one week after the war ended, DuPont announced that its production was to change again, this time to hosiery yarns. DuPont forecast that there would be "enough nylon manufactured to knit 360,000,000 pairs of nylon stockings per year, the equivalent of 11 pairs for every American woman."[110]

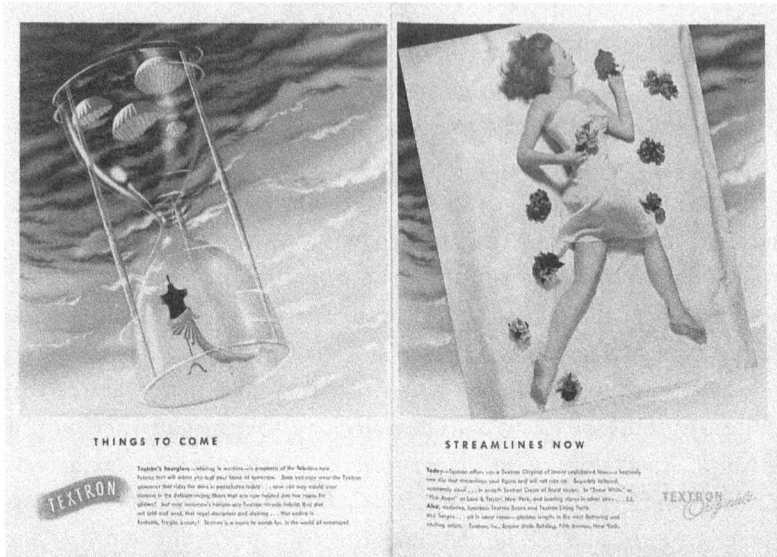

THINGS TO COME

STREAMLINES NOW

1.4 Textron, "Things to Come, Streamlines Now," advertisement, *Vogue*, November 1, 1943, 8–9.

DuPont's reconfiguring of its factories to produce yarn for stockings rather than parachutes was a production feat more easily achieved than, for example, regearing factories to produce white goods, and it further secured the company's image as a provider of "Better Things for Better Living."

In the postwar US, nylons continued to generate buzz in popular culture and the media. Banking on this enduring appeal, in 1945, the Nylon Girls burlesque group advertised their touring show in an African American newspaper, instructing women to bring their husband to a "zippy pageant of sexotic [*sic*] sirens starring Dian Mason New Goddess of Strip Tease."[111] In early 1946, newspapers covered the so-called nylon riots that erupted when the first postwar deliveries appeared in shops.[112] Door-to-door salespeople offered an alternative way to shop for nylons, often to marginalized communities. *The Pittsburgh Courier* featured updates on "The Nylon Man," a popular figure who sold nylons directly to local Black households.[113] By the end of 1946, nylon production had stabilized, and journalists reported that nylon queues were shrinking or even disappearing in some cities.[114] Nylon was in demand and continued to make headlines.

DuPont's prewar discovery of the world's first fully synthetic fiber proved revolutionary. In the postwar United States, having passed the touch test and now freed from military restrictions, intimate "nylons" became commonplace, replacing silk—a position that remains unchanged to this day. Nylon had come a long way from the initial skepticism voiced by cautious members of the public, who were troubled by earlier lethal attempts at "artificial silk," its connections to a leading explosives company, and its suspicious "conjuring" from tar derivatives. The popularity of nylon's soft, sensual materiality in the form of women's hosiery paved the way for the successful domestication and development of many other synthetic materials that revolutionized dress and the fashioned body. Synthetics—once feared by some—gradually entered the domestic sphere, intimately coating the environment, daily interactions, and bodies with materials derived from petrochemical industries. In the postwar period, DuPont began work on the next "miracle" materials. Its labs were abuzz with chemists working on "Orlon" acrylic, "Dacron" polyester, and eventually "Lycra" spandex fiber.[115] New materials produced by an explosives company once called "merchants of death" were generally accepted and less feared, as evidenced by the postwar success of N-Day sales.

2 NYLON AND THE TEST TUBE GIRL / RACIALIZED GENDER NORMS AND PLASTIC FUTURES

In the November 1939 issue of *National Geographic*, assistant editor Frederick Simpich, reporting on DuPont's display at the New York World's Fair, marveled, "At a fashion show, we saw a girl clad from head to foot in artificial materials. Everything she wore was made from synthetic stuffs created by chemists. . . . Only the girl herself was natural—natural flesh and bone wrapped in her own waterproof skin. There she stood, a startling symbol of this new artificial world risen so fast since the [First] World War."[1]

For modern readers familiar with "Space Age" fashion, she hardly looks futuristic but rather resembles a fashionable woman of the late 1930s. Crowned "A Living Symbol of the Chemical Age" (figure 2.1), this white figure, from the tip of her cellophane hat to the tops of her nylon-encased varnished toes, became known as "Miss Chemistry" and the "Test Tube Girl." Dressed in synthetics from head to toe, she embodied the latest scientific developments, including nylon, and welcomed a plastics future.

Nylon, the world's first fully synthetic fiber, was presented in 1939 on the white female figure, highlighting changing attitudes toward synthetic materials and their proximity to the body. In this chapter, I analyze the image of the Test Tube Girl, exploring the racialized and gendered aspects

2.1 "A Living Symbol of the Chemical Age," *National Geographic*, November 1939, 609.

of this performance, as well as related visual and material culture. I chart how she became a "living symbol" of a new age of designing an American fashionable female ideal with science and plastics, both on the skin's surface and beneath it. While Miss Chemistry is one image created by the DuPont Advertising Department for an exhibit, it is indicative of how changing material technologies were promoted and domesticated in relation to the gendered and racialized body, as represented by her and other figures like her—largely envisaged by powerful white men, including journalists, chemists, filmmakers, designers, and advertisers. At the 1939 world fairs and in accompanying imagery, DuPont presented a binary vision in which chemists (male, mind) improved nature (female, body). In depictions like the Test Tube Girl, "the future is plastic"—but apparently only for a select type of white American woman, whom the company considered to be malleable.

Unlike the foundational scholarship on nylon and its presentation, I incorporate critical thinking on the body, race, and gender.[2] I offer an important new critical perspective on the Test Tube Girl by contextualizing her presentation within wider cultural discourse around racialized gender ideals within fashion, streamlined design, and eugenics. Race and gender were also encoded into experimental synthetic objects, such as light peach-colored experimental "wear-test" underwear. These images and objects collectively reinforced and embodied white womanhood as the unmarked norm and ideal.

Drawing on original research using Black publications and trans feminine media, I also show how some individuals and companies actively worked against these biases and binaries in the ensuing decades. These actors developed, designed, produced, and marketed nylons specifically for a wider range of consumers. Their approach to design and marketing questioned unmarked norms of a hegemonic American womanhood. These examples offer a vital counterhistory to nylon and the Test Tube Girl's "normalization of bodies" and mass production that mobilized a standard ideal of womanhood as white, thin, heterosexual, and cisgender. Instead, companies, including Basin Street Nylons and Tana & Mara, envisioned a future of nylons designed to cater to a wider range of consumers beyond such restrictive norms.

Staging Chemistry and Mapping "Wonder Worlds": Race, Gender, and Colonialism

In 1939, DuPont presented its "Wonder World of Chemistry" exhibits at two US world fairs on the East and West Coasts: the Golden Gate International Exposition in San Francisco and the New York World's Fair. In February, the company introduced its nylon stockings to the public at the San Francisco exposition, where its display attracted over two million visitors.[3] DuPont invested particularly heavily in its sprawling New York exhibit, commissioning renowned American industrial designer Walter Dorwin Teague and his team to design the pavilion. Designed to hold over fifty thousand visitors at a time, the pavilion's surface area spanned over 36,222 square feet and cost over $500,000.[4] DuPont recorded a total of 9,734,408 visitors between its opening on April 30 and October 31, 1939, and from May 11 to October 27, 1940.[5] As noted in the previous chapter, these visitor numbers aren't broken down by demographics. However, it is important to note that most visitors were white.[6] Visitors to DuPont's display were welcomed by what the press release called a "100-foot tower typifying the spirit of chemical research," referencing abstract chemistry apparatuses and a neon "Wonder World of Chemistry" sign.[7] It featured "use of color, active liquids in circulation and changing lights to achieve the spectacular effect sought."[8] The massive scale of these attractions and the funding they demanded demonstrates DuPont's dedication to its public-facing image.

Once inside, visitors saw a mural measuring over thirty feet high, displaying chemistry's transformative powers and anthropomorphizing the discipline into a white man (figure 2.2). Light radiates from this central muscular figure, inspired by classical sculpture. A 3D figure modeled in shiny aluminum, "Chemistry" is centered and appears ethereal; standing out from his two-dimensional surroundings, he is fabricated from a different material altogether. Crowned with a laurel wreath, "Chemistry" references Apollo, one of the twelve Olympian gods, the classical Greek and Roman deity of the sun, the god of prophecy and divination, associated with oracles, knowledge, healing, care of herds and flocks, and the protection of the young.[9] DuPont spelled out its intentions as leading contributors to American society in the mural's banner: "BETTER THINGS FOR BETTER LIVING THROUGH CHEMISTRY." Accompanying imagery emphasizes a narrative of scientific progress. On the left, a rural, mountainous, green environment—home to a white family, hunched, barefoot, in drab clothing while offering natural raw materials—is transformed on the right into a thriving group in clean, modern outfits, standing tall and looking toward the future, with an industrial backdrop looming behind them. The message is clear: DuPont's goal is to transform nature and improve families' everyday experiences by removing their heavy burdens and making life brighter, cleaner, and easier. A press release declared that "the figures are painted on a heroic scale" and celebrate "the significance of

2.2 John W. McCoy, *Better Things for Better Living... Through Chemistry*, mural, 1939. Hagley Museum and Library, Wilmington, DE.

the chemist in industry, transportation, agriculture, mining and other fields."[10] Here, the focus is on what chemistry can do to improve the lives of white families through raw materials, rather than specifically on transformation of women.

Plastics were an important part of DuPont's corporate vision. Synthetics presented at its 1939 exhibits promised a utopian future filled with an abundance of exciting, shiny new goods. In the context of the recent end of the Great Depression and the outbreak of World War II in Europe, DuPont promised the American public that a brighter future, featuring these products, was on the horizon, offering a welcome escape in such dark times. Shaped by the company's technocultural narrative, an advertisement announced for "the world of tomorrow... a new world and a new material: NYLON" (figure 2.3). Published in the *New York Herald Tribune*'s "Woman's Forum," it courted a high-income white female readership, inviting them to see nylon at DuPont's world's fair display and join "Mr. Chemist" on a journey to an uncharted future realm. DuPont is represented as the chemist, a mythical patriarchal figure with the power to safely navigate new territories and protect his wards. This depiction

2.3 DuPont advertisement for the 1939 New York World's Fair, *New York Herald Tribune*, October 30, 1939. Science History Institute, Philadelphia.

strategically aimed to appeal to the racialized heteronormative family, thereby humanizing DuPont's image as more than that of a munitions and chemicals manufacturer.

Colonial comparisons between plastics and an uncharted or "untouched" paradise, such as DuPont's 1939 nylon announcement, were frequent in US petrochemical companies' promotional material during this period and continued well into the postwar era.[11] The National Association of Manufacturers echoed this narrative, celebrating chemists as modern "pioneers" and plastics as a "frontier" where industry was "forging ahead through research and science" and offering a "dramatic challenge to those who claim there are no longer any frontiers for Americans."[12]

American whitestream publications also promoted plastics using a similar rhetoric of colonization and the Old West. An example of this is *Fortune* magazine's frequently referenced October 1940 issue, featuring nylon and countless other synthetic materials plotted on a map called "Synthetica: A New Continent of Plastics" (figure 2.4).[13] Historian Jeffrey Meikle notes that "the map of Synthetica firmly rooted plastic in the extractive materials culture of the past."[14] The cartographic approach is different from one focusing on the democratic potential of "organic chemistry's promise to free the human race from geographic accidents of scarcity and supply."[15]

From advertising copy to illustrations, synthetics—still a fairly recent development—were frequently imagined through long-established dualist Western colonial and patriarchal tropes of extractivism and domination over the unknown "other." World War II increasingly clarified the limits of colonial economic systems, fueling the United States' and Germany's desire for a material autarky through synthetic replacements or "ersatz" materials that did not require imported raw materials. The outline of Synthetica's continent appears to reference that of South America and its colonial exploitation of materials such as "ivory, rubber, gutta-percha, coal and other natural resins and substances."[16] In this narrative, plastics—just like natural resources, foreign lands, and othered individuals—could be extracted, ruled over, and transformed through chemistry, creating "Better Things for Better Living . . ." This petrochemical imaginary targeted a predominantly white audience and perpetuated a dualist, technocultural utopian vision of progress through science and domination of nature.

The DuPont Advertising Department understood the importance of appealing to the American whitestream public's politics, imagination, and emotions, promoting its offerings through nationalist sentiment. *A*

2.4 "Synthetica," map by Ortho Plastic Novelties Inc. in *Fortune*, October 1940, 92–93.

New World Through Chemistry (1939), a promotional film produced by DuPont to coincide with its exhibits, compares the company's chemical advancements with those of Europe. This film, available on loan from the DuPont Motion Picture Bureau, was in "active demand by service clubs, educational institutions, church organizations and other groups."[17] It celebrated the idea of American exceptionalism and the company's role in contributing to the country's world-leading position in plastics, boasting, "[Synthetics] made of coal, air, water well mixed with brain. The brains of American scientists who in the brief period between 1914 and the present have brought America scientific achievements from second place, failing Europe, to first place, leading the world. A position it is not likely to soon if ever to relinquish."[18] In DuPont's reordering of the world, "American brains" were framed as a key ingredient in transforming natural matter into nonnatural matter.

As Synthetica illustrates, plastics were a burgeoning field of scientific and industrial research. DuPont's expansion into this realm provided the company with an opportunity to bring its R & D from the lab to the public, which the DPAD and the industry experts it hired believed to be key to the company's commercial success. Teague's pitch for DuPont stressed increasing scientific competition and the need to "[take] the public behind the scenes" of the company's activities. He felt it pertinent to "acquaint the 'man in the street' with [DuPont's] startling contributions to human progress and welfare through chemistry."[19] Displays were designed to be increasingly interactive, including the staged re-creation of a chemical laboratory. Here, "chemists" in lab coats performed experiments and explained how products commonly used in factories, shops, and homes are derived from raw materials and refined.

This was an age of corporate showmanship. Major American businesses, particularly those with interests in antiadministration efforts such as DuPont and General Motors, invested heavily in promoting to the American public their social roles and entrepreneurial achievements in science and industry.[20] The DPAD employed industry leaders of American advertising (Batten, Barton, Durstine & Osborn) and design (Teague), with whom it carefully crafted a "Wonder World of Chemistry" for the public. At the heart of the New York display, Teague positioned the aforementioned laboratory demonstration stage. Raised above the audience on a platform, amid displays of natural resources and the workers who sourced them, "chemists" performed live experiments with different materials. These individuals were mostly white male college students, recruited for the summer. They were trained to give information and record notes on visitor engagement in daily reports shared with the DPAD. Referred to as "lecturers" by the DPAD, they were chosen for their appearance and ability to attract audiences.[21] They had no prior knowledge of DuPont's activity, relying instead on scripts provided by the DPAD. Nevertheless, the show was deemed a success, and on weekends, visitor figures reached over one hundred thousand, with a total of eighty-six staff employed at its peak.[22] The DPAD directed DuPont's largest and most ambitious exhibit, presenting a carefully staged vision of chemistry as a transformative force in everyday life.

Nylon, Miss Mitton, and Gender

DPAD staff responded to lecturers' reports that routinely noted nylon as the most requested product. Original floor plans and reports show that the DPAD had not anticipated nylon's popularity in its exhibit design. However, seeking to maximize the company's public appeal, the DPAD actively

2.5 Miss Chemistry and nylon stockings, 1939. Hagley Museum and Library, Wilmington, DE.

responded to visitor feedback, making changes throughout the display's duration. This included adding a live nylon model, a white woman, whose first recorded performance took place in June 1939. Originally hired as DuPont's receptionist for the New York World's Fair, Katherine Mitton was reinvented as "Miss Chemistry." Her first performance was described in a weekly report: "On the research section, a very successful addition has been made. Miss Katherine Mitton takes part in the nylon demonstration, on a special platform built for the purpose, seven feet above the floor. As Miss Mitton steps up on to the platform, the lecturer calls attention to the fact that she is 'dressed by chemistry from head to toe.'"[23]

Mitton embodied the white "girl next door" and effectively humanized DuPont's products (figure 2.5). At the end of each performance, she would descend from the platform to join the public and answer questions, many of which were posed by women and related to her experience wearing nylons. In this early iteration of the Miss Chemistry show, Mitton served as a mediator between the lab and the gendered dualist power relations that shaped it.

The show generated increased interest in nylon, and Mitton became known as the "Lady of Chemistry." Journalists, DPAD staff, DuPont Style News Service staff, and lecturers interchangeably referred to Mitton and the figure she represented as "Miss Chemistry." This name references beauty culture and "Miss" beauty pageants dedicated to products such as Miss Cotton and Miss Corn. It also connotes the racialized beauty ideals many pageants promoted, such as Miss America, which until 1970 was only open to contestants "of good health and the white race."[24] DuPont's other names for Miss Chemistry included "Princess Plastics," a young heiress to a future of plastic possibilities, and the "Test Tube Girl," further emphasizing her links to scientific processes and technology.[25]

The test tube became a more prominent feature of Miss Chemistry's staging (figure 2.6). In response to Miss Chemistry's popularity, DuPont transformed her presentation into a multisensory musical spectacle. In this increasingly dramatic version, Miss Chemistry stepped out from a giant lit-up plastic test tube; the DPAD instructed her to "parade on top of the demonstration counters, engaging in repartee with the chemist to bring out the chemical origin of her nylon hose, rayon frock, Lucite jewelry and so on through her costume."[26]

By 1940, Teague had become increasingly ambitious with his designs, and he established a new "Industrial Showmanship" department within his firm dedicated to producing commercial spectacles like Miss Chemistry.[27] Visitors to DuPont's constantly updated display were presented with a sleek, celebratory vision of how the company created its products. Unpleasant chemical smells were removed from lab shows.[28] With the lab's messy sensorial complexities neatly edited out, DuPont's R & D was, according to Teague, "staged with care and imagination that should make the best of Hollywood respectful."[29]

The star of the show, Miss Chemistry, was to be presented as one of the company's many products for better living. She was strategically positioned with other "products" at the end of the manufacturing section. Teague instructed that she be fully clothed in synthetics and "made to appear from an apparently empty test tube of 'Lucite' plastic, as if she herself, in addition to her costume and accessories, had been created by chemistry."[30] A press release from a month before the end of the New York fair, apparently inspired by Teague's instructions, remarked, "Where else but the DuPont exhibit could you see a live girl apparently 'created' inside

2.6 Miss Chemistry steps out from the test tube, 1939. Hagley Museum and Library, Wilmington, DE.

a giant transparent test tube?"[31] The DPAD's vision of a white woman in a test tube, dressed from head to toe in synthetics, remained a popular trope for the company and others.[32] On June 15, 1940, almost a year after Mitton first appeared on stage in the "lab," the DuPont Style News Service sent a press release titled "Test Tube Fashions," featuring imagery of a white woman in a test tube, to 568 newspapers.[33]

In this technocultural discourse, boundaries between fashions made from the latest advancements in synthetic fibers and the racialized, gendered body became blurred. In DuPont's promotional materials, Miss Chemistry's physical form was increasingly represented as the creation of a white male chemist (figure 2.7). The accompanying text reads, "During the summer of 1940 at the New York World's Fair . . . DuPont introduced to the public a 'Miss Chemistry of the Future' dressed from head to toe in . . . articles made with nylon yarn. . . . Her underthings and foundation garments all were made with the new yarn. . . . Nothing was said about 'her operation' but she could well have completed the picture with an incision sewn up with nylon surgical suture."[34]

2.7 DuPont, "Nylon: Versatile Product of DuPont Chemistry," booklet, 1941. Hagley Museum and Library, Wilmington, DE.

Now that Miss Chemistry was dressed from head to toe in nylon's fully synthetic fibers, the discourse surrounding her began to more intimately incorporate her body. Unlike earlier versions, which included semisynthetic apparel such as a rayon dress, this Miss Chemistry was dressed almost entirely in nylon, including satin shoes, stockings, and a blue lace dress. Beyond what the audience could see, her undergarments were also made from nylon. DuPont's advertising writers additionally hinted that she may have even had hidden incisions stitched with nylon sutures, drawing attention to a lesser-known nylon application featured in the display.[35] The reference to sutures is opaque, leaving it unclear whether her operation was for medical treatment or cosmetic purposes. Perhaps its vagueness indicates the latter. This insinuation stands in contrast to descriptions of Miss Chemistry's previous iterations, such as the remarks made in November 1939 by journalist Frederick Simpich, quoted at the beginning of this chapter. He marveled at how "only the girl herself was natural—natural flesh and bone wrapped in her own waterproof skin. There she stood, a startling symbol of this new artificial world risen so fast since the [First] World War."[36]

2.8 Cover of *Amazing Stories*, January 1942, featuring "The Test Tube Girl."

By the summer of 1940, Miss Chemistry had become a figure whose entire body was shaped by advancements in plastics, science, technology, and medicine. Her body was increasingly becoming a site of transformation—no longer on the skin's surface, but also beneath it. One of the final scripts accompanying the Miss Chemistry show instructed the male chemist to exclaim, "Is there a new beauty distilled from the vessels and test tubes of chemistry?" further implying that a female body complete with clothing and accessories could be made using science.[37] The vision of Miss Chemistry as "fully synthetic" "distilled beauty" from the test tube fits into the wider contemporary discourse on building, designing, and creating racialized, gendered bodies.

In January 1942, a story titled "The Test Tube Girl" was featured on the cover of *Amazing Stories*, an American science fiction magazine. The image depicted a slender, green female figure suspended in a tube, with white men dressed in futuristic suits gathered around her amidst a lab full of "dials and meters and controls; bubbling liquids in crystal globes and tubes" (figure 2.8).[38] The story's setting, Eugenic Laboratories, is not dissimilar to the aesthetics of DuPont's Wonder World of Chemistry, with its one-hundred-foot-tall lit-up tower of bubbling liquids that welcomed visitors outside, and the giant test tube from which Miss Chemistry emerged. In the dystopian tale of "The Test Tube Girl," "The world was dying, slain by an invisible, silent ray of sterility [shot by Hitler] that had smitten womankind, destroyed her miraculous power to reproduce her race, to perpetuate it."[39] However, straight white men in lab coats who work for the Dowling Institute of Military Technology—potentially an allusion to the Dow Chemical Company—are able to save humankind using science.

The Test Tube Girl and Eugenics

The Test Tube Girl and the white female body as a malleable material of transformation are echoed by "Fashions of the Future." American *Vogue* commissioned Teague and eight other white male industrial designers for this editorial feature in its February 1939 issue that celebrated the New York World's Fair. The fashions also debuted at a New York World's Fair party, were displayed in windows of popular New York shops, and appeared in a *Vogue*-Pathé newsreel. Four of the nine designers went beyond designing clothing and accessories to comment on and construct future white women's bodies for the year 2000; their creations alluded to eugenics, as the designers fantasized about a future female race.

Influential streamline industrial designers Teague, Donald Deskey, and Raymond Loewy speculated that, in the year 2000, women's bodies would be "perfected" and "scientifically beautified" by supposed technological advancements, resulting in transparent clothing as the norm. Freed from restrictive clothing and underwear, Deskey's white models appear active, as if in the throes of a eurhythmic dance (figure 2.9). He claimed, "Medical science will have made her body perfect. She'll never know obesity, emaciation, colds in the head, superfluous hair, or a bad complexion—thanks to a controlled diet, controlled basal metabolism. Her height will be increased, her eyelashes lengthened—with some X-hormone."[40] Deskey reasoned that with the scientific perfection of women's bodies, foundationwear would become redundant: "She'll consider (as I do) corsets and brassieres as unattractive as surgical appliances, and underwear coy."[41]

Teague similarly "forecast near-nudity" (figure 2.10). He designed a revealing transparent robe. The white model's Lucite-heeled feet and other body parts are semi-exposed. Teague noted that in the future, "most women will have beautiful bodies, and the present trend toward nudity will continue at an accelerated pace." In Teague's vision, which appears to have taken inspiration from his work with DuPont, further R & D into "[transparent] materials . . . of chemical origin" would coincide with what he termed "better" bodies, each accentuating and complementing one another.[42] In Teague's hierarchy, plastics R & D and women's improved bodies were given the same importance and value, interchangeably showcasing each other's perfections, just like the Test Tube Girl.

Like his white male colleagues, the engineer and artist George Sakier visualized a white woman for his designs. He wrote, "The woman of the future will be tall and slim and lovely; she will be bred to it—for the delectation of the community and her own happiness."[43] Sakier's preference for tall, slim, "bred" women is underpinned by the same racialized gendered ideals of white Anglo-Saxon womanhood that circulated on the pages of *Vogue*. Raymond Loewy was more direct in referencing this ideology, stating, "Eugenic selection may bring generations so aesthetically correct that such [transparent and revealing] clothes will be in order."[44] *Vogue*'s 1939 World's Fair issue demonstrates that industrial designers applied streamline design to the female body itself. Her body was reduced in size, her flesh minimized, and bodies that did not conform to white, Anglo-Saxon, or Northern European appearances were omitted.

2.9 Anton Bruehl, "Donald Deskey Foresees a Great Emancipation," *Vogue*, February 1939, 73.

2.10 Anton Bruehl, "Walter Dorwin Teague Forecasts Near-Nudity," *Vogue*, February 1939, 78.

Eugenic ideology infiltrated middle-class culture in 1930s America and became embodied within streamline design's aesthetics.[45] Streamline is considered the United States' first mass-produced, large-scale industrial design style. Historian Cristina Cogdell argues that "streamline designers approached products the same way eugenicists approached bodies."[46] Loewy often used white female bodies as a metaphor for streamlined design. He famously referred to Betty Grable's body in his autobiography, reflecting that "[her] liver and kidneys are no doubt adorable, though I would rather have her with skin than without."[47] In 1934, he created a chart titled "Female Dress and Figure," mapping changes in popular female costume and the fashionable female figure.[48] This stands out from his other evolutionary charts showing product design development, such as those of the desk telephone or railcar, in that it includes the human body. Loewy compared the white female figure alongside designed artifacts, implying that the white female body could be designed and sculpted in accordance with the fashionable ideal. The chart denotes that efficiency and improvement in design can be streamlined to an ever-reduced, svelte body. The ideal female figure is increasingly minimized, and unlike his other product designs, all that remains of the future woman is a sliver-thin question mark suspended in midair. In these examples, women's bodies are presented as speculative streamlined products of white male scientific advancement.

The Test Tube Girl and *Vogue*'s streamlined woman of the future was white. Historians have described how individual ethnicity was increasingly effaced in the United States during the interwar, World War II, and postwar periods, in favor of a hegemonic Anglo-Saxon "American" whiteness.[49] Scholars have also shown how cosmetic surgery served as a tool for some Italian Americans, Jewish Americans, and other recent immigrants to alter their appearance and social presentation.[50] Assimilating to American ideals included Anglicized names, facial surgery, skin lightening, and hair straightening, removal, or bleaching, in accordance with the representation of then-dominant American Anglo-Saxon or Northern European ideals. Historian Nell Painter discusses "a keen awareness of the difference between *American* standards of beauty and the bodies of women increasingly being called 'ethnic.'"[51] In the 1930s and 1940s, "ethnic" became a label for groups whose appearances diverged from the tall, slender, athletic look celebrated in the media, while desirable bodies were expected to project a middle- rather than working-class image.[52]

Hegemonic American whiteness relied on the othering of Black, Indigenous, and people of color. This othering often occurred through their absence and exclusion. It can also be seen more directly in a British Pathé newsreel accompanying *Vogue*'s "Fashions of the Future," which was shown in "five thousand theatres before estimated audiences of twelve million people" (figure 2.11).[53] The behind-the-scenes footage opens by informing the viewer that the year is 2000 and immediately afterward shows a Black woman dressing a white woman.[54] The Black woman is outfitted in a maid's uniform—a stark contrast to the futuristic fashions of the white women she is dressing. Her attire signals that she is excluded entry from the future. In this sequence, she must remain subservient in clothing that marks her as being from the 1930s and the recent American past, but not from the future. In this white supremacist vision, the future—and the promises of better living it brings—are accessible only to white people. This narrative is reliant on the othering and continued subjugation of Black, Indigenous, and people of color. Far from a democratic plastic utopia for all Americans, this future replicated established forms of violence and oppression.

Similar imagery can be seen in 1939 feature film *The Women* (figure 2.12). A group of wealthy white women attends an exclusive fashion show at a New York City department store. During intermission, white women dressed in the latest fashions, including futuristic plastic hats, are handed refreshments by a Black woman wearing an apron. The representation of these white women as modern, elite, desirable, and fashionable socialites again relied on the stereotypical representation of what cultural theorist Stuart Hall describes as the "spectacle of the 'other.'" Writing on the power of representation, Hall notes that Black actors were relegated to subservient roles in 1930s mainstream American cinema.[55] The intermission scene is followed by the show's futuristic finale, featuring a laboratory-style set reminiscent of DuPont's Test Tube Girl and her *Vogue* contemporaries. A sequence of white women emerges from a revolving test tube and descends from the stage, much like the Test Tube Girl. The show's final look comes complete with a hood, like Miss Chemistry's (see figure 2.6), and is embellished with plastic jewels. The show's white woman host comments, "You will be able to study the flow of the new line as it responds to the ever-changing flow of the female form divine," alluding to changes in dress design *and* white women's bodies.

2.11 Still from *Eve, AD 2000!*, British Pathé newsreel, 1939.

People of color were largely excluded from the 1939 New York World's Fair. Only a handful of African Americans were employed in manual labor jobs, leading to protests.[56] The fair's Black attendance was low, partly due to economic factors: the cost of admission was inaccessible to many during the Great Depression. Furthermore, as historian Robert Rydell has noted, "Barely welcomed as consumers and certainly not as citizens in these futuristic world's fair cities, African Americans had far less reason than whites to take at face value promises about the 'dawn of a new day.'"[57] Black representation at the fair was limited to entertainment, where musicians and performers were cast in stereotypically racist roles. Rather than being streamlined, Black bodies were othered and depicted as overtly sexual and unruly. In this racist dualism, Black bodies were presented as "undesirable" and "uncivilized," representing the past, while white bodies were modeled as desirable and embodying a "civilized" future.

In DuPont's 1939 displays, white women's bodies were placed in the lab and represented as materials to be transformed for "the better" by

2.12 Still from *The Women*, directed by George Cukor (Metro-Goldwyn-Mayer, 1939).

the supposed scientific superiority of white men. Only whiteness was represented, implying that only these bodies could be improved. A 1939 *Vogue* article titled "Good Form in America" praised an elite class of "American women" for their meticulous efforts toward homogeneous, streamlined figures and thanked "controlling garments" for their contribution: "It's not just an accident of nature and heredity that American women, as a group have the most admirable figures in the world. Giving their due to our long-legged frameworks, our athletic lives, and our conscientious efforts towards sleekness, much credit still belongs to those little persuaders, corsets (that includes *all* controlling foundations: girdles, all-in-ones, brassieres)."[58]

Stretchy synthetics used in "controlling foundations" served to reshape the body into the fashionable, streamlined silhouette of an unmarked, slender, white American ideal. In 1944, *Vogue* enthused to readers that foundationwear, which utilized the latest synthetic developments, provided a useful cosmetic aid with which to combat "the pattern of heredity" and shape your body as you wanted.[59] Here, the pliable materiality of plastics became associated with modeling the feminine body into a racialized biopolitical ideal of standardized, slender, American whiteness.

Redefining Nude: Nylon Futures Beyond Test Tube Girl Pink

In the 1930s United States, the difference between idealized and actual bodies became blurred. This skewing resulted from the coalescence of anthropometric processes, increased standardization, and the demand for fast turnarounds in mass-produced fashion.[60] Historians have shown the influence of popular eugenics on efforts to create standards and define concepts of the "perfectly" "average" and "normal" American.[61] Nuanced linguistic differences between "norm," "type," and "ideal" were erased, resulting in the terms often being used synonymously. The impact of biopolitical efforts to define whiteness as unmarked "normality" is evident in Test Tube Girl imagery and related material culture of this period, such as nylons and foundationwear. In keeping with streamline design's dominant ideology of standardization for mass production, nylon stockings and undergarments are shaped by interconnected discourses of whiteness, sexuality, and gender. Nylon hosiery and undergarments both catered to and were shaped by ideals presented as the standardized norm—that of the slim, "yet-softly-curved," "average" young cis white woman.[62]

Nylon was a major technological innovation, yet it was paradoxically shaped by established industry standards, racialized gender norms, and their influence on one another. This is seen in a rare surviving example of DuPont's nylon experimental "wear-test" underwear from 1938, which are a light beige color in keeping with standard 1930s trade definitions of "nude" (figure 2.13). This color choice was consistent with dominant lingerie trends and is similarly reflected in surviving early wear-test stockings, which are variations of a light peachy or beige color.[63] When it came to creating future foundations, variety was not a priority for chemists or advertising staff at DuPont. Wear-test underwear and stockings were made in limited sizes, catering to an assumed slender white woman, again revealing wider ideological assumptions around body sizing and racialized gender ideals. These objects show how DuPont's experiments with nylon undergarments and hosiery not only centered whiteness but also, arguably, connected a singular configuration of womanhood with "normalcy."

The term "nude" to describe color became prevalent in the 1930s.[64] Since the 1910s, peach- or pink-toned underwear described as "flesh" color was an increasingly popular alternative to white, which had previously dominated

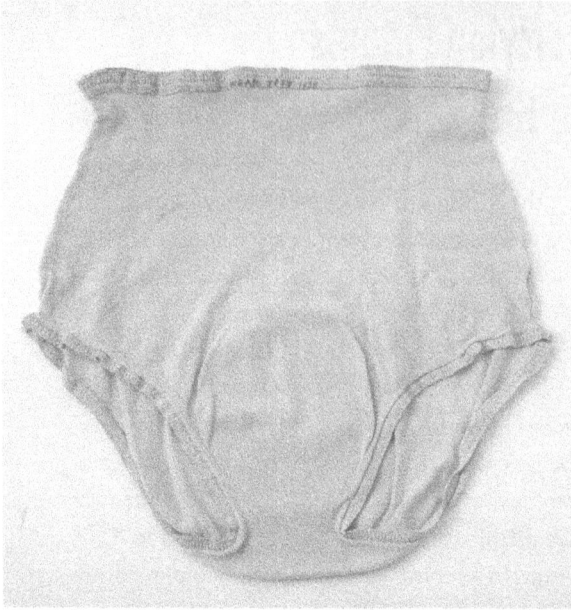

2.13 Nylon wear-test underwear, 1938. Joseph X. Labovsky Nylon Collection, Science History Institute, Philadelphia.

the industry. Light "flesh"-tone brassieres that catered to a restricted audience were worn under sheer blouses to create a "nude" look. By the 1930s, "nude" was frequently used to describe a pinkish-beige hue intended to create the desirable illusion of wearing "nothing" underneath.[65]

Color sample books from the 1930s demonstrate efforts to define this term. Elizabeth Burris-Meyer, colorist and dean of the New York School of Fashion Careers, included "nude" in her 1935 list of "color names in common use in merchandising and advertising," describing it as "natural skin color" of a yellow-red hue.[66] Burris-Meyer's definition excluded darker skin tones, thereby rendering them "unnatural." Similarly, the Textile Color Card Association of the United States labeled a light beige fabric color swatch as "nude" in its 1941 Standard Color Card.[67] These examples of color standardization informed the industry, as evidenced by foundationwear and hosiery. For example, throughout the 1940s, Sears limited most of its foundationwear, such as corsets and brassieres, to light "nude" or "tea rose" colors described as "flesh pink." These examples show how "nude" was presented exclusively as a light pink-beige color, rather than as a palette of different shades designed to reflect ethnic and racial diversity. Such exclusion reveals the systemic racism in which these colors were created and continues to apply to "nude" color descriptors in many contexts today.

The resounding absence of race and skin tones from definitions of nude, as well as nude hosiery and foundationwear copy, is striking. This absence of discussions of race is also evident in the other 1939 imagery I have discussed, such as that of the Test Tube Girl. Historian Julian Carter notes a similar absence of discussions of race within US efforts to "standardize" the body as white throughout the 1930s and even in earlier decades.[68] Whiteness is consistently presented as the unmarked standard, thereby violently erasing Black, Indigenous, and people of color, while constructing and reinforcing whiteness as the invisible norm.

It is important to contextualize the choice of colors offered in these advertisements. At this time, the synthetic organic chemical industry in Germany and the United States had revolutionized the dye and pigment market, enabling it to replicate almost any color.[69] While the science and technology to produce a wide range of pigments had been developed, it largely did not lead to the creation of hues specifically for people of color.

In 1939, when DuPont debuted nylon stockings, textile chemists and colorists were concerned about how established dyes would take to the new material. In response, the American Association of Textile Chemists and Colorists researched the effect of established dyes on nylon. It found that "the first experimental dyeing of hosiery was carried out with the neutral dyeing acid colors generally used for dyeing silk hosiery."[70] Some aspects of nylon dyeing could be improved with minor adjustments, such as optimizing the dyeing temperature. The study of nylon hosiery and dyes concluded, "Our tests indicate that popular hosiery shades can be produced with acetate dyes comparable in fastness to light and washing with pure silk hosiery at present on the market." It also generalized the study to apply to "underwear garments."[71] Here, the American Association of Textile Chemists and Colorists assured its readers that established "popular hosiery" shades—likely pink "nude" and "flesh" color dyes—could be applied to nylons. DuPont promoted nylon as a new material for a new world. But standardized colors and the racialized gender norms they reflected and reinforced were simply grandfathered in and applied to this new technology.

Throughout the 1930s, black lingerie and stockings were associated with sex workers and illicit sex. This othering depended on white, pink, and light beige being presented as the everyday "norm."[72] Historian Jill Fields shows how black lingerie and stockings have a long history of being associated with sexuality and race.[73] Black lingerie, which gradually became more popular in the postwar period, "allowed women, especially white women, to express, and their bodies to convey, the eroticism attributed to black women via a safely contained and removable black skin."[74]

Although most postwar hosiery companies offered a range of opaque colors, they did not explicitly address a variety of skin tones. Instead, color descriptors—such as the Harrison Hosiery guide to six beige and brown tones shown in figure 2.14—assumed a white wearer and listed the season's fashionable shades of clothing with which they should be worn. The practice of standardized color coordination was in keeping with norms developed after World War I by the Textile Color Card Association of the United States. Color stylist Burris-Meyer noted that, before the war, "black, brown, and white were about the only [hosiery] colors."[75] However, after the war, shorter skirts resulted in the emergence of pink flesh-colored stockings. A greater variety of shades were gradually added, leading to a "saturated" stocking market, after which the industry turned to renaming old colors to maintain sales. The Textile Color Card Association, in cooperation with the National Association of Hosiery Manufacturers, coordinated "hosiery colors with the seasonal colors in dress and eliminat[ed] the duplication of color names, resulting in a Standard Hosiery Color Card that contained the most sold staple shades."[76] These standardized color guides, which included light pink or peach-beige colors defined as "flesh" and "nude," were only concerned with matching hosiery colors to womenswear trends and implicitly assumed a white consumer.

The standardized practice of developing hosiery colors to match the seasonal fashions for an unmarked racialized gender norm continued after World War II. Throughout the 1950s, Sears catalogs listed a range of three to five colors, from "light to darker," starting with shades like "royal sand" or "ambertone" and ending with colors like "royal mist"(a grayish taupe) or "royal shadow" (a dark gray or translucent black). This is not to say people of color did not wear them, but the shades do not acknowledge or even mention differences in skin tones. Instead, companies such as Harrison Hosiery and Sears offered a range of shades that were "fashion-right" and designed "to complement every color in your wardrobe."[77] This shows that although a range of dye shades was technologically available to hosiery brands and manufacturers, they chose to develop and promote their colors as suited to an assumed white wearer's varied and colorful wardrobe rather than to a range of skin tones. In turn, some people of color dyed hosiery at home to better match their own skin tones.

Some postwar advertisements for nylons in Black publications explicitly emphasized the range of skin tones available. *The Crisis*, the official

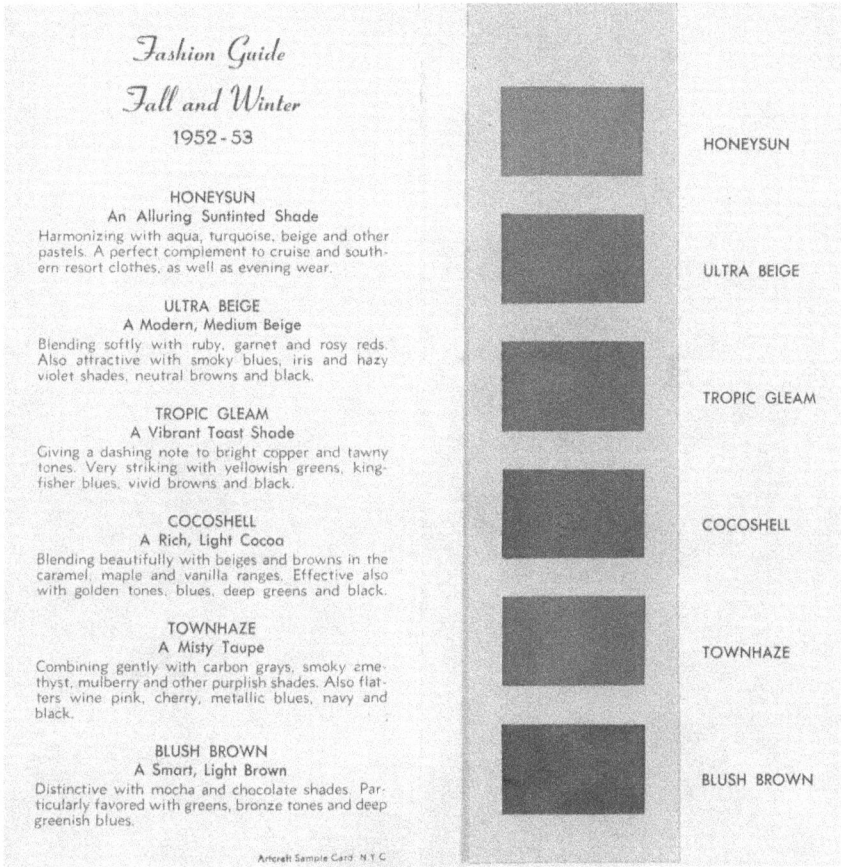

Fashion Guide
Fall and Winter
1952-53

HONEYSUN
An Alluring Suntinted Shade
Harmonizing with aqua, turquoise, beige and other pastels. A perfect complement to cruise and south-ern resort clothes, as well as evening wear.

ULTRA BEIGE
A Modern, Medium Beige
Blending softly with ruby, garnet and rosy reds. Also attractive with smoky blues, iris and hazy violet shades, neutral browns and black.

TROPIC GLEAM
A Vibrant Toast Shade
Giving a dashing note to bright copper and tawny tones. Very striking with yellowish greens, king-fisher blues, vivid browns and black.

COCOSHELL
A Rich, Light Cocoa
Blending beautifully with beiges and browns in the caramel, maple and vanilla ranges. Effective also with golden tones, blues, deep greens and black.

TOWNHAZE
A Misty Taupe
Combining gently with carbon grays, smoky ame-thyst, mulberry and other purplish shades. Also flat-ters wine pink, cherry, metallic blues, navy and black.

BLUSH BROWN
A Smart, Light Brown
Distinctive with mocha and chocolate shades. Par-ticularly favored with greens, bronze tones and deep greenish blues.

Artcraft Sample Card N.Y.C.

HONEYSUN

ULTRA BEIGE

TROPIC GLEAM

COCOSHELL

TOWNHAZE

BLUSH BROWN

2.14 Harrison Hosiery guide, 1952–53. Science History Institute, Philadelphia.

magazine of the National Association for the Advancement of Colored People, ran an ad for DuPont "Beauty Fit" nylons that stated that they "[come] in all skin tones."[78] A 1951 ad for Basin Street Nylons was more direct in its messaging, exclaiming, "NOW DESIGNED ESPECIALLY FOR YOU!" It features a headshot of a smiling Black woman introduced as Jean Williams, a color stylist. It states that she "created our three colors that are definitely right" (figure 2.15), alluding to the fact that other nylon colors did not explicitly cater to Black women. The company, apparently named after Basin Street in New Orleans, also took inspiration for its color no-menclature from US cities famed for African American arts and music: "Harlem—for light complexion; Memphis—for medium complexion; St. Louis—for dark complexion."[79] In addition to offering colors designed

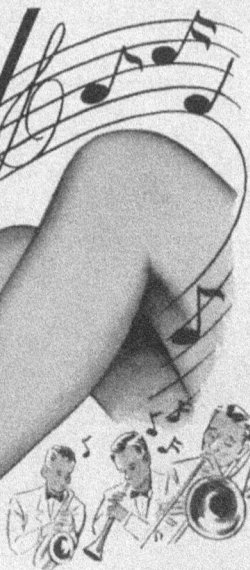

2.15 Basin Street Nylons advertisement in *Tan Confessions*, November 1951, 55. Beinecke Rare Book and Manuscript Library, Yale University, New Haven, CT.

for Black consumers, this advertisement for Basin Street Nylons did not follow industry color standards, which drew on names of foods like "pecan," "toast brown," and "spice brown," referencing colonialism and exoticization.[80]

Information on Basin Street Nylons is limited. In 1951, Sumner Hosiery, a white-owned hosiery company from Gastonia, North Carolina, filed a patent for Basin Street. It is unclear whether Williams was a color stylist employed by Basin Street Nylons or a fictional figure used as a promotional strategy for this ad in *Tan Confessions*.[81] The inclusion of Jean Williams in the advertisement is striking. It reimagines a future for nylons in which Black women are named, promoted to positions of authority, and given the agency to develop products in a wider range of skin tones beyond a single, reductive light shade.

Everyday postwar American imagery, such as Sears catalogs, represented a limited ideal of standardized womanhood that centered on whiteness and maintained the 1930s preference for normative, streamlined bodies (figure 2.16). Historians have shown the ongoing impact of American industry sizing standards, which were created during the broader 1930s racialized pursuit of the "average" American.[82] A repeat image in Sears' nylon section and sizing guide reinforced this ideology.[83] Although supposedly created to illustrate different sizing needs—"petite," "shapely," "classic," "tall"—it bizarrely depicted essentially the same slender, young, white, often blonde woman in the same dress, merely scaled up in height with consecutively longer legs. Sears paired such uniform images with text praising proportions and endless customization: "Each is wearing size 8 1/2 . . . but in the length and width perfect for her."[84] Stocking sizes ranged from 8.5 to 11.5 to fit shoe sizes 3.5 to "10 and over" and explicitly included wearers four foot eight to six foot two. By 1961, Sears expanded its range of sizes, claiming to cater to consumers as tall as six foot three.[85] However, as evidenced by its imagery, Sears—in keeping with whitestream media ideals—continued to illustrate difference by paradoxically depicting a homogeneous, slender white leg.

By the 1970s, the promotion and coverage of nylons in American print media reflected wider cultural shifts and challenged singular ideals of femininity. A decade later, in 1971, advertisements for Big Mama pantyhose appeared in numerous US publications. The advertising and packaging featured three light-skinned, plus-size women and listed sizes for a wider range of consumers.[86] Sugar & Spice also reimagined nylons. This Chicago-based company advertised nylons in *Ebony* in the 1970s. Unlike

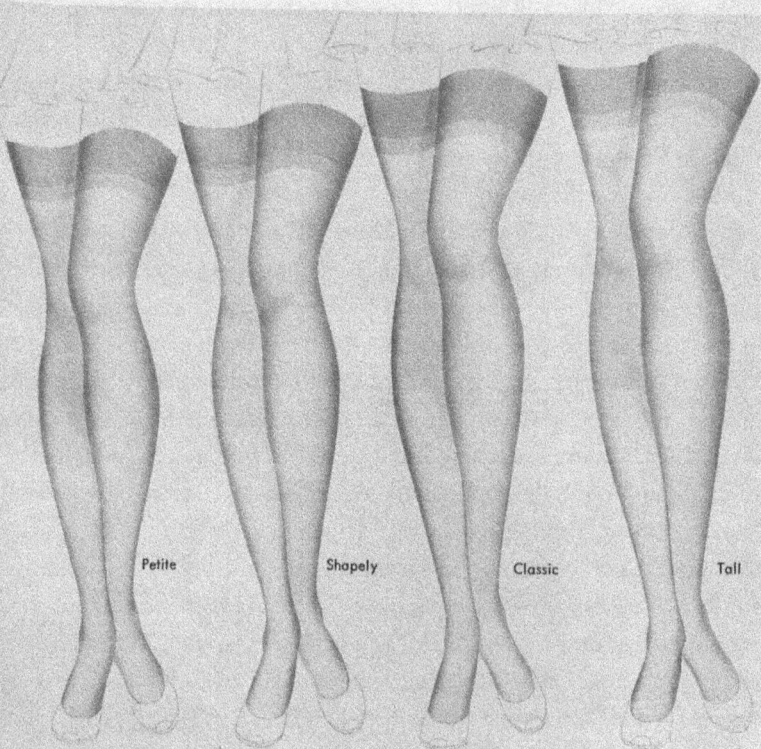

2.16 Advertisement for nylon stockings, Sears catalog, Fall–Winter 1956, 258.

Basin Street, its advertisements did not feature the face of a designer or stylist but instead showcased a group of dark-skinned Black women models and celebrated the message "Black is beautiful."

Nylons were an integral part of dress and identity for many queer, trans feminine, and gender-nonconforming people. However, this has not been considered in histories of nylon. The Tana & Mara advertisement shown in figure 2.17 is a rare example of nylon advertising aimed at queer and trans feminine consumers. It is marketed to wearers who did not fit standard-length stocking sizes and is part of a series published in six early issues of *Female Mimics* from 1963 to 1965. Tana & Mara—a New York–based mail order fetishwear company established in the 1960s—featured extralong nylons in ads for its illustrated catalogs. The "super-length" nylons were listed as available in gray or black, with seams, in sizes 9–11. Striptease artists and stars of the underground fetish publishing scene, Tana Louise and Mara Gaye ran the company together. Louise regularly contributed to *Exotique* as a model and columnist, and Gaye graced the pages of such publications as *Frolic, Dazzle*, and *Cavalcade of Burlesque*. Although many copies of Tana & Mara's catalogs were likely sold based purely on their image content, rather than by prospective apparel purchasers, it is important to note that the company explicitly advertised "super-length" nylons in the trans feminine magazine *Female Mimics*.

Publications in the Digital Transgender Archive provide valuable documentation of readers' varied lived experiences with nylons, including guidance on safe purchasing practices and insights into sizing challenges. In 1963, Betty-Bill wrote about her recent experience of purchasing clothing: "Like most of the girls who write you, I buy most of my clothing from mail-order catalogs." She felt this was "the safest way, since you can pick up the items at the catalog sales desk and the clerks think that you are merely picking up something that has been ordered by your mother or sister."[87] The nylons Betty-Bill collected were "size 11 super long and a light shade."

Throughout the 1960s, nylon sizing remained an issue for many trans feminine, queer, and gender-nonconforming people. Chris Ames—a star of the famous 82 Club and avid reader of Nutrix Co. publications, including *The Art of Female Impersonation*—enthused about the importance of sharing experiences and meeting other girls in person.[88] She complained about regular-length nylons running and snagging on stage from the strain of being worn with a garter belt for longer periods, as they weren't designed for her long legs.[89] Suli Montiel shopped in specialist stores such

2.17 Tana & Mara advertisement in *Female Mimics* 1, no. 3 (1963): 67.

as "Tall Girls Shops," sharing that she could get "sizes that fit me fairly well, without having to have them custom made."[90] Meanwhile, Bobbie Weaver noted, "My nylons are bought in a special store by my sister, because I wear a size 12 extra long."[91] Erica remarked, "I feel I really stand out in a crowd when I request size 12 super long . . . nylon stockings instead of just buying panty hose."[92] While its sizing wasn't as expansive as specialist "tall girls" stores, Tana & Mara filled a gap in the market for customers who did not fit the standardized sizes, preferred stockings over hose, or did not feel comfortable purchasing in person from a store.

Some trans feminine readers expressed a preference for darker shades of nylon. Tana & Mara's advertisement lists nylons in gray and black—color descriptors that featured less frequently on the light-color hosiery pages of mass-market catalogs like Sears. The pages of *Letters from Female Impersonators* attest to the popularity of darker shades of nylons and readers' experiences with wearing them.[93] Dorothy observed, "My legs look much slimmer in dark nylons than in any other shade."[94] Similarly, Betty-Bill preferred a darker tint for her stockings and customized her pair with Tintex, "[dying] two pairs deep brown"—a design process also practiced by African American fashion designer Zelda Wynn Valdes, who dyed hosiery to match the skin tones of the dancers with whom she worked.[95] Wilma complained, "The light nylons did not show up as well as I would have liked, so I put on a pair of dark hose with seams and these showed up much better in the next photo."[96] Jakki J. shared, "I love to wear black nylons because they look very sexy and they hide the hair on my legs rather well."[97] Meanwhile, another reader noted that light nylons were ideal for showing off "well-shaven legs."[98]

The materiality and sensation of wearing nylons was another popular topic. Some contributors to trans feminine publications described the pleasurable sensation of skin encased in nylon, its materiality, and the admiration it inspired for its "glossiness." Joyce cited nylons as a motivating factor in crossing social boundaries of gender: "I began by my desire to want to wear luxuriously silky garments made of materials not common to male attire (such as nylon, satin, and silk)."[99] Nylons were so central to some trans feminine individual's gender presentation that Lorraine Channing, art editor of *Turnabout: A Magazine for Transvestitism*, suggested running an article titled "Silk vs. Nylon—Roundtable topic discussed by prominent members of the Fetishist Division of the American Society of Transvestites."[100]

Nylon's appeal was enduring. By 1971, the mainstream fashion for stockings was in decline, and Erica concluded that mail order was her preferred way to shop. She complained about new seamless mesh knit hose— "To me this is just so much gauze around my legs"—and invited readers to share their experiences of wearing them.[101] Unlike the Test Tube Girl, who embodied a narrow heteronormative, cisgender white vision of the future, nylons' uses and meanings were shaped and reimagined by the individuals who wore them.

DuPont's representation of nylon on the Test Tube Girl was entangled in discourses on the creation of norms, standards, and ideals of a hegemonic white American womanhood. Powerful white cisgender men, including industrial designers and advertising executives, imagined an imminent future in which one could scientifically control and aesthetically design the perfect female body, inside and out. The absence of race within much of the discourse around imagery—such as Miss Chemistry, *Vogue*'s "future breed of women," nylon wear-test items, and standardized "nude" color descriptions—reinforces white womanhood as the unmarked norm. Bodies that did not fit this narrow, slender, heteronormative white ideal of gender were marked as nonnormative. However, the creation of such standards and the ideals they embody did not reflect reality. Yet the concept of norms and their modeling of ideals undeniably affect societal structures, lived experiences, and perceptions of the self.

The legacy of the creation of "nude" colors that standardized an unmarked norm of womanhood as white can still be seen today in hosiery, underwear, and apparel branding. However, this has not gone unchallenged and there are increased efforts to address this, particularly among such Black-owned businesses as Nude Barre. It is important to historicize the emergence of standard practices, such as singular definitions of nude and limited sizing, and their efforts to normalize one-dimensional definitions of gender. By deconstructing these practices and looking at less-documented alternatives, one can uncover and address the embedded power relations within standardization. Nylons by companies like Basin Street Nylons and Tana & Mara, as well as people's lived experiences using them, provide an important and previously untold counterhistory of nylons. These polyvocal narratives show how individuals and communities resisted the creation of racialized norms of gender and sexuality while actively shaping their own nylon futures.

FOAM / PADDING

3 SOFT POWER / PLASTIC FOAMS, DESIGN, AND POSTWAR BODIES

Shortly after VE Day in 1945, a group of US plastics experts entered the laboratory of Otto Bayer at IG Farben's Bayer Corporation plant in Leverkusen, Germany.[1] They were amazed by what they found: new types of plastic foams they had never seen before. The foams had been developed in a series of experiments to produce alternative synthetics to DuPont's nylon.[2] The US government and military had sent this group of white men, all of whom worked for US-based chemical companies, to collect information on plastics technology for transfer to the United States. Enchanted by the lightness and fluffiness of Bayer's foams, they documented them in a series of reports and photographs that conveyed both their haptic properties and the enthusiasm of those who interacted with them. The reports recommended that plastic foams such as polyurethane be returned stateside.

In the postwar United States, polyurethane foam (also known as urethane foam) became the plastic foam material of choice for padding. It found applications in transportation, furniture, and upholstery, as well as in building insulation. In addition to apparel and shoes, polyurethane foam was used more intimately in foundationwear to pad the curves of bombshells. Today, polyurethane foams continue to pad and cushion human bodies. They can be found in office chairs, sports shoes, shapewear, mouse pads, car upholstery, wheelchair seats, couches, and mattresses. Yet, little is known about these plastic foams' military-industrial origins. This chapter traces the history from a series of World War II US military intelligence reports that recommended the postwar transfer of plastic foams from Germany. Using a materials-centered approach, the chapter examines how, in the postwar United States, polyurethane foam's materiality and soft appeal to the body moved from the lab to domestic spaces, particularly furniture.

To investigate why and how actors working for the US military and chemical companies transferred foam plastics from Germany to the US postwar consumer market, industrial, scientific, military, corporeal, and domestic environments are explored via archival materials. I examine polyurethane foam in the context of Cold War soft power politics, paying particular attention to how gender and race shaped the presentation and marketing of its haptic properties. During this period, the commercial uses of polyurethane foams mushroomed, enveloping postwar bodies in everyday encounters. In environments ranging from public spaces to private residences, from morning routines to nighttime relaxation, and spanning transportation seating, living room upholstery, mattresses, and foundation garments, human flesh could come into contact with foams at every conceivable encounter. However, as this chapter will demonstrate, actors—including chemical company and advertising executives, as well as designers and journalists—often associated foams' soft fleshy properties with women's bodies.

Tracing Polyurethane Foam's Provenance and Its Archival Legacy

Polyurethane foam is a type of plastic foam. Plastic foams can include foam rubber, initially made from whipped natural rubber or latex and later gradually replaced by synthetic alternatives derived from petroleum by-products. In 1956, the US Society of the Plastics Industry, probably wanting to clear up the confusion, announced its adoption of "the name 'urethane foam' for what used to be known as isocyanate, polyurethane, or polyester foam."[3] This new type of synthetic foam technology was first developed by Otto Bayer in 1937 at IG Farben Laboratories—a subdivision of the Bayer Corporation—and can be traced back to nineteenth-century German isocyanate chemistry. Bayer (no relation to the company's founder, Friedrich Bayer) is considered the originator of the polyure-thanes industry. He later reflected that his research and development in polyurethanes stemmed from IG Farben's "urgent problem of creating something similar or better, independent of [Carothers's nylon] DuPont patents."[4] Foam plastics advancements emerged as a reaction to—and in pursuit of—an "ersatz" (replacement) for nylon, the first fully synthetic fiber developed in the United States. Bayer developed Perlon, a type of synthetic fiber, for parachute fabrics. While experimenting with the syn-thesis of novel materials not covered by DuPont's nylon patents, he real-ized that new materials with unique properties could be created.[5] Bayer simultaneously discovered that certain polyols—for example, esters and ureas—could be used to produce foamed urethane polymers.[6] The pres-ence and type of additional ingredients in the polymerization formula determined whether the resulting polymers were solid or foamed, varying from liquid elastomers to rigid solids.[7]

In the years leading up to World War II, research and development in plastic foams was not only a key component of an international exchange of knowledge and patents but also part of a larger push for synthetic rub-ber replacements. Bayer chemists discovered that rigid urethane foams offered comparatively "high strength and a wide range of elasticity."[8] Du-Pont, attracted by the foams' commercial potential, was issued a series of basic US patents from 1937 to 1939 based on Bayer's work, detailing the reaction processes between the key ingredients.[9] After 1939, trade between the United States and Germany ended, halting all research exchanges on plastic foam developments. During World War II, German aircraft, tanks, and submarines used rigid urethane for adhesives and fillers and soft Porophor N foam variations for upholstery and packaging.

Although DuPont purchased a series of basic patents just before the war, the Anglophone historiography of plastic foams notes that the poly-urethane group of polymers later piqued the interest of plastics and chem-icals groups in the United States. This renewed interest followed further developments in Germany, which were finally unveiled to US industry representatives via military intelligence.[10] These scholars posit that it was only in 1945, after the Allies entered Germany, that American scientific teams could carry out military intelligence reports on German wartime scientific and technological developments, including foam advancements and applications.[11] In these histories, references to these important reports are vague, lacking citations or authors' names.[12] However, in my research, I was able to identify and locate these reports.

The German Plastics Industry Quartermaster Report

The political aspect of polyurethane foam becomes clear in the German plastics industry Quartermaster report, produced under the direction of Brigadier General Georges F. Doriot, who worked on military research, development, and planning. Under the auspices of the US Army Quartermaster Corps, a team of experts was sent to Germany immediately after VE Day in 1945 to collect data on plastics. Its initial mission was to "investigate any technical developments in the field of plastics which might have immediate application to the plastics research program of the QM Corps in connection with the war in the Pacific."[13] John M. DeBell (DeBell and Richardson, Springfield, MA) led the team, joined by W. C. Goggin (Dow Chemical Company, Midland, MI) and Walter Gloor (Hercules Powder Company, Parlin, NJ). Over three months, they visited and reported on the operations of forty individual plants in all four occupied zones in Germany (British, French, American, and Soviet). With headquarters set up in Frankfurt, the group traveled ten thousand miles in tanks. The team returned from missions every few weeks, "loaded with samples and bedraggled personnel."[14]

Four interlinking factors drove the plastics team's intelligence mission. First, General Doriot, who headed the Quartermaster Intelligence Agency's Procurement Division, had an active interest in materials. He believed that improved R & D in this area would create a more efficient and technologically advanced military, capable of operating on all terrains.

The second main driver for the reports was rivalry in the chemical industry, particularly in dyes and plastics, which had long been a point of competition between the United States and Germany. The 1917 Trading with the Enemy Act triggered a concentrated effort among American businesses, the federal government, and universities to work toward American independence from foreign chemicals.[15] In the United States, Europe's technological advances remained a concern for decades, especially Germany's position as a global leader in the production and R&D of synthetics before, during, and after World War II.[16] The American military-industrial complex was eager to overtake European—and specifically German—advancements in synthetics, securing a position for the United States as a world leader in scientific and technical achievements, including plastics.

Third, it was hoped that the commercial potential of any unknown plastics advancements could contribute toward reparations. Rebuilding Germany's factories and reinstating industrial infrastructure was key to supporting US military and industrial R & D interests generally, as well as working toward repaying reparation costs. As director of the Military Planning Division for the Quartermaster General, Doriot was keen to provide industry with access to military intelligence on plastics: "On the matter of giving away information we should make very sure that all companies, large, small, medium, individually, anyone who might want it can have it."[17] An important concern at this point was the equitable and rapid distribution of the German plastics reports among US industry.[18]

Fourth, there was a heightened sense of urgency when it came to collecting and documenting plastics, as the Procurement Division pushed to do so before Soviet forces. Correspondence from the Gloor Papers at the Science History Institute shows that there was frustration and worry that the Soviet Union was not sharing its technological findings in relation to plastics.[19]

In August 1945, the team presented its observations in a technical intelligence report directly to Doriot and his Procurement Division, stating that the German plastics industry during World War II was largely influenced by two factors. First, "the enormous wartime demand for substitute articles derivable from the limited raw material available." Raw materials were restricted due to excessive wartime production demands, the destabilization of colonial trade, and the end of most international trade with Germany. Faced with these limitations, Bayer's chemists turned to materials they could source more easily, such as derivatives of acetylene and coal. Second, the German plastics industry was shaped by "the predominant position of the well-organized and well-staffed research divisions of the I. G. Farbenindustrie A. G."[20]

During World War II, Germany's plastic industry strategically focused on the development of chemical processes rather than mechanical developments to circumvent the limited wartime access to materials. Polyurethane foam was one of the new synthetic materials created through a chemical reaction at a time when access to established materials for building conventional factory equipment was scarce. As a result, plastic developments created via chemical processes were particularly well suited to knowledge extraction, as they did not require bulky mechanical equipment.

Plastics advisers employed by the US government argued that, in the context of reparations, extracting information—such as polyurethane foam processes—was preferable to removing equipment. One adviser noted, "If we take equipment out of a German plant, the plant is unable to produce, but if we take information only, the Germans are not deprived of the use of this information. In this way we obtain reparations without greatly impoverishing Germany's economy."[21] Plastics, like other technological and material developments, could serve as currency for reparations and help stabilize Germany's economy. DeBell, the team leader, reasoned that if left intact, Germany's established plastics production capacity could help alleviate American supply shortages and reduce foreign relief costs to American taxpayers.[22] Experts were keen to identify "responsible men in Germany who can work out proper policies [on plastics], have them accepted by Washington and London, and then carry them out."[23]

While their initial report noted that comparing the wartime progress of the German and American plastics industries was challenging, the experts identified some clear differences and potential military developments that could be transferred to the United States. They observed, "[Germany has] new chemical types which we consider to be most interesting for the future."[24] One of these areas was plastic foams: "[German] chemical research has been impressive, and has produced fast polymerization of synthetic rubber, two new tough, high-melting or high-curing polymers related to our polyamides (caprolactam and the isocyanates)."[25] Polyurethane polymers, a type of isocyanate, represented an entirely new development and were not yet available on the US market.[26] DeBell, Goggin, and Gloor recognized the great value of isocyanates for US military and civilian applications. Isocyanate and polyurethane developments were predominantly located at Bayer's laboratory and the IG Farben Leverkusen plant, both within the British occupied zone. For these locations, they recommended installing a commission directed by "a very competent chemical development organizer and executive, familiar with plastics and rubber." The commission would extract the information, and the laboratories would then begin working on applications to support the essential civilian economy and certain military nonweapon needs.[27]

At this time, Otto Bayer's isocyanate discoveries had not yet reached full plant-scale production. However, his pilot-stage production, which showed great promise, was earmarked by the US plastics experts as "profoundly significant, since this new chemical family has likelihood of wide application in synthetic fibers, high temperature thermoplastics, protective coats, leathers, foams and adhesives."[28] Indeed, polyurethanes became one of the largest and most versatile polymer families, still widely used nearly a century later. Otto Bayer, they observed, had a large laboratory and staff, and although no work was underway when the experts visited him on June 12 and 13, repairs were already in progress. The report noted, "[Dr. Bayer] should shortly be in position to take on investigations in American interests particularly in utilizing isocyanates, or . . . special foams," and that his "splendid laboratory . . . could well be put to work on assigned problems in the interest of the United States."[29] Additionally, they believed that "wide industrial dissemination in the United States of information on the work of [Otto Bayer's I. G. Leverkusen] laboratories" would lead to increased competence and production in this new area.[30] Through these two channels, Bayer's advancements were immediately utilized for US applications.

Bayer developed Moltopren, a soft type of polyester-based polyurethane foam, which was described by the US plastics experts as "a new light and foamy plastic, which is used as a construction element and insulating material."[31] It was "under investigation for airplane structures and unsinkable airplanes" for use in aviation wings and stabilizers.[32] Additionally, Moltopren's insulation properties were harnessed in submarines for soundproofing and in tanks for heat.[33] Porophor N—another chemical product developed by Bayer that was used to create plastic foams—was used for floatation, marine storage compartments, and assault boats; meanwhile, "for peace uses, auto cushions, insulation, and bathroom mats are indicated."[34] At this point, no mention was made of foams for apparel or domestic furniture. However, the body and its interactions with foam certainly factored into the team's reporting on these new plastic foams.

Soft plastic foams' materiality and responsiveness to touch plays an important role in the German plastics industry Quartermaster report. The document, accompanied by 346 photographs, offers a unique and perhaps unlikely record of the importance of the materiality and haptic qualities of plastic foams, which made them so inviting to touch and engage with physically. Most of these photographs document equipment, manufacturing plants, and heavily bombed industrial sites. In contrast, the images of the different plastic foams stand out due to the visible enthusiasm of the subjects and their phenomenological engagement with these materials. In these photographs, we see US investigators and German chemists interacting with one another and the materials. Cigarette cartons are used throughout the report to indicate scale, indicating the haphazard nature of the reports. It is only with the foam images, however, that we see human touch being recorded to indicate materiality. This is evident in a series of images from the report that are worth analyzing in more detail.

In the upper part of figure 3.1, a light-skinned hand reaches in, gripping a white circular object laid out on a dark surface—possibly wooden floorboards—and confidently squeezes the item between thumb and fingers. The thumb pushes down, almost sinking and vanishing into one of the seven circular holes; the fleshy, white, marshmallow-like material invites touch. It is portrayed as easily malleable but also offering resistance; it can be compressed yet has the ability to return to its original molded shape, a theme that was popular in the US promotion of plastic foams.

In the lower image of figure 3.1, there is a group of five white men holding up an assortment of polycarbonate–urethane (PCU) foam samples. Unlike the other individuals depicted, John DeBell (second from left) is excitedly smiling at the camera. He stands slightly behind two other men, leaning into the space between them, his right hand lightly holding a small gray ring made from plastic foam. With his little finger lifted, the object appears so light that he does not need more fingers to grip it. In his left hand, he raises a larger white circle—the stowage rack from the upper image. The other men convey a range of facial expressions; they seem less confident, standing somewhat awkwardly behind their assortment of foam items, as if to protect themselves from the camera's gaze. They are wearing suits rather than uniforms, and one is dressed in a lab coat; it is likely that these are German chemists and engineers working at Continental Gummi Werke AG, Hanover. DeBell's enthusiasm while presenting these objects to the camera (presumably the photographer is either Goggin or Gloor) is striking. He appears eager to share his findings from the plant—a new type of material not seen before in the United States, so light that one barely has to lift a finger to raise it and simultaneously so satisfying to touch and squeeze.

In the report, Otto Bayer is the only German chemist named in a photograph caption. In figure 3.2, Bayer proudly looks directly into the camera, smiling as he holds up two foam items. The item in his left hand is delicately balanced between his fingers, further emphasizing its lightness. There are some items on the table behind Bayer; however, Bayer and the foam items he is holding have been selected as the focus of this photograph. Before World War II, Otto Bayer had participated in international exchanges on plastics. Bayer confidently interacts with the photographer; perhaps he was pleased to finally share his developments with American plastics experts and resume his engagement with international plastics research.

PCU porophor foamed submarine galley stowage rack (Report 20c)

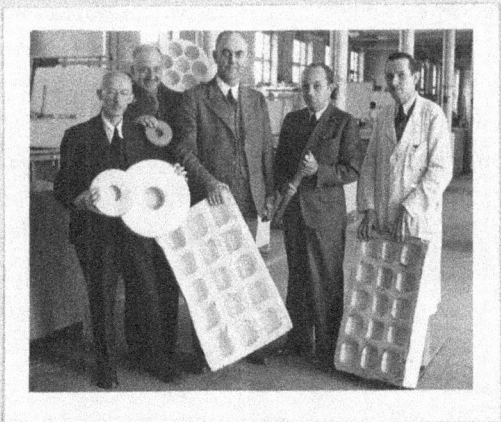

Variety of porophor-foamed PCU samples (Report 24d)

3.1 Photographs of PCU Porophor N–foamed products, ca. June 1945, in "Quartermaster Report: Technical Intelligence Reports; The German Plastics Industry." Walter E. Gloor Papers, Science History Institute, Philadelphia.

In figure 3.3, Bayer's "physical testing lab" appears with foams in the foreground. A white man dressed in a suit and tie stands behind a funnel-like shape, his right hand resting on the side of it as he smiles at the camera. Unlike many of the images of other (nonfoam) plastics locations in the report, the lab looks bright, tidy, and clean. In contrast to the photo of Continental Gummi Werke AG, Hanover (figure 3.1), the report highlights Bayer's cooperative staff, who appear happy to share their research with the investigators. The report also notes the well-functioning condition of the labs.

The complete military intelligence Gloor and his colleagues returned with encompassed written reports, photographs, and material samples. Gloor was particularly keen on sharing samples so that American chemists and industrialists could handle and touch the materials.[35] However, the sharing of plastics samples was logistically challenging, and much to Gloor's disappointment, his dream of a tactile plastics sample library was never realized.[36] Despite this, the report's photographs documented human interactions with plastic foams and helped convey their novel materiality.

Using plastics trade and industry publications, a complex network of US agents in government, military, and industrial positions rapidly distributed the German plastics intelligence that had been collected by American investigators between May 20 and September 1, 1945. In October 1945, just one month after initial technical intelligence extraction had been completed, the trade publication *Modern Plastics* ran the first of two review articles reporting on plastics developments in Germany during World War II.[37] These articles were authored by its technical editor, Gordon Kline, chief of plastics at the National Bureau of Standards, a physical sciences laboratory within the US Department of Commerce. Kline's report further celebrated polyurethanes as an exceptionally dynamic group of materials previously unknown to the United States, providing an overview of technical information, applications, and economics. In addition to technical representatives from the Quartermaster Corps and the National Bureau of Standards, the publication involved members of the Ordinance Department, the Chemical Warfare Service, the Technical Intelligence Industrial Committee, and other groups, such as the Field Information Agency, Technical (FIAT). This range of contributors underscores the intricate involvement of numerous stakeholders in the transfer of German plastics.

In 1946, DeBell, Goggin, and Gloor's *German Plastics Practice: A Record, Rewritten and Amplified, from the Quartermaster Reports* soon

3.2 "Dr. Baeyer [*sic*], Porophor foamed rubber," ca. June 1945, in "Quartermaster Report: Technical Intelligence Reports; The German Plastics Industry." Walter E. Gloor Papers, Science History Institute, Philadelphia.

3.3 "Leverkusen, [Dr. Bayer's] Physical Testing Lab, foams in the foreground," ca. June 1945, in "Quartermaster Report: Technical Intelligence Reports; The German Plastics Industry." Walter E. Gloor Papers, Science History Institute, Philadelphia.

followed Kline's article. *German Plastics Practice* provided far greater technical detail than Kline's introductory articles on the findings. This 554-page, indexed publication sold for four dollars (roughly sixty-nine dollars today) and included all the images from the original Quartermaster reports.[38] Legal disclaimers preceded the US-based publication of German industry intelligence, distancing the US government from potential copyright infringements. For example, Kline's 1945 articles stressed that the US Department of Commerce was "merely distributing technical information" from Germany and that it "should be made available to all United States citizens interested in it," but that caution should be exercised regarding patent laws.[39]

Major air, space, and weaponry technologies within the military-industrial complex, particularly Wernher von Braun's research on space exploration, were essentially Nazi technologies transferred to the United States from Germany via Operation Paperclip, a large-scale secret intelligence program.[40] The extraction of German technological expertise through information, equipment, scientists, and engineers also contributed to the successful emergence of the US military-industrial complex and R & D culture that was to shape postwar US technology and everyday life via "spin-offs."

Distributing plastics intelligence to plastics R & D labs across the country ensured the stimulation of activity in an area the Quartermaster Corps was actively interested in, and it requested that companies alert the service to any promising new materials that could be useful to the US military.[41] The US Department of Commerce's involvement in the technological extraction and transfer of intelligence about Otto Bayer's polyurethane foam inventions demonstrates how plastics were part of the postwar political push by the United States to secure its position as a scientific and industrial world leader. US intelligence that had formerly been classified and was deemed less sensitive, such as that collected by Gloor, Goggin, and DeBell, was theoretically accessible to "all." Plastics experts noted that the US Department of Commerce's publishing of this data meant that it was "available not only to the US but to the whole world, and certainly the Russians are taking full advantage of it."[42] Urethane foams and other German plastics advancements, as documented in the German plastics Quartermaster reports, were no longer secret and were now available internationally. Thus, polyurethane foams began their journey into postwar domestic life and more intimate everyday interactions with the body.

The Plastics Race: Demonstrations of Soft Power

After the war, the US Air Force and American industry took an interest in polyurethane foams. Defense research and development contracts, primarily interested in the application of rigid foams for aviation, tanks, and submarines, were placed with several organizations associated with the military-industrial complex, including DuPont, Monsanto, Princeton University, Goodyear Aircraft, and Lockheed Aircraft.[43] A competitive postwar plastics race of military and civilian applications ensued.

The first polyurethane products to be commercially available on the US market were rigid foams. They were developed by Lockheed's chemical engineers, who patented a foamed-in-place rigid polyurethane technique.[44] In the postwar United States, the press and plastics companies made production comparisons with Europe, adopting a narrative of US exceptionalism to extol the superiority of American products. Plastic foams were not excluded. Industry and press reports used militarized language and references to the United States overtaking European competition.[45] By 1956, according to an unsigned article in *Industrial and Engineering Chemistry*, "the scores of products presently being manufactured from urethane foams in [the United States] represent only a fraction of European applications"; however, projected outputs indicated that the United States would soon "surpass Europe."[46]

In 1951, owing to its continued research and development in Germany, Bayer offered the first commercially available urethane protective coatings, followed by polyester-based flexible urethane foams in 1952.[47] DuPont and Monsanto had been producing pilot plant amounts of diisocyanates by 1950 in the United States, and other companies, including National Aniline and Chemical Company and B. F. Goodrich, were quick to follow, consequently initiating and expanding their R & D programs.[48] One company, however, was to take the lead in polyurethane foam technology supply and licensing in the United States. In 1954, after lengthy negotiations, Farbenfabriken Bayer AG and the Monsanto Company tapped into the surge of interest in polyurethane technology by establishing a joint transnational company, Mobay Chemical Company, which would facilitate the introduction of these German technologies to the United States.[49]

The technological and ideological postwar battle of the Cold War arms and space races has been well covered in scholarship.[50] Scholars have argued that there was a third race—one of "living standards"—fueled by the soft power of design and the rivalry between the "West" and the "East."[51] Polyurethane foams, while not considered in these studies, were certainly part of this race. Drawing on evidence from the Quartermaster reports, I argue that there was also a "plastics race." US government agencies were focused on intelligence related to plastics, particularly new polyurethane foams, to stimulate the national chemical industry. This is an essential aspect of polyurethane foam's origin story, demonstrating how the material was embroiled in an ideological, Cold War scramble for technological domination and interrelated displays of soft power. Sociopolitical constructions of postwar American racialized gender norms were at the heart of soft power displays of technological progress and consumerism. As with earlier displays of fluffy polyurethane foams, somatic engagement would remain a central theme in Mobay's promotional materials.

There is something truly spectacular and performative about the active materiality of polyurethane foam, which lent itself well to displays. Chemists at Mobay and Bayer's companies utilized this quality in promotional activities, both in the plant and lab and at public exhibitions. A Mobay employee recalled the early days of experimenting at the pilot plant in St. Louis during the 1950s. Staff would open the converted garage to curious locals, who would see the production process for the first time: "a nozzle squirting a liquid down on a moving belt; . . . the liquid turn[s] milky as carbon dioxide was formed within it; . . . the liquid mass rather suddenly begins to foam up, increasing in volume many times."[52] Wide-eyed visitors would enthuse about the expanding bubbling mass, "It's like Bromo Seltzer!"[53] Here, plastics chemistry came to life, rapidly expanding, and drew in a general audience. Polyurethane foam's raw, vital, thriving mass certainly was eye-catching and performative, making it ideal for entertaining displays that celebrated the wonders of Big Science on a smaller scale.

In the postwar era, white men in lab coats or smart suits performed such displays on varying scales. Figure 3.4 shows Otto Bayer smiling enthusiastically as he stands behind a table adorned with beakers, the largest of which contains foam material swelling over and angling toward him. Unlike actor "chemists" in DuPont's 1939 US world's fair displays, Bayer performed to an industry or university audience.[54] The mass bubbles over the confines of the container, mushrooming into a chemical cloud not entirely dissimilar to carefully monitored US imagery of mushroom clouds looming over Bikini Atoll, Hiroshima, and Nagasaki.[55]

Similar to other plastics, foam was promoted as a magical material with infinite possibilities. Unlike the photograph of Bayer, figure 3.5 (ca. 1950–59) was shot in a studio and staged. It depicts a white man dressed in a lab coat. He stands behind a table, with an assortment of immaculate glass chemistry accessories at his disposal. These are pristine and untouched: He works confidently and knows exactly what to do. Surrounded by abstract shapes suspended in midair, he has a creased brow and a focused expression. His left arm reaches up to pour the fluid contents of a metal bucket into a transparent conical shape, secured by a stand and steadied with his right hand. The caption reads, "The chemical being poured and the foams surrounding the scientist . . . are made from Du-Pont's new organic isocyanates. Customers combine [these] with other chemicals to produce urethane foams, which can be soft or rigid, heavy, or light and have potential uses in many industries."[56] The DuPont company circulated this image to promote the infinite transformations that "men of science and industry" had at their fingertips, thanks to this new and highly innovative material. An instant conjuring of shapes is shown, concealing its complex technological production processes, while the test tubes, pipettes, and beakers remain unsullied. Finished shapes appear as if by magic, suspended in time and floating in space. This image promotes polyurethane foam as malleable, controllable, low-cost, and instant.

As French theorist Roland Barthes observed in his 1957 essay on plastics, "So more than a substance, plastic is the very idea of its infinite transformation."[57] This photograph connotes that foam could be shaped into any form. It depicts a white man in a lab coat, responsible for creating a new "miracle" material that is no longer reliant on natural imports. In keeping with capitalist models, labor processes are disguised, rendered neither messy nor complex: Chemistry utensils are left untouched, finished objects appear as if by magic, the product of a sole white male creator.

In 1957, *Life* magazine featured a two-part photo series named "Man's New World." The second part, titled "Tomorrow's Life Today," celebrated the latest technological developments on the consumer market, focusing on domestic spaces and "such modern wizardry as plastics."[58] Photographs documented a bright future of colorful plastic living, including Nylon Airhouses—inflatable structures made from US Rubber's Fiberthin, a vinyl-covered nylon. Another highlight was Monsanto's iconic experimental

3.4 (*left*) Otto Bayer performing a foam experiment with polyurethane during a lecture in Göttingen, Germany, 1952. Bayer AG.

3.5 (*below*) DuPont, "Uses of Organic Isocyanates," ca. 1950–1959. Hagley Museum and Library, Wilmington, DE.

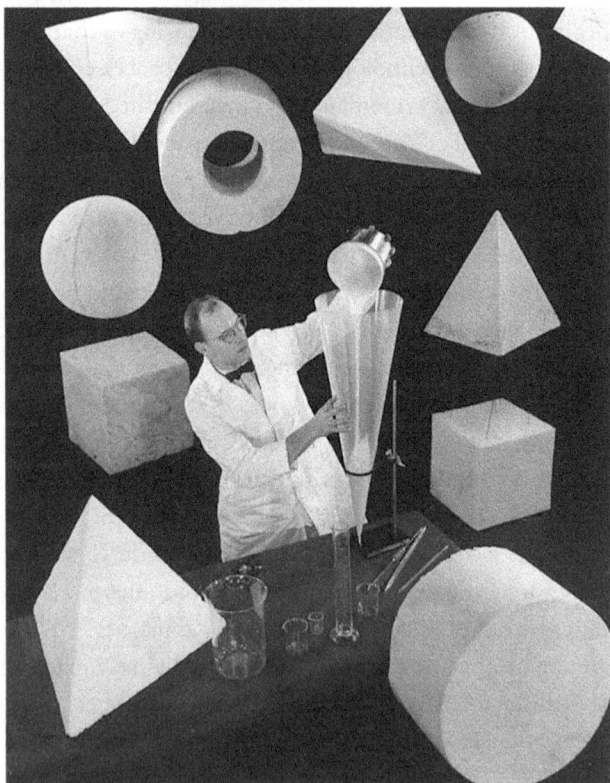

"House of the Future," made entirely of plastic and held together by rigid urethane foam, on display at Disneyland's Tomorrowland.

Also featured in "Tomorrow's Life Today" was a spread titled "New Plastics for Interior" (figure 3.6). The accompanying photograph, staged on a minimalist white set, features abstract shapes in bright colors that was described as "foam furniture created for a stylized interior-of-the future by the Mobay Chemical Company." At half the weight of competing foam rubber materials, these modern polyurethane pieces were proudly displayed as improvements to living standards. An intergenerational group of slim, light-skinned models—signaling a nuclear, heteronormative white family—appears to rest comfortably on an assortment of giant primary-colored polyurethane foam shapes. The group is dressed in tight black leotards—those worn by the male models cover their legs and shoulders, while the female models are dressed in a shorter more revealing version.

Most of the group watches intently as a young boy in the foreground pours a beaker of red liquid into a bowl balanced on circular, electric blue slabs of foam, essentially reenacting Bayer's foam displays. The accompanying caption reads, "Boy in foreground is demonstrating action of foam which rises, yeastlike, 30 times its size." In this future vision of the American home, furniture has been stripped down to ultralight, basic shapes in primary colors that can seem to be magically conjured from the latest materials, which mushroom from chemicals mixed in scientific beakers. However, despite polyurethane foam's ability to unlock novel approaches to design, *Life*'s foam future imaginary ultimately reinforces established heteronormative, racialized gender roles—it is the young white boy who has the means and vision for building the future in "Man's New World."

Chemical companies promoted polyurethane foams to manufacturers as a material that could increase profits and efficiency by reducing the costs of design, labor, production, materials, and transportation.[59] For example, a premolded foam cushion or piece of furniture requires a different approach to craftsmanship and assembly than a hand-upholstered, less streamlined design stuffed with nonsynthetic fibers like wool, cotton, or hay. Instead, polyurethane foams could be premolded to a standardized shape, a concept that chemical companies celebrated with promotional imagery evoking an illusory alchemy of petrocapitalism. Barthes reflected, "Plastic . . . is less a thing than the trace of a movement . . . and as the movement here is infinite, transforming the original crystals into a multitude of more and more startling objects, plastic is, all told, a spectacle to be deciphered: the very spectacle of its end-products."[60] In the postwar United

3.6 "Foam Furniture" in "Tomorrow's Life Today—Man's New World: Part II," *Life*, November 11, 1957, 138.

States, capitalist imagery of seemingly instant soft foam shapes signified absolute abundance and offered new design possibilities—apparently magically conjured from air, polyurethane could be molded into any shape, size, or density required.

Soft power was also an element in the ideological battle of the Cold War.[61] There are four interconnected tenets central to postwar political displays of American soft power: the domestic, racialized gender, technological advancements, and material consumption. These principles are also key to polyurethane foam's commercial promotion and successful peacetime transfer to American suburbia. Popular social commentator David Riesman's satirical essay "The Nylon War" (1951) describes an imaginary US Cold War tactic (Operation Abundance) that involved bombing the Soviet Union with American consumer goods, including the dropping of "200,000 pairs of nylons." Riesman explained that the reasoning behind this operation was simple: "That if allowed to sample the riches of America, the Russian people would not long tolerate masters who gave them tanks and spies instead of vacuum cleaners and beauty parlors. The Russian rulers would thereupon be forced to turn out consumers' goods or face discontent on an increasing scale."[62] In Riesman's satire, labor-saving devices and glamour for women proved a winning combination in the fight against Communism. Similarly, in Stanley Kubrick's atomic film parody *Dr. Strangelove* (1964), an American bomber aircrew flying over the Soviet Union is famously handed survival kits for their landing, packed with rubles, food, medical supplies, condoms, lipsticks, and nylons.

Incidentally, the export of a white American feminine ideal as shaped by the latest synthetics was supported by US government action. Ida Rosenthal, founder of Maidenform, was the only woman and underwear manufacturer invited to join an exchange between the United States and the Soviet Union as part of the 1962–63 US-USSR Exchange Agreement.[63] Journalists enthusiastically covered Rosenthal's trip, likening her foam-padded bullet bra designs to a powerful arsenal capable of seducing even the most devout Communists with the allure of capitalism. Upon her return, Rosenthal lamented the inferior quality of synthetics in the Soviet Union: they "claim to be their own, but who knows."[64] Emphasizing the superiority of the United States' synthetic advancements, Rosenthal pledged to export Maidenform's products across the Iron Curtain. Nationalist narratives of plastics advancements, consumerism, and a glamorous, racialized gender ideal were interconnected. White American women shaped by the latest synthetic foundations were central to Cold War displays of soft power.

The Cold War period saw a range of consumer goods strategically paraded by the United States in a spectacular standoff display between the East and the West.[65] American consumer goods were also deployed in the Americanization of postwar Europe and Japan. Design historians have argued that modern design and products were used to export and promote American ideology during the Cold War, when they were "assigned an ambassadorial role in Europe" and displayed in exhibitions, including MoMA's iconic "50 Years of American Art."[66]

The Kitchen Debate of July 1959 is a well-documented example of Cold War ideological combat, fought out as words soared over objects on display in an exhibition.[67] On July 24, American Vice President Richard M. Nixon and Soviet Premier Nikita S. Khrushchev toured the American national exhibit at the Moscow Fair. At one point, they stopped in front of General Electric's radiant yellow kitchen display—a domestic space shut off from the public—and debated their respective ideological values. The Kitchen Debate demonstrates the political importance assigned to the domestic by Nixon, who stressed that the ideal of the suburban home equipped with modern appliances and a white nuclear family was the foundation of the United States' global superiority.[68]

The "model" home was not just the domain of the white middle-class, heteronormative male breadwinner and female homemaker, but also a vast assortment of consumer goods, "represent[ing] the essence of American freedom."[69] Essential to this problematic understanding of "liberty" was what Nixon referred to as the "housewives' choice": the availability of numerous manufacturers and labor-saving devices.[70] In this context, freedom became equated with spending power, material consumption, and the variety of goods available for purchase. Heteronormativity, consumerism, social mobility, and the domestic were powerful tools in American postwar constructions of freedom and peacetime prosperity. Imagery in whitestream media "reinforced an idea that postwar American suburbia just happened to be white, when in fact racial segregation was legislated through federal laws and private development practices that privileged white homebuyers exclusively."[71] In this carefully constructed and contained American suburban landscape, there was growing affluence among the white middle-class.

In the postwar United States, many African Americans prioritized spending on furnishing their interiors in the latest fashions. The average number of objects in American postwar homes increased, afforded by changes in spending power, manufacturing methods, a multiplicity of design movements and styles, plastics, and consumer habits. In a widely read 1949 *Ebony* editorial, John Harold Johnson, founder of Johnson Publishing, noted that Black consumers invested in the home to create a haven from daily discrimination and violence. For Johnson, equality through consumption differed from mere imitation and was an important aspect of African American pride and identity.[72]

While the wages of Black families were markedly lower than those of white households, African Americans continued to spend a higher proportion of their income on furniture and home furnishings.[73] A 1963 *Ebony* report on the furniture market noted that in the 1950s and into the 1960s, Black consumers "made up the most dynamic purchasing power in the home furnishings industry." The report noted that this was an "astounding" statistic, considering that African Americans had "accomplished this rating despite gross economic discrimination."[74] In cities beyond the segregated South, such as Chicago, access to property and home funding through mortgages improved for some African Americans, which had a knock-on effect on demand for home furnishings, where polyurethane foams continued to thrive.

Postwar American racialized gender ideals were central to highly politicized displays of American soft power. After working on the US home front in the absence of men, middle-class women were encouraged to return to the home. The ideal American suburban home—portrayed as a showpiece of affluence and public display, an "exhibitionist myth"—was stealthily used as what design historian Gregg Castillo terms a "Trojan House" in the Cold War battle of conflicting material ideologies.[75] Here, as design historian Beatriz Colomina observes, "The housewife had become a soldier on the home front; the kitchen, the command post from where she not only controlled the domain of her living space but was purported to defend the nation."[76] The default white, heteronormative housewife was presented as a domestic embodiment of American capitalist values.

The bombshell, scholars argue, is the antithesis of the housewife.[77] In *Pornotopia: An Essay on Playboy's Architecture and Biopolitics*, Paul B. Preciado discusses that in the postwar United States, female bodies outside the domestic space were represented as "disruptive" sexual forces that needed to be disarmed and redomesticated in the private realm of the home.[78] Engaging with postwar histories of domesticity, Preciado writes that US government entities perceived "the increasing presence of women within the public sphere during and right after the war as a civil danger and a sign of sexual disorder, relating paid work and economic independence to promiscuity and prostitution."[79] The independent female body outside the home became further politicized and denounced. It was increasingly represented as dangerous and capable of mass destruction, a threat to the racial and sexual purity of the white, heteropatriarchal American nuclear family. In this context, the self-supporting woman was understood as unpatriotic and "in some way un-American."[80] Political bodies such as the American Social Hygiene Association and the Civil Defense Association sought to make white middle-class women return to the home.

Simultaneously, many US postwar homes were increasingly filled with alluring new technology, including the latest in plastics, along with so-called labor-saving devices. Historians of design and technology have extensively discussed the politicized role of gendered consumer goods and technologies during this era.[81] In the Kitchen Debate, Nixon famously praised postwar advances in domestic technology, such as the washing machine, for emancipating the housewife from domestic chores, thereby highlighting their ideological function in racialized capitalist gender roles. However, social inequality remained clear when comparing the percentage of washing machine ownership between Black (49 percent) and white US households (76.2 percent) in 1963.[82] The most potent weapons in Riesman's imaginary American arsenal were gendered consumer items, such as nylons, vacuum cleaners, and beauty products. Historian Susan E. Reid reflects that "a 'universal feminine' desire to be a leisured consumer and to beautify herself and her home was presumed to transcend the Cold War ideological divide."[83] Western observers constructed Communist Bloc women as "deprived, dowdy and work-worn"—the antithesis of the white American middle-class housewife.[84] While Eastern Bloc women were imagined as rough and tough, the ideal American housewife was presented as polished, soft, accommodating, glamorous, and embodying a consumerist hyperfemininity of the capitalist West.

The Power of Selling Softness: Mobay and American Domestic Markets

Plastic foam companies such as Mobay identified that urethane foams were a lucrative area of investment and tapped into a burgeoning postwar interiors sector. House furnishings accounted for the largest market for flexible urethane foams, with 60 million pounds consumed in 1960—a figure projected to reach 840 million pounds by 1975.[85] For example, by 1965, Englander, a subsidiary of the Union Carbide Corporation, advertised its gendered "Princess" mattress, with an asterisk next to the word *foam* to specify that it was made from urethane, which had replaced most of the earlier types of foam rubber.[86]

The second largest market was the automobile industry, followed by "marine and flotation, textiles and coverings (including footwear, luggage, wall and window coverings, etc.), and floorings."[87] Indeed, foams spilled out of the domestic space and into the car, with American car advertisements of this period positioning seats made from plastic foams as a luxury selling point: "big extras at no extra cost."[88] Besides its use in home furnishings, urethane foam was particularly appealing to the aviation industry. By 1956, cored flexible foams were being successfully applied to US airplane seat cushioning, where their weight savings alone reduced costs.[89]

At the time, the material was understood as boasting both longevity and lightness, providing ideal ventilation and seating properties.

Mobay Chemical Company capitalized on the softness of polyurethane foam and centered its promotional materials on the substance's inviting tactility. Promotional materials, including the booklet *Put the "Soft Sell" of Urethane Foam into Your Furniture Sales Story*, emphasized the somatic joy of touching and squeezing foam to its readership of "salesmen."[90] It included a summary page of polyurethane foam's main selling points, with six photographs (figure 3.7) featuring light-skinned hands engaging with the material. The accompanying copy listed urethane foam's benefits: "TOUGH!," "STRONG!," "CHEMICALLY STABLE!," "LIGHT WEIGHT!," "SOFT!," and "VERSATILE!" The summary page encouraged salespeople to replicate these words and actions in a sales pitch aimed at female consumers.

Chemical companies also enlisted renowned design professionals to promote their materials. In an ad published in *Interior Design*, Mobay featured influential designer Freda Diamond approving the softness of its foam. She is shown in a professional setting, behind a desk, dressed smartly in a jacket and holding a sheet of urethane foam. With her light-skinned fingers spread across the soft, marshmallow-like material, she looks into the camera. The words "truly a designer's material says Freda Diamond" are printed at her eye level.[91] In this image, a white woman design professional is used to signal authority and approval of a new material for a middle-class female consumer.

Chemical companies frequently employed white women's bodies to domesticate their foam products. Depicting and appealing to middle-class "homemaker" consumers was an especially important part of their promotional activities. Goodyear, which specialized in latex foam rubbers, ran a series of ads throughout the 1950s that depicted a white woman's hand sinking into foam.[92] The accompanying text described the fleshy active materiality of foam: "Press your fingers down on Airfoam and it gently gives, cradling them buoyantly. Lift your hand and AIRFOAM instantly plumps back to full-rounded shape."[93] In these ads, Goodyear chose to represent the manicured left hand of a white woman wearing a ring on her fourth finger, signaling that she is married. The image of the woman's hand is sensuous, with her red varnished fingers sinking into the foam, encouraging the viewer to imagine the tactile sensation of the soft, squishy foam. However, the gendered sensuality of this image and text is carefully contained and redirected toward racialized, heteronormative understandings

here's why furniture designers and manufacturers

choose . . . **URETHANE FOAM**

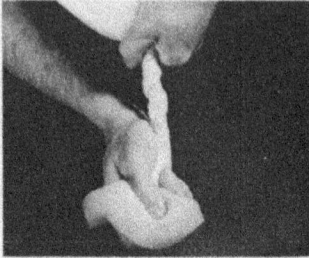

TOUGH! Tear strength up to 8 psi; tensile strength up to 40 psi; may be stretched to 500% of original length without parting.

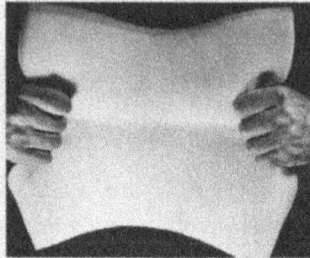

STRONG! Urethane foams have been flexed at 30 cps at 30-80% deflection for 3,000,000 cycles with minimum property loss.

CHEMICALLY STABLE! Urethane foams are unharmed by dilute acids, alkalies, detergents, water or dry cleaning solvents.

LIGHT WEIGHT! At 2 lbs. pcf, properties of urethane foams are superior to other foams weighing twice as much.

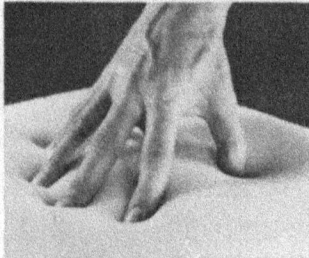

SOFT! An entirely new experience in cushioning effect. You will have to *feel* urethane foams to understand the growing preference for these materials in cushioning markets.

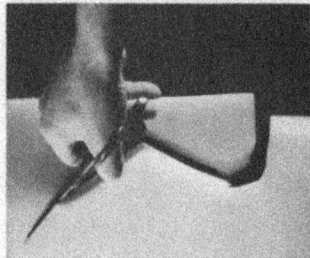

VERSATILE! Urethane foams may be sliced, sawed, sewn, stapled, glued, sprayed or foamed in place, die-cut, flocked, laminated, tufted, contour-cut, with standard fabricating equipment.

3.7 *Put the "Soft Sell" of Urethane Foam into Your Furniture Sales Story* (Pittsburgh: Mobay Chemical Company, 1959). Freda Diamond Collection, Archives Center, National Museum of American History, Smithsonian Institution, Washington, DC.

of domesticity. As with the Hays Motion Picture Production Code that regulated most films by major studios in the United States from 1934 to 1968, "the treatment of bedrooms must be governed by good taste and delicacy."[94] In this Goodyear image, as in many other whitestream representations of the time, the sensuality of touch in the bedroom is depicted as exclusively reserved for white, heterosexual married life. Gender roles were being reshaped in the postwar period, but advertisements like these reinforced established norms and redomesticated white, middle-class American women. Some advertisements likened the inviting softness of white women's bodies to that of pliable foams used to furnish the home. Ultimately, in such imagery, white middle-class women were frequently positioned as accommodating objects of desire and domesticity belonging firmly in the family home.

Mobay's sales executives were keen to distance the company's urethane foams from other established plastic foams, such as Goodyear's. Instead, Mobay claimed its urethane foams offered an "entirely new experience" that was exciting, modern, and cutting-edge.[95] Mobay instructed salespeople to promote polyurethane foam as a synthetic material designed by chemical engineers with a particular function in mind. The company was concerned about distinguishing urethane foam from earlier foam rubbers made from natural materials, which had been prone to porousness and crumbling. Copywriters strategically chose to place urethane foam within an established plastics context. They notably referred to nylon as a gendered example that salespeople could use to relate Mobay's foams to another synthetic, one that had supposedly already improved the everyday lives of many women: "Perhaps best-known to your homemaker customer, is their replacement of silk (a natural fiber) with nylon (a synthetic) in women's hosiery." A quick reference to safe, soft, and sensuous nylons, worn intimately on the body, could hopefully assuage concerns about plastics and their proximity to people.

In the postwar United States, advertisers used industrial and peacetime reconversion as a seal of quality and domestic safety.[96] Chemical company sales advisers consulting for Mobay reasoned that if polyurethane could pass safety tests for aviation and automobiles, it would surely meet the strict standards of their target consumer: "Mrs. Homemaker." The company instructed salespeople to assure prospective clients that their foams were inert and "one of the cleanest and safest materials ever created by man or Nature." They argued that "safety" and "purity" were a few of the reasons why "[urethane foam] is being selected by automotive and aircraft firms for seating, insulation, and safety padding."[97] In the context of the Cold War United States, urethane foam's military-industrial track record could offer comforting yet also technoscientific protective padding to the domestic body, which became increasingly enveloped in plastics.

Comfort and softness are central to a gendered and padded icon of postwar modern design that references the uterus: Eero Saarinen's Womb Chair. Saarinen designed this organic-shaped chair, upholstered in plastic foam, for Knoll in 1946.[98] Florence Knoll famously asked Saarinen to create a comforting, cocoon-like chair for her to retreat into "like a big basket of pillows that I could curl up in."[99] She was so enamored with the result that she initially named Knoll chair no. 70 the Curling Chair but later called it the Womb Chair. It was frequently depicted as upholstered in the company's signature bright red color, which simultaneously emphasized its anatomical nomenclature.[100] Hans Knoll allegedly refused to refer to the chair by its popular name and unsuccessfully implored for a more conventional alternative. Saarinen responded, "I have been thinking and thinking about a printable name for that chair, but my mind keeps turning to those which are more biological than less biological."[101]

In a period of US history commonly referred to as the baby boom, the Womb Chair's generously padded curves provided comfort and support to those who could afford it. Its popular name firmly resituated and reinforced women's reproductive organs as supportive infrastructure and objects of comfort in the home. Scholars have explored the imagery of the Womb Chair, connecting it to Cold War anxiety, suburbia, modernism, gender, and the construction of racial whiteness.[102] Additionally, the Womb Chair became a "[stalwart] of US touring exhibitions of the 1950s," serving an ambassadorial role in Europe.[103] Modern design and objects such as the Womb Chair played a significant role in shaping racialized, classed, and gendered identities. Moreover, designers, journalists, advertisers, and other actors employed polyurethane foam's soft materiality to construct postwar domesticity and racialized gendered identities.

In addition, polyurethane foam was used in mass-market furniture across a range of styles and marketed to a broad spectrum of consumers in the postwar United States. While most of Mobay's promotional materials focused on modernist design, the company was keen to show that polyurethane foams were not just reserved for modern furniture. Mobay also employed "designeering" (a combination of the words *design* and *engineering*) professionals to detail how this material could enhance more traditional forms, such as the classic wing chair, commonly found in homes and retail spaces across the United States.[104] While imagery such as "Tomorrow's Life Today" (figure 3.6) reimagined foam furniture in a white, gallerylike space, in reality, most working-class people living in urban environments in the postwar United States faced cramped living conditions, particularly when the influx of returning troops led to a housing shortage. Polyurethane foam's lightweight properties also made it ideal for use in convertible furniture, such as Castro Convertibles, whose couches featured a lightweight "feather-lift" mechanism "to conquer living space" and maximize floor space for people living in crowded apartments.

In the postwar period, *Jet* and *Ebony* showcased soft, upholstered, streamlined sofa designs featuring curved shapes. For many African Americans, the home was a space to enjoy freedom of movement, design, and consumer goods at a time when many spaces remained segregated.[105] It was also a space in which to entertain and build community. Foam's lightweight quality and its ability to be shaped into modern, organic, flowing forms—described as "graceful s-shaped, bumper-ended" and "sleek" and credited with freeing up space while "[adding] beauty and an illusion of spaciousness to the modern home"—appeared to be more worthy of comment in *Ebony* and *Jet* than the material itself (figure 3.8).[106]

Foam furniture coverage in *Ebony* and *Jet* focused on the design of the homeware itself and only mentioned the materials in passing, if at all. Whitestream publications like *Life* overtly celebrated plastics as "technical triumphs to shape daily lives" in photo series such as "Man's New World" and ran ads from chemical companies for foam plastics and mattresses. Goodyear did not advertise foams in *Ebony* and *Jet*; instead, the company advertised its car tires, an object for which touch is not a priority. Unlike *Life* or *Interior World, Ebony* and *Jet* did not run lengthy features on plastics futures. It was not until the late 1960s and early 1970s that foam chemical companies like Monsanto started gaining space in *Ebony*, not for reporting or advertising their plastic materials for the home, but for recruitment purposes only.[107] When it came to coverage of plastic foams in Black publications, the focus was on pleasing the eye. A modern piece of foam-upholstered "silhouette furniture" was admired for its "slender willowy effect as graceful as a fashion model's outline[;] the pieces are simple, yet artistically designed and strongly constructed."[108] This commentary focuses on the overall design aesthetics and the joy of seeing and experiencing them, rather than on technoscientific developments in plastics.

Designers and advertisers often used gendered language that referenced beauty culture and the body to describe urethane foam. Bernard

3.8 S-shaped sofa in "Modern Living: New Ideas for Sectional Furniture," *Jet*, June 9, 1955, 38.

Castro advertised his Castro Convertible designs as "slimline" and "Sleeping Beauties."[109] Freda Diamond similarly praised urethane foam's "slimline" qualities and its ability to afford designers "greater flexibility in the achievement of sleek, trim, graceful, upholstered furniture designs."[110] Another designer admired the way that polyurethane foam could achieve "intricate curves" and be "shaped to any contour."[111] In some descriptions, Mobay's "new look" foam took on fleshlike qualities: "cushions remain plump and lively" but lacked the "excessive buoyancy and annoying 'jiggling' effect characteristic of other types of foams." Here, Mobay presented urethane foam as pliable and soft, yet its softness was not "out of control," like that of gendered flabby flesh or hyperactive bouncing foams: It moved and reacted exactly as directed. Mobay's foam is superior; absolute control is exerted over foam's materiality. The company pursued furniture designers by asking, "What other material at your command can be controlled in density to fit its function; provided to you in any shape or form; or molded in place?"[112] In promotional materials like these, urethane foam cushioning avoids excessive buoyancy: The "designeer" retains full control, adjusting the material to the ideal softness and without the "annoying" uncontained wobbling of competing foams.[113] Here, foam is presented as a gendered material of science, exuding Modernist design precision, discipline, and control.

Strikingly, chemical companies like Dow and designers such as Gaetano Pesce later experimented with seating by dressing curvy polyurethane foam shapes in snug-fitting, stretchy nylon and Lycra skins.[114] These "garments" could be stretched taut and zipped shut to contain wobbly foams, similar to the shaping technology of a smoothing girdle. This was particularly significant in Pesce's 1969 UP 5 chair design (figure 3.9), which he modeled on the curves of the Swedish actress and iconic white blonde bombshell Anita Ekberg, famed for her role in Federico Fellini's *La Dolce Vita* (1960).[115] As with the Womb Chair, Pesce's design was upholstered in red material. The chair came vacuum-packed; when opened, it instantly mushroomed to its fully actualized, final state. Design historian Jane Pavitt writes, "For Pesce, plastic was design made flesh. His material explorations were rooted in an interest in the visceral and sensual power of objects. Pesce's objects had life-cycles—they reflected and recorded the process of ageing and decay of the human body."[116] In these later examples, polyurethane foam's fleshiness was even more closely linked to postwar US concepts of the bombshell and her associated plump, curvaceous corporeality.

Polyurethane foam could be shaped into any form; its seamless, foamed-in production was particularly pleasing to the eye, aligning with modern, streamlined visions of excellent "craftsmanship." Its production methods seemed instantaneous, as liquid was poured or injected into a pre-molded shape, enabling the creation of fashionable, smooth, organic forms. Polyurethane foam was informing approaches to design, and design was affecting the latest petrochemical company developments in foams. The lightness and supple pliability of polyurethane foam enabled postwar designers to experiment with form—creating sleek modernist lines—and function, as it was purportedly easier to clean. Foam's materiality was touted as immediate: Shapes could be produced quickly to meet the demands of designers and prototypes, as well as consumers themselves, making it an ideal material to expand from military and industry into burgeoning US postwar domestic spaces. The "soft power" of polyurethane foam embodied US postwar political values of technological advancement, the domestic, material consumption, and heteronormative, racialized gender ideals.

As documented in military reports, plastic foams like polyurethane offered a unique range of material properties that invited touch and corporeal

3.9 Gaetano Pesce, B&B Italia, UP seating series, 1969.

interaction. This materiality also served as its impetus, from the lab to the living room, ensuring its postwar transfer to the United States and successful introduction to the domestic market. The politics of plastic foams' commercial application in the designed environment of the post-war United States is entangled with discourse on soft power and modern design. Chemical companies like Mobay presented polyurethane foams to designers and consumers as a scientifically engineered, ideal, pliable, and safe material with which to furnish the modern middle-class American home. Key actors central to the promotion of plastic foams, including chemical company advertising staff and industrial designers, harnessed these sensual properties for foam's postwar domestic promotion and applications by presenting a carefully constructed ideal of white American womanhood and heteronormativity.

4 OUTWARD, UPWARD, AND INWARD / IMPLANTING FOAM FOUNDATIONWEAR IN THE POSTWAR UNITED STATES

This chapter follows and contextualizes the impact of foam plastics on the shaping of the body, focusing on the mid-1940s to the mid-1960s. Building on the previous chapter's exploration of urethane foam's military-industrial origins and materiality, it charts the movement of foam across the surface of the body, through padded designs on the skin's surface, to its implantation within. The chapter's structure follows a material and thematic trajectory—outward, upward, inward—rather than a chronological one, illustrating examples within these three areas of application.

Foam plastics moved outward from some of their original military-industrial applications. Polyurethane foam, for example, pushed flesh outward from the body and into the fashionable silhouette—from conical missile bras to accentuated hips and buttocks. The chapter begins with a section on Hollywood and the entertainment industries to contextualize foam falsies and implants within the "star system," as well as wider US celebrity and consumer culture.

Foam plastics moved upward in whitestream media, as actors—including copywriters working for foundationwear and chemical companies, cosmetic surgeons, and journalists—claimed that foam plastics were an ideal tool for self-transformation. Queer, trans feminine, and gender-nonconforming people, largely excluded from this discourse, also engaged in DIY and self-fashioning practices with foam plastics.

Foam plastics moved inward as cosmetic and plastic surgeons implanted foam-based objects, resembling fashionable falsies, to permanently shape the body. In this final section, I examine the relationship between racialized and gendered ideals of the conical breast in fashion and medicine.

A wide range of actors used foam plastics for body contouring, both externally with padding devices and foundationwear and internally with implants. In some instances, foam technologies of the body upheld and conformed to American postwar norms of femininity and womanhood; in others, they resisted and subverted them. Changes in technology informed the fashionable conical bustline aesthetic that remained popular in the United States throughout the 1950s and 1960s, while the silhouette, in turn, influenced the development of foam plastics and related technologies such as cosmetic surgery. Actors in medicine and cosmetic surgery pathologized racialized gendered bodies, in keeping with the fashion for a conical bust. By incorporating a polyvocal intersectional narrative and analyzing the connections between seemingly disparate fields, I show that what medical professionals presented as healthy and aspirational was shaped by changes in technology and fashion. Makers and users crafted their own aesthetic interventions from foam plastics, enabling them to enact agency over the body and shape it as they desired. However, this agency was often influenced by intersecting oppressions and resulted in access inequities.

Foam Moves Outward: The Entertainment Industry and Falsies

The history of the "bombshell" and the contouring of her body, particularly her conical bustline, are enmeshed with the entertainment industry, glamour, and celebrity culture.[1] Foundationwear brands, such as Hollywood-Maxwell or Frederick's of Hollywood, capitalized on the allure of silver screen sirens to market their products by referencing the American film industry's capital in their names. The conical "whirlpool" stitched bra is often associated with the postwar era, when it became known as the "bullet" or "missile" bra; however, it made its initial appearance in the 1930s. In 1934, a year after the release of the film *Blonde Bombshell* and prior to its later use of foam padding, Hollywood-Maxwell launched its signature whirlpool stitched bra for a conical bustline. Advertising emphasized its links to Hollywood studios. Similarly, Maidenform's best-selling design, the "Chansonette"—patented in 1937 and launched in 1938—was promoted as "delightfully young in line" and promised to "[give] bosoms the new pointed roundness."[2] It arrived a year after the white actress Lana Turner starred as the "innocent girl next door" in *They Won't Forget* (1937), gaining the nickname the "Sweater Girl" for her tight-fitting top and pointed bustline.

The fashion for a conical bustline began to emerge in the 1930s but was later halted due to rationing. The synthetic materials that lingerie designers were using to achieve such a silhouette were soon needed for military applications, preventing further experimentation and production. In the postwar United States, once materials were free from rationing and made commercially available, foundationwear brands such as Maidenform could offer a range of conical designs for achieving a fuller, uplifted bustline. The popularity of Maidenform's Chansonette design continued until the late 1950s, and its iconic "I dreamed . . ." advertising campaign ran until 1969.[3] Indeed, as objects, images, and other primary sources discussed in this chapter will demonstrate, the Sweater Girl conical bustline remained a fashionable silhouette well into the 1960s. A journalist writing for *Ebony* magazine noted that, in the postwar United States, the look returned to popularity during this period, making repeat appearances on its front cover and in other Black publications, including *Jet* (figure 4.1).[4]

Celebrity culture, as presented by Hollywood, encouraged spectators to experiment with different looks, roles, and identities modeled after their silver screen counterparts. Lana Turner's conical silhouette offered a demure, innocent "Sweater Girl next door" look. Meanwhile, Jane Russell's smoldering femme fatale image in *The Outlaw* (1943)—in which she was famously rumored to wear a cantilevered bra specially designed by the film's producer, former engineer Howard Hughes—was more overtly sexualized and met with censorship. Rita Hayworth's title character in the film *Gilda* (1946) is another example of the lusty vamp figure that was so popular during this period. Platinum blondes, such as Marilyn Monroe in *Gentlemen Prefer Blondes* (1953) and Jayne Mansfield in *The Girl Can't Help It* (1956), offered a more vivacious, "va-va-voom" vision of the curvaceous bombshell.

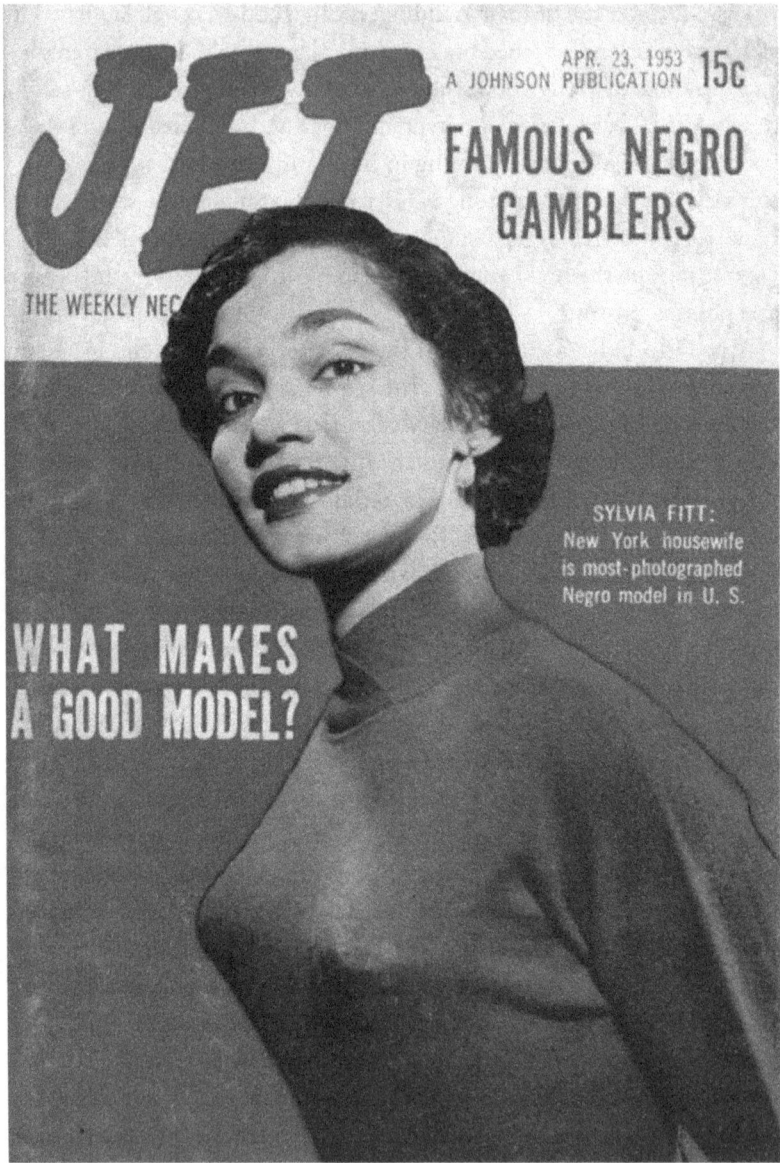

4.1 Model Sylvia Fitt on the cover of *Jet* magazine, April 23, 1953.

Although Hollywood's film industry exerted relatively little influence on shaping body image ideals for African Americans, its promise of glamour and stardom had widespread appeal; in reality, however, young starlets of color were frequently subjected to exclusion, colorism, and roles confined to negative racial stereotype.[5] African American magazines such as *Jet, Ebony, Sepia, Tan Confessions*, and *Our World* reported on Hollywood's double standards for aspiring Black actresses. Journalists observed that Hollywood glamour in the postwar period was "taboo" for talented Black women like Lena Horne, whose MGM contract restricted her to singing parts only.[6]

Because representations of Black actresses in Hollywood were incredibly limited, the most popular celebrities within Black communities tended to be singers. Historian Maxine Leeds Craig interviewed Black women about their postwar US experiences, and "many said there were no famous beauties with whom they could identify."[7] Lena Horne, Dorothy Dandridge, and Diahann Carroll were the most frequently named women in her interviews who were admired for their beauty. However, these figures represented a light-skinned vision of Black womanhood, with "sharp" white European features and "good hair."[8]

Given the lack of opportunities for women of color in the US film industry, African American magazines encouraged their readership to pursue modeling and thereby expand the concept of celebrity outside Hollywood. Unlike whitestream fashion magazines, these periodicals promoted a less lean, more curvaceous figure. This body ideal was also seen in the Black models who won beauty and pin-up contests, as well as those who were cast as chorus girls.[9] Throughout the 1950s and 1960s, Black professionals working in the entertainment and media industries used the postwar market as a way to "rewrite the visual rhetoric of Black bodies in commercial representation" from racialized stereotypes to "the middle ground of commodified sexuality."[10] Historian Elspeth Brown notes that this was in keeping with a "midcentury version of glamour" that harnessed and "sanitized" sexuality to sell goods.[11]

African American magazines featured fashion stories and pin-up centerfold shots, as well as advice for readers on how best to match foundationwear made from the latest synthetic developments to new dress trends. Some publications featured more sexualized advertisements and imagery than others. *Ebony*, modeled on *Life*, printed lingerie ads that tended to be more reserved. Meanwhile, *Jet*, a weekly magazine aimed at a younger Black demographic, ran ads for the mail order company Frederick's of Hollywood, which was famed for more extreme conical brassiere designs, highly padded foam foundations, and overtly sexualized marketing. In comparison, *Tan Confessions*, known for its more sexual content, ran more spots for extreme pointed brassieres, girdles, and transparent lingerie using the latest in synthetics.[12]

The materials available to foundationwear manufacturers were a major determining factor in design and production. Foam plastics' yielding materiality played a key role in the proliferation of foundationwear items used to achieve a curvaceous look. Padded foam details could be produced quickly and at low cost, further securing their part in giving shape and support to structured hourglass figures. Such changes in plastics technology shaped the new silhouettes and the competitively priced shaping devices they required. Foam pads could be made simply by pouring urethane foam solution into designed molds to create identical forms and new options for "pointed roundness." Magazines and advertisers claimed that a foundationwear wardrobe was needed to keep up with ever-evolving styles and necklines, and urethane foams' short production times fulfilled this task.[13]

The fashion for bras and girdles padded with plastic foams after World War II is certainly linked to findings in the US Army Quartermaster reports on German plastics, as well as the subsequent increase in the production and availability of plastic foams in America. News reports and press releases from foundationwear brands, furniture manufacturers, and chemical companies reveal that synthetic materials and foams, which had briefly been introduced to female consumers before the war, were improved on through wartime usage. These materials were now increasingly available to the postwar US consumer market.

In 1946, within a year of the end of World War II, Maidenform made use of materials developed during wartime to launch two new versions of its Masquerade bust pads, which had first been introduced in 1936 (figure 4.2). The company noted that it was unable to supply these popular products because military demands "cut off the supply of rubber."[14] Maidenform assured its buyers that "now that rubber has been finding its way back into civilian goods, lighter, airier foam rubber bust pads are being introduced ... in a new shape and ... sizes. They are hollow cone-shaped pieces of foam rubber encased in Rayon Satin envelopes." Early 1940s Sears catalogs also offered "Flatterettes," two types of conical "bust pads."[15] However, with more foam plastics available, competition rapidly increased in the postwar era. The Corset and Brassiere Association

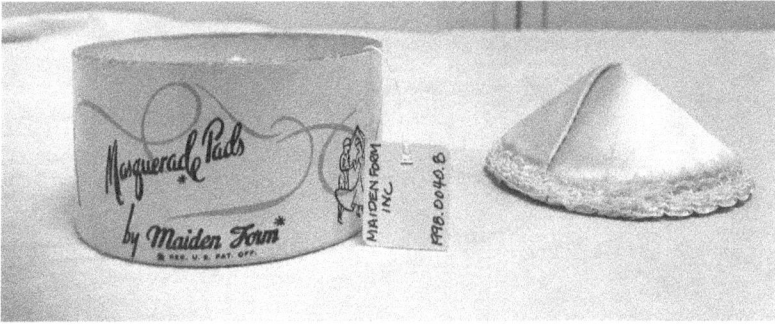

4.2 Maidenform Masquerade falsies and packaging, ca. 1948. Photograph by author. Division of Home and Community Life, National Museum of American History, Smithsonian Institution, Washington, DC.

of America estimated that over five million pairs of falsies were sold in 1946, "with the purchasers including some of Hollywood's most glamorous damsels."[16]

Foundationwear companies used the latest foam plastics to experiment with form. When it came to falsies, this included more realistic designs. In 1946, Maidenform marketed the second version of its new "Masquerade" forms as "light, airy, comfortable and, above all, convincing."[17] The Smithsonian's National Museum of American History, which holds the Maidenform collection, has no examples of this second type of falsie, and to date, I have been unable to find them in any other US museum. However, I would expect them to look like the designs shown in figure 4.3, which also include nipples. Maidenform claimed that changes in technology during World War II meant that it was now able to offer "exact replicas" of "natural bosom contours," complete with nipples.[18] Despite these technological innovations, the default, light peach color of these designs and others, including Sears's Flatterettes, reinforced white women's bodies as the unmarked norm.

Maidenform was not the only brand to offer a range of products using the latest in plastics innovations. By 1951, the Sears catalog offered twenty-two variations of this bust-transforming product in various designs, including "aerated" foam rubber cones, teardrop urethane pads, and synthetic molds, complete with erect nipples and specially designed rayon satin covers.[19] Although the designs of these objects varied in shape and size, their colors did not; options remained limited to "white" and "flesh pink," thereby ignoring Black, Indigenous, and people of color.

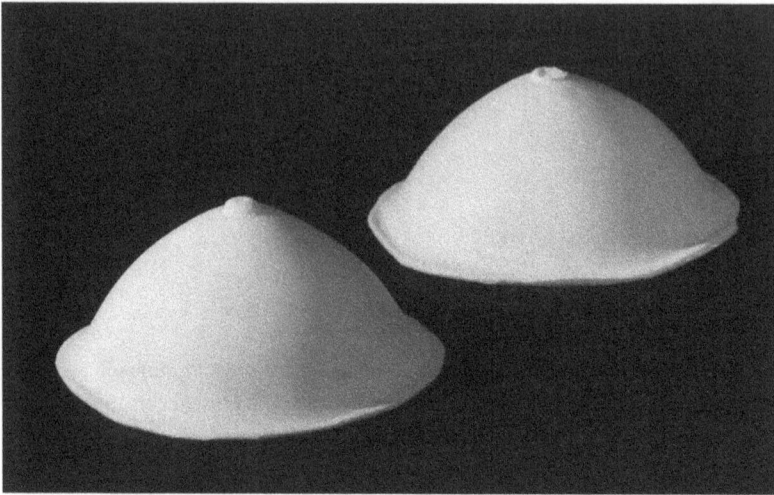

4.3 "Shape-Ups" foam rubber bust pads, 1950s, Collection of Kristina Haugland. Photograph by Jack Musgrove.

Frederick's of Hollywood, founded in 1947 by returning GI Frederick Mellinger, was clearly influenced by the full-bosomed ideal of Alberto Vargas pin-ups, which he would have been surrounded by on the war front.[20] The mail-order catalog company, which advertised in a range of publications, including *Jet*, *Charm*, and *Esquire*, was notorious for its overt sexuality, embodied in its aggressively shaped and pointed designs, as well as copy phrases like "bait him" and "keep him."[21] Unlike the Sears falsies listed in 1951, Frederick's offered pads in "black" and "white" colors; however it remains unclear whether these terms refer to skin tone. The fact that Frederick's offered falsies in black set them apart from other brands and suppliers at the time, which appeared to have almost exclusively stocked falsies in white and pink "flesh" tones—an issue that persists among many foundationwear companies today.[22]

Catalogs like Sears and Frederick's could offer consumers anonymity, which functioned on multiple levels. Lingerie specialist Frederick's was able to offer more risqué designs, which some customers may have felt embarrassed to try on or be seen purchasing in a store. Furthermore, catalogs could reach customers throughout the United States, in both rural and urban areas. Accordingly, shopping from outlets like Frederick's could appeal to many marginalized consumers, including but not limited to African Americans living under Jim Crow racial segregation, as well as queer, trans, disabled, and people of color, who faced continued accessibility issues, discrimination, violence, and profiling in retail spaces across the country.

In their advertisements, lingerie brands established in earlier decades, such as Hollywood-Maxwell—which claimed to sell exclusively its original whirlpool "Holly-Ette" conical bra to Paramount Pictures—had already shown the enduring appeal of movie glamour to consumers.[23] A 1947 mail-order ad for "Hollywood's famous Liftee" bra promised *Ebony* readers "sweater girl glamour," thanks to a design "made . . . to meet Filmland's demand for enchantingly beautiful breast lines."[24] However, it offered these designs only in "white" and "tea rose" colors and featured the image of a white model. Frederick's also employed the language of Hollywood and the entertainment industries to sell its body-shaping designs: Bra names included "Showbiz," "Understudy," and "Rising Star." Similarly, in keeping with earlier promotional copy that pushed the idea of owning multiple foundationwear items to suit a varied wardrobe, Mellinger enthused, "Gone are the days when one bra goes with everything! A bra wardrobe is an invisible but invaluable necessity! Here are some behind the seams secrets from Hollywood!"[25] Consumers were urged to invest in multiple designs to sculpt their bodies according to the fashion-shaping requirements of their outfits. Promotional copy, such as that of Frederick's, promised that participation in the practice of temporary body contouring would bring the glamour, status, attention, and social mobility associated with the entertainment industries. "Movie-star cleavage" could be achieved with an assortment of "in-up" push-ups.[26] Foam-padded foundations like these enabled the wearer to accentuate and build curves where they saw fit, as well as experiment with more exaggerated silhouettes.

Foam's soft and pliable materiality—sometimes integrated into foundationwear and at other times offered as separate external pads to enhance corsetry—enabled its users to take control of the aesthetic shaping of their bodies. African American publications during this period promoted multiple beauty ideals, unlike the whitestream fashion press, which tended to encourage a more "streamlined," slender physique and be more critical of women's bodies.[27] In a *Jet* magazine interview, Josephine Baker enthused over the gowns in her wardrobe. "They have everything," she said, pointing to the integrated corsetry. The journalist remarked on Baker's trim physique and the contouring effect of these designs—"These [built-in corsets] subtract inches from her already slim waistline"—and how she was "coyly shy about the foam rubber falsies (also built-in) which inflate her rather slim upper structure."[28]

4.4 Miss Leonard's foam falsies, 1950s. Photograph by author. Shirley May Jones Exotic Dancer Collection, Division of Culture and the Arts, National Museum of American History, Smithsonian Institution, Washington, DC.

In addition to being staples in many women's everyday and evening wardrobes, plastic foam falsies were also integral to the professional stage wardrobes of strippers and other sex workers during this period. The Smithsonian's National Museum of American History holds the costume collection of Shirley May Jones, a Cincinnati-based white exotic dancer. Performing under the stage name "Miss Leonard," she toured the United States in the 1940s and 1950s, accompanied by an eye-catching wardrobe that included rhinestones, Lucite-heeled peep-toes, feathered fans, and a couple of well-worn plastic foam falsies (figure 4.4).

The photographs in Leonard's collection show her carefully posed or, in a few instances, caught in motion mid-performance. This blurred sense of movement and the body's relationship to dress is also inscribed within her frayed and fringed wardrobe. The marks left in Leonard's foam set shown in figure 4.4 reveal the signs of prolonged wear, remnants of her life on- and offstage. The skin-like texture on the surface appears puckered, wrinkled with age, wear, and friction from corporeal impact. Parts of the foam have crumbled away, revealing a dense, fleshy inner material. There is some discoloration inside the cups, likely caused by perspiration from wear and the aging of the material. The hollow shells remain incredibly lightweight and springy. These foam falsies would have served their purpose well in enhancing the bustline, worn with Leonard's many different brassieres and lending additional volume to her feather fan tease at a time when topless dancing (i.e., without "pasties") was strictly prohibited.

Leonard was not alone in her use of falsies. Other examples include the Cotton Club's Tondaleyo, who later became an exotic dancer; Betty

"China Doll" Dickerson of the all-Black performance venue Savannah Club; Latinx dancers Tongolele and Ethel Rojo; Bunny Lake, a light-skinned trans feminine performer known as a "blonde bombshell"; Asian American drag queen Toni Lee of the Jewel Box Revue; and Black female impersonator Kicks Wilde of New York City's Crazy Horse. Burlesque star Hedy Jo Star, a white transsexual woman, shared her experience of wearing custom falsies on stage: "My breasts were developed, but I was hardly a Jayne Mansfield. In fact, I'd always worn falsies as a dancer. They weren't ordinary falsies. I'd had them specially tailored for me and they looked very natural. So natural that in over eighteen years of dancing no one had ever guessed I wore them."[29]

Foam Moved Upward: Queering Falsies

In the postwar United States, falsies and foam padding were an important part of trans feminine and queer life, both on- and offstage. In the 1950s and 1960s, trans community publications with longer print runs first emerged, reflecting wider sociocultural changes and a relaxation of obscenity standards. The pages of trans feminine periodicals like *Female Impersonators on Parade* (1960), *Lavender and Lace* (1964), *Female Mimics* (1963), *Female Impersonators* (1969), and *Trenns* magazine (1969) attest to the range of approaches and experiences that people in these communities had with self-fashioning the body, including the use of falsies and plastics. The falsies depicted and described include several mass-produced items—premade foam pads with nipples, smooth, fabric-covered foam pads, padded bras—as well as creative adaptations, such as repurposing everyday objects (e.g., a powder puff), homemade prostheses crafted from materials such as nylon and birdseed, and the latest plastic developments.

Although foundationwear companies like Frederick's of Hollywood and Maidenform offered ready-made solutions, a DIY approach was particularly popular when it came to trans feminine self-fashioning. Connie, a white self-identified transvestite, noted "I'm a little flat in the behind for a girl, but that's easy enough to pad out with sponge rubber, which I sewed into the seat of a panty girdle."[30] She disliked the feel of plastic foam padding, remarking that it "[does not] stay at the right body temperature." Instead, she preferred filling leakproof plastic sandwich bags with water: "A cup and a half to two cups has the same weight, inertia and bounce as real flesh."[31] Unlike store-bought falsies or realistic-looking prostheses, falsies made from everyday items were less conspicuous in domestic settings for wearers who were concerned about privacy.

Gender-nonconforming and trans feminine people experimented with materials at home, often researching and developing plastic prostheses over many years. One such plastics experimenter is Pudgy Roberts, a white self-described professional female impersonator who ran the Crazy Horse in New York City's Greenwich Village. Roberts, who wore a homemade, "personally designed one piece falsie" made from foam and affixed with Max Factor spirit gum, was rumored to "[hold] the distinct honor of being the first mimic to be featured in burlesque."[32] Roberts's homemade falsie, described as a crowd-pleaser, drew many compliments. In 1968, Roberts noted that although many people were "fascinated by the illusory breasts I wear . . . the materials and the process is my trade secret; a gimmick I'm not about to reveal to my competition."[33] Perhaps this just acted as a teaser, since Roberts had, in fact, already revealed this trade secret a year earlier in a self-authored publication titled *Female Impersonator's Handbook*.

Roberts spent many years perfecting a method for creating a breast prosthesis from a material called "Art-Foam." His handbook detailed a step-by-step guide to making custom foam falsies using affordable synthetic materials available from most dime stores.[34] A seasoned expert in DIY techniques for making "life-like" falsies and "inventing [other] devices to aid the profession," Roberts, who referred to himself as "the first activist of the field of female impersonation," shared his expertise in the hope that it could improve readers' employment chances in show business.[35] At a time when US laws against trans and queer communities actively policed bodies, the stage and show business industry—including burlesque— could offer greater acceptance, employment, and community for some queer, trans, and gender-nonconforming people. Roberts felt that offering support to his community was important to his practice. He collected and documented materials on the history of female impersonation for future generations. Additionally, Roberts mailed out free information on "crossdressing and related subjects" internationally and volunteered as "the only crossdressing contact on New York's Gay Hotline" for over twelve years.[36]

Synthetics research and development were also an essential part of San Francisco–based Sally Douglas's set of instructions, "the result of intensive experimentation over a period of several years."[37] In *New Trenns* magazine, a trans feminine publication with a focus on "off-beat fashion," Douglas, a white self-identified bisexual transsexual woman, detailed an incredibly complex, six-step guide—over twelve pages—on how to make and wear "realistic breast prosthetics" for "Ladies and Female Impersonators."[38]

Unlike Roberts's practice of sourcing readily available materials, Douglas's DIY approach engaged more directly with industrial plastics R & D. Douglas listed the key ingredients as polyurethane foams from Upjohn's Chemicals Plastics Research Division, and silicone rubber and mold release from Dow Corning. By including contact details for both companies, Douglas encouraged readers to order supplies directly. Douglas identified these materials as ideal for making DIY breast prosthetics due to their texture and ability to be "shaped in the normal household environment."[39] She prioritized material safety, noting that her chosen materials "are not irritating to the skin when fully cured."

Blending colors was an essential part of DIY falsie-making for both Roberts and Douglas. Unlike companies such as Maidenform, Roberts noted that when it came to making falsies, "there has to be communication [with the wearer] for size, skin coloring etc."[40] Roberts felt that plastic foam presented itself as the ideal material with which to shape the body, its ability to be dyed to match the wearer's skin tone being one of many benefits: "Why I chose [plastic foam] is because it is flexible, it can be formed easily, it is light, and it is the most effective. It is a thin foam material and by using ordinary clothes dye you can dye this material to match the coloring of your skin."[41] Douglas clearly recognized how skin tones vary greatly, stating that "if care is exercised in the color blending and testing, exact matches to the skin may be achieved."[42]

At a time when mass-market foundationwear companies did not explicitly cater to trans and queer consumers, Douglas and Roberts empowered users to become makers and create customized technologies of the self. While their respective approaches and positionalities differed, both encouraged readers to make their own personalized bust forms, building trans feminine communities of practice and knowledge exchange for crafting customized DIY technologies of embodiment. These maker communities operated outside restrictive societal gender norms and economies.

Publications including *Female Mimics*, *Lavender and Lace* (figure 4.5), and *Female Impersonators* featured trans, queer, and gender-nonconforming people of different races and ethnicities. Photographs often depicted them performing together, posing for pictures, or attending events. However, *The Queen* (1968) shows that Black trans feminine performers were frustrated by the bias of drag contest judges, who tended to be white and show a preference for white performers. In this documentary,

LAVENDER & LACE

Female Impersonators..at Play

4.5 Cover of *Lavender and Lace*, 1964.

Crystal LaBeija, a Black trans woman who was crowned Miss Manhattan in 1967, spoke out. LeBeija was frustrated by the preference given to white contestants and consequently set up her own Black drag ball. Notably, most trans feminine and gender-nonconforming people depicted in *Female Mimics* and *Female Impersonators* are white. This is particularly true within *Female Impersonators*' reoccurring "Greatest Stars" feature, which mainly celebrated light-skinned models, despite the more diverse readership suggested by readers' letters and photos.

African American publications in the postwar United States featured coverage of Black trans feminine communities and individuals. In the 1950s, *Jet* magazine ran a series of articles on interracial drag balls, also referred to as "female impersonator balls," noting that almost every large US Black community had at least one nightclub dedicated to drag.[43] Annual drag balls were large-scale events attracting crowds of thousands, such as the 2,500 participants, "garbed in girdles," at the 1952 annual Thanksgiving Ball held at New York's Rockland Palace.[44] In 1957, Bronzeville Mayor John E. Lewis judged Chicago's nineteenth annual Finnie's Masquerade Ball.[45] Miss Major, a Black trans woman and activist who participated in drag balls as a teenager in 1950s Chicago, reflected, "We had the balls then, where we could go out and dress up. You had to keep your eyes open, had to watch your back, but you learned how to deal with that, how to relax into it, and how to have a good time."[46] A few years later, Major joined the famous Jewel Box Revue touring company of female impersonators as a performer.[47]

Mother Camp, Esther Newton's mid-1960s study of female impersonators in the Chicago area, provides an important insight into the role of falsies. She noted, "The 'breasts' especially seem to symbolize the entire feminine sartorial system and role. This is shown not only by the very common device of removing them in order to break the illusion, but in the command, 'tits up!' meaning, 'get into the role,' or 'get into feminine character.'"[48]

In Newton's study, falsies are presented as central to this scene, yet this important context has not been considered in histories of foundationwear.[49] Publications such as *Female Impersonators on Parade* described how some performers removed their falsies—termed a "reveal"—to a "flabbergasted" audience.[50] In these descriptions, the performer's "realness" as a drag performer, and not as a cis woman, had to be proven by removing foam falsies to signal queer embodiment to the audience. Acts like these "challenged viewers' trust in gender as a visually verifiable trait."[51]

Some performers featured in trans feminine publications preferred not to participate in such displays. Pudgy Roberts commented that this action was "overdone" and "no longer a novelty." As Roberts observed, the use of falsies and other technologies of embodiment was not an exclusive "secret" reserved for trans and queer performers. Roberts argued that such objects were ubiquitous in American everyday life, stating, "Many women pad their natural assets this way or that. False eyelashes, phony bosoms, contraptions that plump up scrawny hips and derrières and what not."[52] As Judith Butler famously noted decades later, after reflecting on Newton's *Mother Camp*, "There is no original or primary gender a drag imitates, but gender is a kind of imitation for which there is no original."[53] Roberts echoed this sentiment, saying that "in her own way, every woman is an impersonator of her own sex."[54] For Roberts, there was no need to authenticate drag by removing gendered items at the end of a performance. For many individuals, the "realness" of gender is questioned and redefined in such acts of reimagining the self. Furthermore, these affective gestures complicate power structures around gender and interrogate who gets to decide the "realness" of gender.

Publications like *Female Mimics* frequently featured models in an exaggerated pose that showed them inserting or removing falsies. Figure 4.6 depicts Tommy, a reader of *Female Mimics* whose photograph was published in the magazine. Figure 4.7 shows Kicks Wilde, a performer at Pudgy Roberts's Crazy Horse. Figure 4.8 features Hanna Crystal from Puerto Rico, who starred at the famous 82 Club. Analysis of the repeated gesture in these and other similar images suggests that, as Newton observed, the act of inserting falsies was transformative. Captured by the camera, this gesture signaled a crossing of social gender boundaries. Descriptions of trans feminine people's lived experiences with falsies also often cited these objects as a key means of fashioning the self. For example, Aleshia Brevard, a white transsexual woman and actress, reflected, "I had yet to develop the proud bustline to which I aspired. In the meantime, there was always Frederick's of Hollywood. My brassiere featured a compartment that could be inflated to the appropriate size. If your contouring left something to be desired-huff, puff, and blow up your chest."[55] Falsies and other foam padded prosthetics were (and still are) an essential part of trans feminine and queer embodiment. They gave the wearer, who was often also the maker and designer, agency over the shaping of their body.

DIY approaches were vital tools of empowerment for trans feminine communities. While trans visibility had increased in the postwar United

4.6 (*right*) Photograph from reader's letter by Tommy, San Francisco, in *Female Mimics* 1, no. 3 (1963): 63.

4.7 (*below left*) Kicks Wilde in *Female Mimics* 1, no. 10 (1967): 2.

4.8 (*below right*) Hanna Crystal with a rare black falsie, *Female Mimics* 1, no. 10 (1967): 49.

States, most trans individuals still lacked gender-affirming care within formal medical structures. Bodies perceived to be trans feminine were subject to ongoing systemic state and interpersonal violence. Performers at Finocchio's—a famous female impersonator venue in San Francisco's North Beach neighborhood, where Brevard once performed—were locked into transmisogynist unionized contracts that prevented them from accessing hormones and gender-affirming surgeries.[56] The venue also refused to hire trans feminine individuals who had received and were receiving such care.

Some models who had posed with falsies later medically transitioned, such as popular performers Holli White from Hawaii and Hanna Crystal. Trans feminine publications also reported that some trans women, such as Coccinelle, a famous white transsexual Parisian showgirl, preferred to no longer wear falsies.[57] Newton noted that "a very significant proportion of the impersonators, and especially the street impersonators, have used or are using hormone shots or plastic inserts to create artificial breasts and change the shape of their bodies."[58] This development was "strongly deplored by the stage impersonators who say that the whole point of female impersonation depends on maleness."[59] However, other performers were more supportive of one another. For example, Hanna Crystal, who traveled abroad for gender-affirming surgery, reportedly received encouragement from fellow performers.[60] When a group of transphobic performers wrote a hateful letter aimed at trans women who had medically transitioned, *Female Mimics* published its response, stating, "We are surprised that you resent knowing about some of your fellow entertainers who have faced serious problems with tremendous courage."[61]

Falsies have been used by many different communities and marketed to fulfill different purposes. "Cuties" bust pads were first produced by US-based G. M. Poix in the 1930s, targeting prepubescent and adult customers who desired a fuller bustline. The company also produced "the first patented bra attachment to enclose mastectomy pads."[62] Mastectomy patients were often frustrated by regular falsies. In 1947, one such patient complained that a regular falsie, described as a "pincushion like affair to be shoved into the empty side of my old brassiere," did not fit properly or resemble a real breast closely enough in texture or form.[63] Unsatisfied with regular fashion falsies made from fabric, she resorted to the services of a specialist surgical corsetiere, who provided a sponge rubber prosthesis, known as a surgical bust form, which more closely resembled a breast and, by doing so, restored her confidence.

Ready-made surgical bust forms were also available to purchase from nonspecialist suppliers such as Sears. In its catalogs, Sears listed bust pads explicitly designed for mastectomy patients alongside bust-enhancing pads for a "beauty lift." It appears that foam falsies and prosthetic breasts often resembled one another in form, texture, size, color (or lack thereof), and weight, in their efforts to replicate the characteristics of the human breast. Prosthetics historian Kristin Gardner argues that this similarity is particularly evident when looking at the patent classification suggested by the US Patent Office, which did not differentiate between bust-enhancing forms and prostheses designed to replace an amputated breast.[64] Hence, almost all the patented solutions could serve either purpose, albeit with minor adjustments on a case-by-case basis.

Companies marketing breast forms as medical devices emphasized foam's ability to adhere closely to the body and absorb heat, thereby warming to body temperature. Advertising copy for Camp's Tru Life custom-made plastic foam breast forms claimed that the material's ability to cling to the body "created a natural feeling, a unity that makes the wearer forget that a form is being worn."[65] This concept of unity was marketed as a way to "alleviate anxiety," and the nonsurgical forms were described as "plastic surgery without plastic surgery."[66] Unlike earlier breast pads, improved postwar foams could be molded more precisely to create seamless designs (figure 4.9).

In the postwar United States, trans feminine and gender-nonconforming people also used breast forms marketed for medical purposes to enhance their curves. For example, in a booklet for Virginia Prince's "Society of the Second Self," a heterosexual cross-dressing group, mastectomy inserts are suggested as useful for readers desiring a fuller bust. These inserts are larger than most other breast pads, with a side section that fits under the arm where the lymph glands have been surgically removed, making them particularly desirable for wearers with a fuller figure.[67] Surgical forms were notably sold at a higher price point than falsies, and not everyone could afford them. Their inclusion in the Society of the Second Self's reading materials reflects the more privileged status of the group's members. This was a predominantly white, middle-class collective that utilized their resources to create an exclusive community where they could express themselves privately without fear of losing their jobs and social status.[68]

A wide range of makers and wearers embraced plastic foam's smooth properties as a technology for feminine self-fashioning. A newspaper reporting on the "Big Bust Boom" claimed that in 1958 the foundationwear

INDIVIDUAL CARE IN CREATING EACH *Tru Life*® BY CAMP
REFLECTS THE QUALITY OF A CUSTOM-MADE BREAST FORM

The effects of mastectomy (breast surgery) are not permanently destructive. When the operation is seen in proper
perspective and the patient knows excellent breast forms,
such as Tru-Life, are available, anxiety about physical
appearance in front of family and friends is alleviated.

Below are photos of the Tru-Life manufacturing process;
they show why Tru-Life is the finest prosthesis available.

HAND CEMENTING ■ Shell is
powdered, placed in an open
face mold, cement is applied
and the shaped foam base is
hand cemented in position. It
is then cured for 3 days.

MOLDING ■ A special formula
of plastisol liquid is metered
into the molds and rotationally
molded to create seamless
breast form shells.

FILLING PROCESS ■ Specially
formulated non-evaporating
liquid is dispensed automat-
ically into each shell size in
exactly desired quantities.

INSPECTION ■ Skilled per-
sonnel hand inspect each shell
to make certain they meet established high quality standards.

FINAL AIR EXTRACTION ■
Trained employee extracts all
remaining air via vacuum process. The nipple opening is then
cemented to seal the liquid in
a vacuum.

4.9 Detail from Camp "Tru Life" booklet, 1968. Walter Spohn Collection, Division
of Medicine and Science, National Museum of American History, Smithsonian
Institution, Washington, DC.

industry had used over twelve million pounds of foam rubber.[69] Throughout the 1950s and 1960s, Frederick's catalogs promoted the latest in plastic
foam developments as key to the company's "in-up push-ups." Plastic foams
offered the perfect lift for varying degrees of pointed cleavage, pushing
"any bust—the small—the natural—the wide" ... "inwards, upwards and
outwards."[70] Frederick's catalogs often employed a technocultural rhetoric
when extolling the innovative, corrective, and transformative advantages
of these materials: "Mr. Frederick's alone knows the scientific formula
for pads that lift you up as they push you in! ... 'A' and 'B' bosoms blossom
to new dimensions! Suddenly, 'C' and 'D' busts are higher, firmer, younger
than Springtime!"[71] Here, consumers were offered a bra with an undisclosed

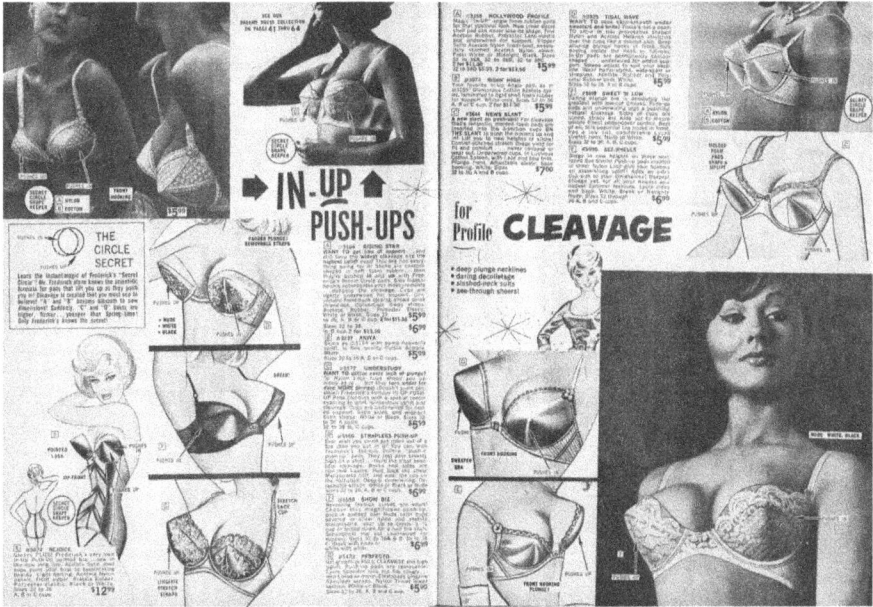

4.10 Frederick's of Hollywood catalog 19, no. 87 (1965): 16–17. Hagley Museum and Library, Wilmington, DE.

scientific formula that unlocked the secret to attaining the desirable hourglass shape (figure 4.10).

Foundationwear companies frequently employed pathologized language to promote padded designs. "Defective" bodies that did not meet the fashionable ideal could be easily corrected with Maidenform's plastic foam pads, "designed to compensate for nature's deficiencies," or, as Frederick's put it more bluntly, "Where nature leaves off . . . ADD WITH PADS!"[72] In addition to bust pads, these companies offered plastic foam padding for buttocks, hips, thighs, and calves. For instance, in 1948, Maidenform filed a patent for a "hip pad" designed to be worn with a girdle. This piece was developed to overcome an "undesirable deficiency or depression" to the buttocks and hips that commonly occurred when wearing a girdle, which was designed to compress the body.[73] Some advertising even included before and after photographs, seemingly referencing pre- and post-surgery photos of two white figures (figure 4.11).

The fleshiness of foam plastics was another selling point for foundationwear companies. Companies claimed these materials, now more

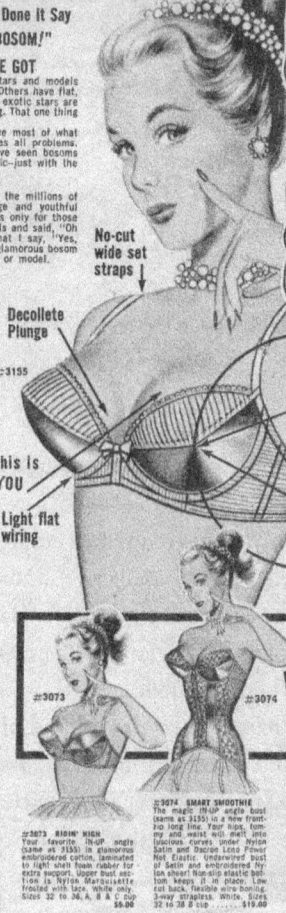

4.11 Frederick's of Hollywood advertisement, 1957.

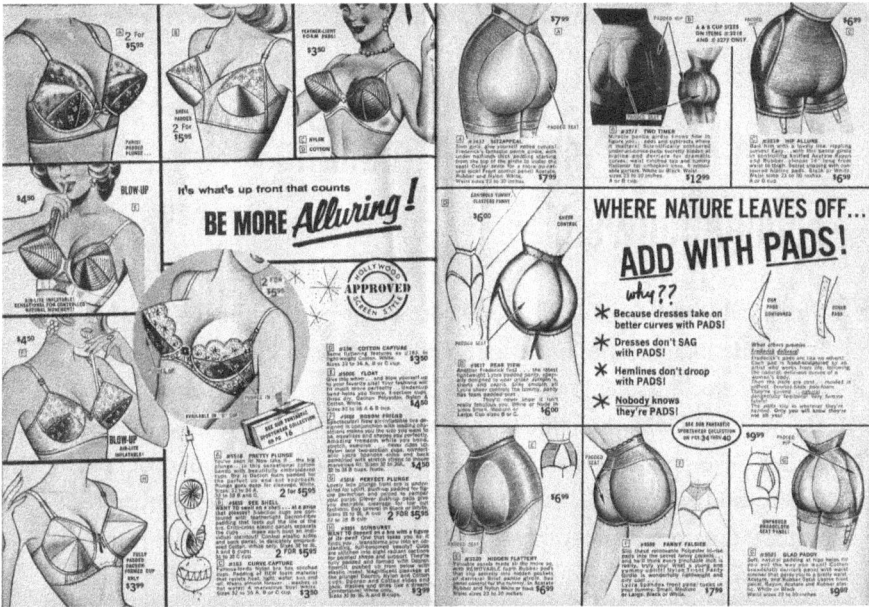

4.12 Frederick's of Hollywood catalog 19, no. 87 (1965): 20–21. Hagley Museum and Library, Wilmington, DE.

realistic than ever before, made the emulation of "perfect" sculptural bodies a reality. A Maidenform advertisement described a pair of falsies as "shaped into exact replicas of natural bosom contours . . . such as one sees frequently in idealistic statuary but not as frequently, alas in actual life!"[74] Referencing sculpture, the text conflated an ideal with a norm or standard for the body.[75] In its pursuit of realistic feminine padding, Frederick's claimed to have employed artisanal expertise (figure 4.12): "Each pad is hand-sculptured by an artist who works from life, following the natural delicious curves of a woman's body. Then the pads are cast . . . molded in softest, bounce-back poly-foam. They're curved . . . *natural* . . . delightfully feminine! Very femme fatale!"[76]

Here, an artist was able to "sculpt" boosting pads, imbuing the foam with agency. The body is reimagined as a plastic sculptural material; foam becomes flesh, and flesh becomes foam. The copy concludes, "The pads slip in wherever they're needed. Only you will know they're not all you!" Presented as a realistic, flesh-like substitute and an immediate way of contouring the body, foam empowered the wearer to use them "wherever they're needed"—thereby enabling them to take control over the shaping of their figure. However, another method allowed these objects to be permanently integrated into the body.

Foam Moves Inward: Implanting Foundationwear

The much-publicized Hollywood cosmetic surgeon Robert Alan Franklyn wrote in a number of his publications that "with all the success the medical world had enjoyed in utilizing plastics all over the body, I could see no reason why the same kind of material couldn't be used to expand under-developed breasts."[77] Here, Franklyn implied that pathologically flawed bodies could be "fixed" with plastics.[78] Changes in medical technology due to advances in wartime plastic surgery, as well as improvements in surgical procedures and post-operative healing, meant that there was now less risk of infection.[79] Furthermore, emerging plastic materials, such as polyurethane foam, could be sterilized, making them ideal for implantation in the body and reducing the risk of infection and foreign body reactions. As will be demonstrated, the similarities between plastic foam falsies and breast implants used in cosmetic surgery are striking in terms of form, design, aesthetics, and materials. They both served a similar function; however, one was temporary and the other, permanent.

In 1948, it was reported that the multimillion-dollar US falsie business was "periled" by plastic surgery, which "removed the need for falsies."[80] Indeed, in 1957, a journalist for *Time* described the respected plastic surgeon Milton Edgerton as using "a form of built-in falsies made of Ivalon sponge" in his operations.[81] Cosmetic surgeons were keen to offer their services as a way of making more permanent the contouring effect that accessories such as bullet bras and falsies could only temporarily provide. They promised that this under-the-skin procedure made the detection of provisional "fakery" impossible, removing the need for padded bras and falsies altogether.

Books by cosmetic surgeons, such as *The Breast Beautiful* (1947) and *On Developing Bosom Beauty* (1959), began with histories of the breast, women's fashion, and the bra, encouraging readers to question the bra's efficiency as a tool to achieve a healthy and aesthetic ideal, before promising that there were no health risks associated with reduction and augmentation surgery.[82] These books also included sections on "racial characteristics of the breast."[83] It is important to note that these histories of the breast presented a racialized taxonomy, in which "youthful" white European "conical" or "hemispheric" breasts were presented as aesthetically superior to those of women of color, particularly Black and Indigenous women. The white plastic surgeons and sexologists who wrote these racist studies often described the breasts of women from Africa and India in offensive, grotesque, and abject terms, suggesting they were inferior, uncivilized, and undesirable. In contrast, they argued that conical, uplifted breasts—particularly in white women—symbolized civilization, desirability, purity, and virginal status before breastfeeding.[84] For cosmetic surgeons and other medical actors, the construction of desirable, "healthy" gendered bodies relied on racialized binaries and pseudoscientific practices of othering.[85]

Promotional and news copy frequently described lingerie designers, health instructors, and surgeons as sculptors or artists of the female form. Comparisons to white nudes, Venus, and classic European masterpieces of art were recurring themes in foundationwear advertising. They referenced racialized ideals of the human figure and beauty tied to strict gender norms and standardized proportion. An ad for "Edith Lances Sculptural Bras" (figure 4.13) displayed a sculpted white torso, with the head and arms missing, on an abstract plinth. The image focuses on the breasts, which could be sculpted to perfection thanks to these custom-made bras "for lovelier bosom contours." There is no need to display the actual product it promotes. The message is clear: The bra is a tool with which to sculpt the body into one's own masterpiece.

Within this omnipresent celebration of youth, the racialized, gendered body—particularly the bust—was presented as an idealized, malleable surface. Lances's comparison of the female body to sculpture even caught the attention of plastic surgeon Herbert Conway, who added the company to his file on prostheses. Conway noted the "identical custom-made plastic form for breast," its cost, and the corresponding phone number.[86] He also kept a cutting of a classified ad for "Nubrest" featuring "individually sculptured" postmastectomy artificial breasts.[87] These cuttings could indicate that Conway connected the design and manufacture of an external object to its potential for internal implantation in cosmetic surgery. Surgeons may have archived information on these external breast prostheses for patients who had mastectomies as part of their research on potential manufacturers and materials intended either for implantation beneath the skin's surface or for use as form guides.[88]

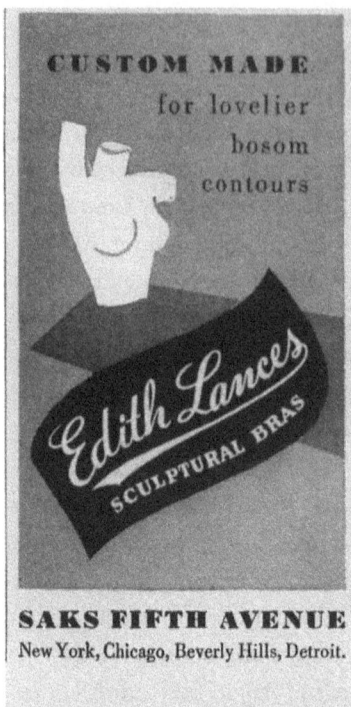

4.13 Edith Lances sculptural bras advertisement, *Vogue*, November 1944, 90.

These examples all point to a pattern of viewing and describing the racialized, gendered body as sculptural on the surface, which was now seen as having the potential for permanent augmentation. Other notes Conway filed under "prostheses" include contact details for an exterminator turned sculptor who claimed to have recently started specializing in prosthetics.[89] Conway may have collected this information to assist him in carving and designing individual implants. The business card, with "exterminators" crossed out and "prosthetics" penciled in, is indicative of the haphazard approach to the manufacture of prosthetics before the Medical Device Regulation Act of 1976.

In his popular book *On Developing Bosom Beauty* (1959), Franklyn presented a sequence of images depicting his "ten-minute simple Breast-plasty surgery," which, by 1976, he claimed to have performed over five thousand times and which, he asserted, had inspired over a million surgeries.[90] The first image was captioned, "Author sculpturing model bust in clay to serve as guide to surgery," implying that a sculpture or representation of the body was used as a guide in a medical procedure to augment the bust.[91] Franklyn also detailed a calculation for what he termed the

4.14 American Society of Plastic and Reconstructive Surgery logo, 1953.

"Breast Quotient . . . a scientific ratio of breast volume to total volume" that combined body measurements to find the ideal bust proportions.[92] Archival papers show that plastic surgeons attended sculpture courses and collected information to improve their work with prosthetics.[93] Indeed, the logo of the American Society of Plastic and Reconstructive Surgery, established in 1931, featured an image of the *Venus de Milo* within a circle, with a gloved hand holding a surgical instrument overlaid (figure 4.14).[94]

When asked by a journalist for advice on identifying the right doctor to perform cosmetic surgery, plastic surgeons—such as New York–based Herbert Conway—recommended consulting a general medical practitioner, a local medical society, or even the AMA for an approved list of competent surgeons. He also stressed the need for the surgeon to have an artistic eye: "The surgeon also must be a psychologist and an artist. Portrait painting and life modelling and sculpturing are studied by already well trained, experienced cosmetic surgeons; for the . . . reshaping of the . . . breasts calls for an understanding of aesthetics, as well as of conscious and unconscious motives of patients."[95]

As in Franklyn's image, the patient's breasts could be remodeled by the expert hand of a surgeon "artist" using foam. Surgeons often asserted their authority in selecting the "proper proportions" for a patient, gatekeeping what they felt was appropriate sizing and feminine gender presentation. Franklyn even referred to himself as a designer, commenting, "We must judge the patient not only as a living organism, but as a design."[96] If the patient is assessed as a design, this would imply that the body is viewed by the surgeon as a malleable material.

As discussed in the first section of this chapter, fashion designs such as the enduring American Sweater Girl look, having been put briefly on

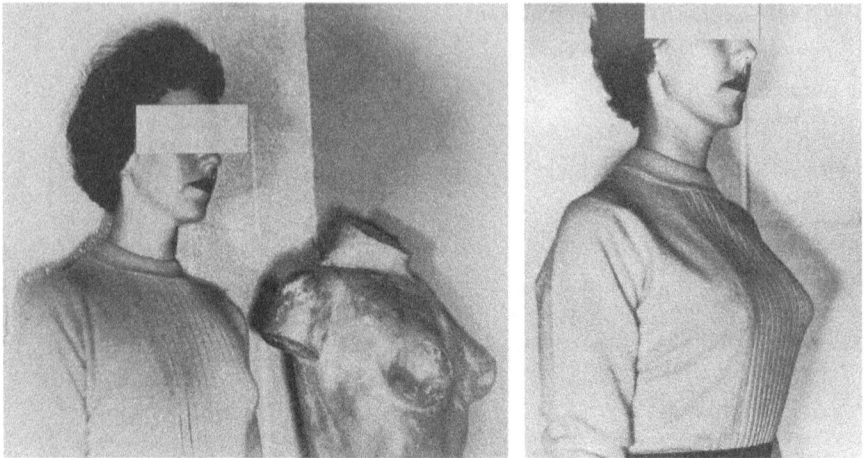

4.15 "Sweater Before Surgery, Sweater After Surgery," in Franklyn, *On Developing Bosom Beauty*, 112–13.

hold during World War II due to materials rationing, remained popular throughout the 1950s and into the early 1960s. This style, shaped by foam, was regarded as a "healthy" and "desirable" female ideal. Franklyn chose to include pre- and post-surgery photographs of patient "case studies" in his books, such as "before" and "after" profile shots of light-skinned patients sporting the fashionable Sweater Girl look (figure 4.15). It is noteworthy that these images appeared in *On Developing Bosom Beauty* (1959) and *Beauty Surgeon* (1960) but are missing from *Augmentation Mammaplasty* (1976), probably because the conical ideal embodied by the Sweater Girl fashion had become outdated.

While most images in *On Developing Bosom Beauty* present nude side views of patients, *Beauty Surgeon* (1960) includes only a sweater-clad "before and after" image of a patient (as this more general publication contains no nudity). *Augmentation Mammaplasty*, however, presents a greater selection of views, including ones from the front. The most extreme conically shaped breasts (figure 4.16) have been removed, while those with more of a "pointed roundness" aesthetic remain.

By 1976, the selection of "after" photos covers a wider range of breast sizes, including rounder shapes (figure 4.17). Franklyn had replaced his Surgifoam plastic material with smooth silicone implants, and the change in shape was evident.[97] These images suggest that breast augmentation surgery, like bra design, was affected by changes in materials R & D and

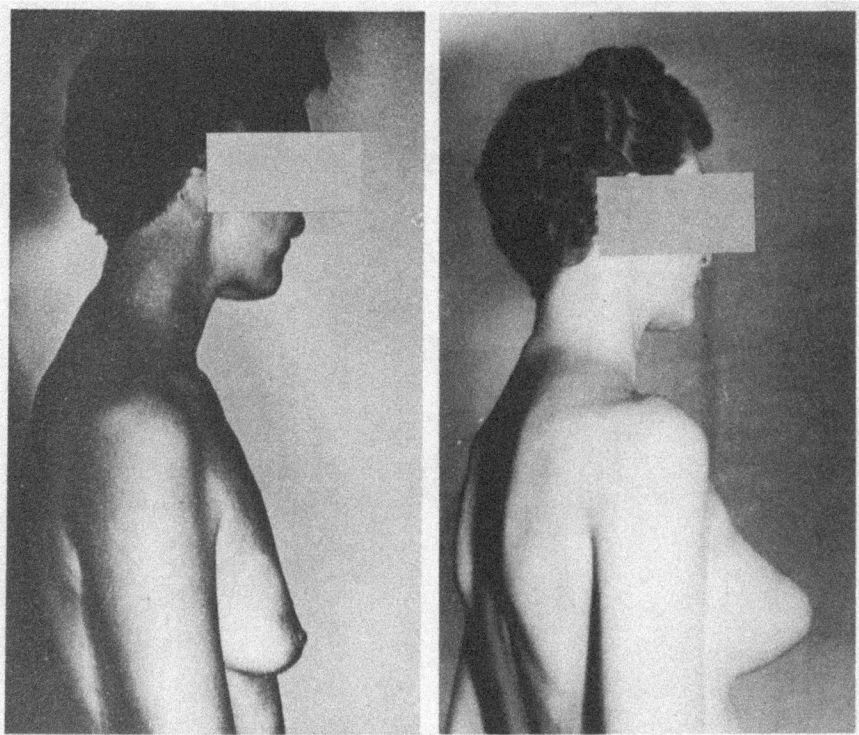

4.16 "Before Surgery, After Surgery," in Franklyn, *On Developing Bosom Beauty*, 96–97.

fashion. Available materials informed the design of implants and prosthetics, thereby influencing both the representation of the fashionable and "healthy" ideal breast and its embodiment in the flesh.

Anne Hollander's *Seeing Through Clothes* (1978) examines the relationship between fashion and depictions of cultural corporeal ideals, as represented in painted and sculptural nudes. She observes, "Although clothes may appear to reduce the grand truth of unclad natural humanity simply by being contingent, specific and intrinsically bound to style, they are recognizably the only thing that gives—has always given—that shapeless and meaningless nakedness its comprehensible form."[98]

Hollander, an art historian who focuses on European artists, argues that in this context, the ideal, disrobed body is culturally shaped by prevailing fashions in clothing at any given time. Flesh appears sculpted by exterior garments, even when they are absent. I would like to

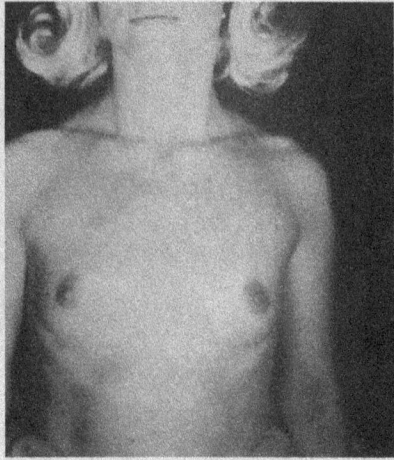

Before surgery

After surgery

4.17 "Before Surgery, After Surgery," in Franklyn, *Augmentation Mammaplasty*, 90.

take Hollander's argument a step further, incorporating sociologist Joanne Entwistle's concept of the "fashioned body" to argue that the pathologized, racialized, gendered nude body in cosmetic surgery is also shaped by the fashions and technologies of its era.[99] This is particularly evident in Franklyn's work, as well as that of his peers such as Milton Edgerton, Herbert Conway, Jerome Pierce Webster, W. John Pangman, and Robert M. Wallace. Another peer and important figure is Elsa K. La Roe, a New York–based cosmetic surgeon who wrote *The Breast Beautiful* (1947), a guide to achieving the perfect bustline with surgery and/or hormone treatments. Hollander's theory applies to the previously discussed photograph of Franklyn posing as a sculptor, as well as to foundationwear companies' promotion of products designed to sculpt the body into the conical fashionable ideal.

Journalists and some surgeons reported that cosmetic breast surgery removed the need for falsies and padded bras. Audrey Minor, writing in *Confidential* magazine, commented in 1954, "Patients for this form of living sculpture [breast augmentation] raid their dressing table drawers the minute they're out of the hospital—to throw away those slip-and-slide falsies and padded bras, once and for all."[100] Similar patient reactions were also recorded in the notes of plastic surgeons, as well as in their publications in the medical and general press.[101] Prospective patients also noted their desire to have a body that could be permanently contoured in line with external shaping devices. Hedy Jo Star said, "I liked the shape of [my specially tailored] falsies so much that I asked the doctor if she could shape my breast implants like them."[102] Another example is an African American female prospective patient who wrote to Jerome Pierce Webster in 1948 describing the psychological pain she experienced from having hips she felt were disproportionate to her waist.[103] She felt that they were "malformed," making it "impossible" for her to wear a "bathing suit, shorts or anything revealing." She wrote, "I have to wear a very heavy girdle in order to appear somewhat natural looking in my clothes and even then, it is quite noticeable." Having listed her attempts at exercise and "dieting to the point of making myself ill," she felt that surgery was the only solution. Her letter is accompanied by in-text illustrations depicting both the appearance of her hips and the body she envisioned achieving through surgery. She hoped that the procedure would enable her to remodel her silhouette to her desired shape.

In the 1950s, prospective patients who thought their breasts did not meet a standardized ideal sought advice from surgeons and the AMA.[104]

Many cited their psychological anguish at feeling excluded from a society where, as Franklyn put it, "the girl who can wear a low-cut gown and display a lovely, clean valley running between two full-grown, well-proportioned mounds of flesh . . . is the girl who will qualify as the belle of the ball in every phase of modern American life."[105] Some journalists presented foam implants as a safe, quick, and easy solution for achieving a fuller bust and gaining the social mobility this supposedly brought with it.[106]

As other scholars have noted, in the postwar United States there was little reporting from African Americans who had pursued cosmetic surgery—unlike the robust coverage found in the aforementioned publications—and media aimed at Black readers generally did not cover cosmetic surgery until the 1970s.[107] An exception is "Plastic Surgery," published in 1949 in *Ebony*.[108] This article discusses George J. B. Weiss, a plastic surgeon who established a clinic in Harlem in 1942 and advertised his services in Black publications. Weiss mainly performed "nose-narrowing" and "lip-thinning." However, the article also mentions a Black patient who asked Weiss to "overhaul [her] breasts." According to the piece, in 1949, there weren't any Black plastic surgeons qualified by the American Board of Plastic Surgery, "although there is at least one Chicago doctor who will probably take the board exam soon." The article noted that because Black people were "barred from the services of many top-flight plastic surgeons," some turned to dangerous quacks. Additionally, the cost of cosmetic surgery and bust augmentation in the United States made it unattainable to most American people of color throughout the 1950s and 1960s.[109] Decades later, an *Ebony* article noted that, among African American and Asian American communities, "there is the feeling that altering the nose, lips or other facial features is a betrayal of one's racial heritage."[110] By the late 1970s, there still were not many Black plastic surgeons working, which further restricted access in an already-unequitable medical system.

Foam-padded items, such as Maidenform's Masquerade falsies and bras, could provide only a temporary solution for generating societal approval and acceptance. As cosmetic surgeons such as Franklyn were keen to point out, the "truth" behind these "little deceivers" would eventually be exposed in more intimate moments. Franklyn's publications featured countless case studies that capitalized on fear of rejection and exposure.[111] In one such study, Franklyn narrated the case of a Black female patient who underwent his Breastplasty surgery to make permanent the affects her padded foundationwear could only temporarily achieve.[112] In another case study, he describes the tale of "Carol," a white female patient who sought out his services because she was afraid of "unmasking" herself as a falsie-wearing "fake" on her wedding night. He accommodated her request by giving her "a thirty-eight-inch bosom every bit as big as the brassiere she had worn into the office."[113] According to Franklyn, once Carol had her surgery, "the Counterfeit Bride became a *real* bride the following week."[114]

In contrast, there were other women who were less concerned about what others thought was "real" or "fake." In Russ Meyer's "shockumentary" *Mondo Topless* (1966) on the West Coast "topless craze," the white go-go dancer Darlene Grey comments, "People will say ah, she's phony, that's all foam rubber. [*referring to breasts; laughs*] I go, 'Honey, believe what you want, I don't care!'"[115] Meyer's sexploitation genre movie includes taped interviews with women on the 1960s West Coast go-go circuit, offering a rare opportunity to hear these women's lived experiences as sex workers.

Medical studies on breast augmentation, such as those of Edgerton and Franklyn, did not explicitly identify trans feminine or gender-nonconforming people among their patients. However, during this period, trans women and gender-nonconforming and queer individuals also permanently shaped their curves with plastics. An example is Charlotte McLeod, a white transsexual American woman whose medical transition was covered widely by the US whitestream press. McLeod traveled from the United States to Denmark in 1953 to undergo gender-affirming surgery, following in the footsteps of Christine Jorgensen, who, in 1952, was the first to be reported as having undergone this surgery. McLeod is said to have had an operation that used plastic foam to "round out her figure into a more womanly shape" and achieve her "38–25–36" measurements.[116] She is counted among La Roe's trans women patients who received her foam breast implant surgery in New York.[117]

After receiving gender-affirming surgery, Hedy Jo Star also sought out the services of La Roe, "one of the world's outstanding plastic surgeons and a specialist in breast reconstruction."[118] A journalist noted that Star, who worked as a stripper, wanted her breasts enlarged with foam, especially since she had not taken hormones prior to her operation. La Roe commented that for trans feminine patients like McLeod, she agreed to "help with her bosom" only after "I examined her and found she was indeed a woman." The journalist explained, "One reason Dr. La Roe examines applicants for mammary enlargement is that the medical profession now frowns on any such treatment for those who are still classified as males, regardless of how soon they may expect or hope to be changed."

This discriminatory and intrusive practice meant that, in theory, only those who could "pass" or had undergone previous gender-affirming treatments within formal medical structures were eligible to request her breast enlargement services.

In the postwar years, women such as Abby Sinclair spent upward of $10,000 on surgery and travel to access gender-affirming care.[119] Sinclair claimed that in addition to traveling to Casablanca to see Georges Burou, a famous surgeon specializing in gender-affirming surgery, she paid $5,000 for the operation itself, $2,000 for electrolysis, and $1,500 for hormones. Despite discriminatory and invasive gatekeeping practices, publications written by and for trans feminine people demonstrate that some individuals who had not had gender-affirming surgeries were still able to undergo breast augmentation procedures using plastic foams. Furthermore, while the treatments were expensive—"about $500 and up"—individuals still sought out and underwent breast augmentation surgery.[120]

Access to gender-affirming care within formal medical structures was further limited for trans people of color. Black trans feminine individuals were frequently criminalized, and their representation in whitestream US publications was equally marginalized. Coverage of Black trans individuals in African American publications at the time offered a different portrayal of trans women and gender-nonconforming people, contrasting with the white heteronormative domestic womanhood of Jorgensen's "good transsexual" presented in whitestream publications.[121] In 1953, *Jet* magazine ran a series of articles on Carlett Brown, a Black professional female impersonator who planned to undergo the same gender-affirming surgery in Europe that Jorgensen had.[122] However, unlike Jorgensen, Brown was arrested for cross-dressing and publicly spoke out against the systemic violence and transmisogyny female impersonators encountered.[123] Delisa Newton, a Black trans woman, shared her experience of gender-affirming surgery with *Sepia*, describing the racism she experienced within the medical system—in her words, "Because I am a Negro it took twice as long to get the [sex change] operation"—and advocated for greater access to gender-affirming care.[124] Unlike McLeod and Star's coverage, these articles do not mention bust, hip, or buttock augmentation procedures using plastic foam, further highlighting access disparities.

Seeking to maximize his commercial success, Franklyn, in an act of informational asymmetry, positioned his Breastplasty surgery as a safe and simple means of achieving a fuller bustline. Surgical diagrams such as those in figure 4.18, illustrating the insertion of a cone-shaped implant under the flesh, closely resemble diagrams and patents for foam inserts in bras (as shown in figure 4.19). Franklyn's accompanying 1959 description of his Surgifoam implant as "polyester plastic foam . . . sheathed in nylon" also mirrors the materials prospective clients would be familiar with from brassieres, falsies, and foundationwear design.[125] In this rhetoric, the body can be "taken in" or "padded out" like a garment, using the very same materials.

Franklyn was keen to stress the ease of his operation. In his description of Breastplasty, slipping a Surgifoam implant through what he describes as a "very small" incision in the fold under the breast is not dissimilar from the language used in figures 4.12 and 4.18, in which Frederick's promises that "the pads slip in wherever they're needed." Franklyn's textual and visual description detracts from the invasive nature of the procedure itself. Instead, it implies the ease and action of compressing foam into the small incisions in a bra to push foam inserts through for optional padding. His language mirrors that of the promotional copy used to describe foam's agentic properties in foundationwear and its ability to shape the body, both on its surface and now from within.

When Franklyn introduced his Breastplasty surgery to the public in 1953, he did not specify the material he used, vaguely describing it as "a soft plastic foam substance—actually a nylon-like resilient material impregnated with penicillin," refusing to reveal the specifics of his scientific secret, much like another foam plastics entrepreneur: Frederick Mellinger.[126] In the following chapter, I examine in more detail the composition and sourcing of Surgifoam and other similar materials. Since Franklyn did not mention the complications of his Breastplasty surgery in his publications, which largely courted a general audience, it is useful to contextualize his surgery within established medical literature of the time.

A number of studies by US medical professionals working with similar synthetic sponges in the 1950s offer early accounts of foam plastics' interactions within living tissue.[127] Initially, some studies of Ivalon, a type of polyvinyl sponge, reasoned that living tissue's tendency to inhabit these foreign structures was a promising sign, demonstrating biological acceptance of the synthetic material.[128] According to these researchers, inert matter became a living extension of the human body. Foreign body reactions and the expulsion of these new plastic materials were not as immediate as those experienced historically with materials such as ivory or silk. In some cases, patients were reportedly pleased with the immediate results of early polyvinyl implants.[129] Issues included difficulty with infection and drainage, which were supposedly fixed with improved sterilization.

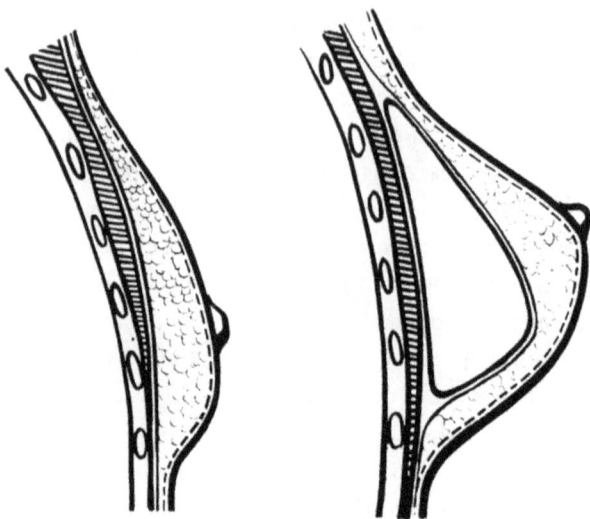

Schematic drawing showing flat chest before and after insertion of plastic foam, remolding breast into pleasant contour without interfering with glandular tissue.

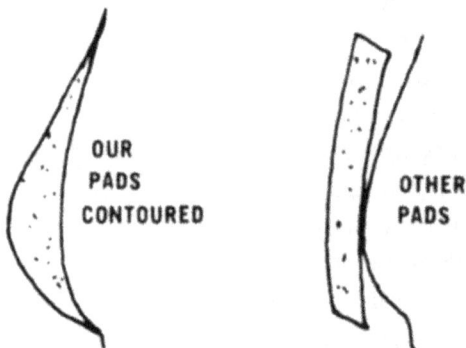

OUR PADS CONTOURED

OTHER PADS

What others promise . . .
Frederick delivers!
Frederick's pads are like no others!
Each pad is **hand-sculptured** by an
artist who works from life, following
the natural, delicious curves of a
woman's body.
Then the pads are cast . . . molded in
softest, bounce-back poly-foam.
They're curved . . . _natural_ . . .
delightfully feminine! Very femme
fatale!
The pads slip in wherever they're
needed. **Only you will know they're
not all you!**

4.18 (*above*) "Schematic Drawing Showing Flat Chest Before and After Insertion of Plastic Foam, Remolding Breast into Pleasant Contour Without Interfering with Glandular Tissue," in Franklyn, *On Developing Bosom Beauty*, 82.

4.19 (*left*) Frederick's of Hollywood falsies, 1963.

Some medical practitioners repeatedly warned of the potential complications associated with synthetic materials, as their long-term effects remained unclear. Earlier animal studies had already indicated that a wide range of implanted synthetic materials—including polyethylene, silastic, nylon, Teflon, and polyvinyl chloride—posed significant carcinogenic risks.[130] Since the medical device industry remained largely unregulated, it appears that the risks of plastics implanted in the human body for prolonged periods did not become clear on a larger scale until later. Issues arising from breast augmentation using Ivalon included loss of sensation in the nipple, inability to breastfeed, and shrinking as well as painful hardening of implants, which for some patients eventually required amputation. While it was hoped that alternative foamed plastics could provide a safer solution, similar problems were also observed with these materials.

Plastic foams, a political material used by the military and industry, continued to develop in the postwar period through different processes of material subversion. Plastic foams' flexibility and softness appealed to a wide range of makers and users, who used them to reshape and reimagine the body. In some cases, this upheld and conformed to racialized, gendered body ideals, and for others, it complicated and resisted them.

Foam's soft materiality shaped fashionable Sweater Girl looks and a plethora of curvaceous bombshells with intersecting identities. Trans feminine, queer, and gender-nonconforming people engaged in technological innovation and processes of adapting existing objects to shape, reimagine, and experience the body. These DIY approaches empowered some individuals within US trans feminine networks at a time when access to gender-affirming care remained limited and cross-dressing laws continued to publicly police bodies and enforce the gender binary.

Cosmetic surgeons compared their role to that of sculptors and crafted foam implants inspired by postwar bombshells, whose looks were simultaneously shaped by the available materials. As implant technologies and R & D changed, so, too, did the desirable shape. This chapter has shown how the design of falsies, prosthetics, and implants shaped conceptions of the ideal body—as influenced by race, gender, and medical discourses, and presented as both healthy and desirable—in medical and popular culture contexts. The same materials that were used to shape the body externally were now being used to shape it internally.

Foam moved outward, away from its original military and industrial applications and into the fashion, beauty, and entertainment industries. It moved upward, enabling many different users to shape their bodies as they wanted. And, finally, it moved inward, appealing to medical practitioners who admired foam's malleability and, in theory, allowed individuals to permanently shape their bodies. Overall, by closely tracing foam's movement across different social groups, this chapter has shown how changes in material, scientific, and medical technologies; fashion; consumer culture; and attitudes toward race, gender, and sexuality led to the shaping of a multitude of bombshell figures.

5 BOMBSHELLS, BOMBERS, AND BUMPERS / PLASTIC FOAM SOURCING AND WOMEN'S BODIES

In *The Mechanical Bride: Folklore of Industrial Man* (1951), writer and cultural critic Marshall McLuhan reflected extensively on the pattern of "sex, technology and death" he observed around him in US popular culture, which he said "[constituted] the mechanical bride."[1] Coining provocative terms such as the "Love-Goddess Assembly Line," he argued that the female body, which he likened to the Ford Model T, had been mechanized and dismembered, her parts easily replaceable and upgradeable by new plastics.[2] McLuhan used metaphors borrowed from the language of cars and industry to comment on the objectification of women's bodies in mass culture, linking this to what he saw as the standardization of homogeneous beauty ideals in the entertainment industry, such as in Hollywood casting and on "mechanical" chorus lines.

As noted by McLuhan in his midcentury studies of the "interfusion of sex and technology," an atomic bomb tested in Bikini Atoll in 1946 was named "Gilda" in honor of the actress Rita Hayworth and her femme fatale role in the contemporary eponymous film. That same year, French designer Louis Réard unveiled his highly revealing two-piece swimsuit—the bikini—explosively named after the atomic testing site. As discussed in greater detail in this book's introduction, numerous comparisons have been made in popular culture between weaponry and the curvaceous ideals for women's bodies—often termed "bombshell"— before, during, and after World War II. In the postwar period, breasts were referred to as "bazookas," presumably inspired by the shell fired from a bazooka antitank rocket weapon, which was first used under this name during World War II.[3] In 1958, when the white model Fran Frost was crowned Miss Bomarc in a beauty pageant held to celebrate the supersonic Bomarc antiaircraft missile, journalists referred to her as the "blonde missile."[4] Bras that molded breasts into the fashionable pointed peak were referred to as "bullet," "torpedo," "missile," and even "rocket" shaped.[5] In addition, falsies and padded bras were often referred to as "strategic" or described using other military-influenced terms in newspapers.[6]

Furthermore, during World War II, US servicemen painted female pin-ups and nudes on to the sides of bombers, figuratively placing women's bodies in close proximity to weaponry (figure 5.1) and in some cases also comparing them to it (figure 5.2). As will be shown in the chapter, similar comparisons were made by other actors in this period, including journalists, advertising executives and copywriters, plastic surgeons, and designers, including those working on cars and furniture. However, this chapter will also show how women's bodies were becoming corporeally interlinked with industrial and military materials R & D in more than a symbolic way, via advancements in plastics and medical technologies.

McLuhan's commentary on postwar US advertising provides a useful cultural example of how Atomic Age actors in popular culture viewed and presented, as well as feared and revered, American women's bodies as sites of self-authorship or "auto-design" via changes in technology. Scholars of cosmetic surgery have explored the profession's cultural and political contexts in the postwar United States.[7] After World War II, the discipline flourished due to medical advancements, increased spending power, a postwar nationalist sense of "technological prowess" through Big Science, and cultural confidence.[8] The face and body could be increasingly shaped acquisitively and materially, creating new ways of self-making.[9] Key actors, including surgeons and journalists, equated cosmetic surgery with social mobility and self-actualization.

5.1 P-51 Mustang pursuit fighter, 1944, on display at the Palm Springs Air Museum, California. Photograph by author.

5.2 Twin-engine B-26 Marauder, "Twin Engine Queenie II," 1944. American Air Museum in Britain, Duxford.

Cosmetic surgery is rooted in consumer culture and situated within the meritocratic ideals and freedom promised by the "American Dream," and its relationship to interconnected discourses around identity, including beauty, race, class, gender, and sexuality. Changes in media representation affected cosmetic surgery's reception in popular culture. Within the historiography of cosmetic surgery, there is a general consensus that, during the US postwar period, the media celebrated medical materials as offshoots of military R & D.[10] These life-saving and prolonging technologies included stainless steel, dialysis equipment, plastics, sulfonamide antibiotics, and organ transplant banks.[11] Newspapers, magazines, radio, television programs, and documentary films often covered these accounts with enthusiasm, promising that human bodies could be fixed, repaired, replaced, and improved, employing such catchy headlines as "Spare Parts for the Human Machine."[12] Medical science came to embody a postwar utopian idea of progress without conflict. However, as this chapter will demonstrate, these advancements were deeply embedded in military and industrial technologies and networks, where Big Science "weaponry" was transformed into "livingry."[13]

The aesthetic contouring of racialized, gendered bodies with plastics, like that of the curvaceous atomic bombshells, is a story of material conquest and subversion. This chapter focuses on the role of plastic foams, centering on an anecdote told by a famous American cosmetic surgeon that circulated in the US whitestream press during the postwar period. Robert Alan Franklyn claimed that at the end of World War II he sat in the squishy seat of a captured German bomber. A decade later, Franklyn was known as a highly publicized Hollywood cosmetic surgeon. In a number of books and articles, Franklyn explained throughout his career, he was inspired by his experience of sitting in this wartime upholstery, drawn to its unique spongy plastic materiality. He used this story in numerous publications to frame his research for the development of "Surgifoam," the material that he developed for his "Breastplasty" operation to cosmetically augment the bustline. Franklyn claimed that Surgifoam was different from the other plastic foams that respected plastic surgeons were using for similar operations.

Franklyn spoke openly of his sourcing of foam from a captured World War II German bomber, effectively linking his practice to the material remnants of war.[14] Indeed, the origin story of Surgifoam continues to circulate in scholarship today. However, to date, it has not attracted an in-depth critical investigation that focuses on its military-industrial context.[15] It is important to interrogate this transfer of material and the associations with war and industry that developed through polyurethane foam's application to women's bodies. This analysis highlights the significance of foam's use within a wider postwar US culture of objectifying feminine bodies.

The chapter draws heavily on the work of two plastic surgeons (Robert Alan Franklyn and Herbert Conway) to address the provenance of plastic foam and its impact on the shaping of women's bodies through cosmetic surgery. These comparative case studies provide in-depth examples of material conquest and subversion. They show, in detail, how foam's pliable materiality informed its application in the surgical contouring of women's bodies. They also provide greater knowledge of the provenance and sourcing of this material, exposing the complex postwar power structures that produced it and how these affected women's bodies.

It is important to begin with Franklyn for a number of reasons: he was the more visible of the two, publishing self-help books, communicating with journalists, and actively trying to sell his services through Yellow Pages listings and ads in pseudomedical journals.[16] He was particularly important during this period because he posed a problem for the American Medical Association. Franklyn was not the only one sourcing his foams from industry; well-respected doctors were doing the same. However, unlike Franklyn, they did not speak publicly about it. One such surgeon engaged in plastic foam implant work was Herbert Conway, my second case study. Based on the East Coast, Conway was recognized as a leading international plastic surgeon by 1936. He established and ran the Division of Plastic and Reconstructive Surgery at New York Hospital–Cornell Medical Center, a major American medical institution. Conway is representative of a small number of active, journal-publishing AMA-approved US plastic surgeons working on cosmetic breast augmentation during this period. Unlike Franklyn, Conway's work and papers have not attracted scholarship on the history of breast augmentation. He published in medical journals but did not divulge the industrial origins of his foams in his articles. However, his archival papers show that he sourced plastic foams for implantation in women's bodies directly via car and furniture suppliers, revealing complex national and international networks of plastics and their cosmetic use.

Franklyn and the Origins of "Surgifoam"

How did the foams used in everyday designed objects—including transportation upholstery, furniture, clothing, and foundationwear—differ from those implanted in human bodies, if at all? Were women being padded from the same materials that surrounded them in their daily interactions with the designed world? In this section, I investigate the extent to which the much-publicized cosmetic surgeon Franklyn's attitude toward material sourcing and transparency differed from that of his well-respected, AMA-approved plastic surgery colleagues who published in medical journals that were less accessible to the public.

First, it is important to understand the difference in scale and approach between Franklyn's communication of his research and that of established plastic surgeons. Franklyn published ten books in the United States between 1956 and 1979 on the subject of cosmetic and plastic surgery, and two in Italy in 1976. These books, most of which were published by trade presses and in multiple editions, were commercially popular, widely available in mainstream bookshops, advertised in newspapers, and circulated in public libraries across the United States. In his publications, Franklyn addressed a female reader, guiding her through steps to modify her body via beauty, diet, and exercise regimes, as well as his cosmetic surgery procedures. He also included compelling stories of women's "before and after" moments, raising hopes for self-authorship, self-actualization, social mobility, acceptance, and fame.[17]

All of Franklyn's books promoted cosmetic surgery; however, their titles communicate this in different ways and to varying degrees. US public libraries have classified Franklyn's books in the 600s of the Dewey Decimal System, thereby broadly situating them within "Technology (Applied Sciences)." Titles with less obvious links to cosmetic surgery, such as *The Art of Staying Young* (1964) and *Instant Beauty* (1967), are classified in 646—"Sewing, Clothing, Management of Personal and Family Life"—revealing how these systems are inherently gendered. Meanwhile, publications with more direct links to surgery are classified as 617, "Surgery and Related Medical Specialties," including *Beauty Surgeon* (1960) and *The Clinical Atlas of Cosmetic Plastic Surgery: A Teaching Manual* (1976). *Augmentation Mammaplasty* (1976) is classified in 618, "Gynecology and Other Medical Specialties." Unlike the latter two books, the more general texts were competitively priced between fifty cents and $1.45 and were commercially published in the United States as self-help guides, reaching a wide audience.

A number of these books were published by Frederick Fell, a New York–based self-help guide publisher. The most popular was *The Art of Staying Young*, with multiple editions published between 1964 and 1968. This indicates that Franklyn's publications were lucrative, warranting repeated investment and new editions. A second edition of *Beauty Surgeon* (1960) was published in paperback by Pyramid Royal Press, in a "new series of books designed to bring women everywhere the latest and most authoritative information on vital topics as Love and Marriage, House and Home, Health and Beauty."[18] In all these publications marketed to a general readership, Franklyn promoted his cosmetic surgery services, stressing how they improved women's lives, thereby further distancing himself from established medical structures and norms in the United States, which frowned on such overt advertisement of medical services.

Franklyn claimed to have completed 2,170 Breastplasty augmentation operations using plastic foam between 1950 and 1957.[19] In 1956, he claimed to perform an average ten to fifteen surgeries a week, and by 1967, Franklyn's operations had allegedly exceeded ten thousand.[20] His services were in demand, and a feature on his extensive racing horse collection in *Sports Illustrated* flaunted his wealth. The article included a description of his Viennese castle–style mansion, as well as "The Beauty Pavilion," a "spectacularly attractive round building" designed for him by renowned modernist architect Oscar Niemeyer on Hollywood's iconic Sunset Boulevard.[21]

In addition to being featured in whitestream media, Franklyn published articles in US and international medical publications. The American journals in which Franklyn published, such as *Southern General Practitioner of Medicine* and *General Practice: The Medical Journal of the West*, are less known and have since folded. The International Academy of Cosmetic Surgery, based in Rome, attested to his status as "director of the first licensed teaching institution in California for Plastic Surgery—The Plastic Surgery Academy Institute."[22] Franklyn also featured in international publications, such as Germany's oldest running journal on surgery, *Zentralblatt für Chirurgie*, which, after World War II, was run under Soviet occupation from East Berlin in the German Democratic Republic. Notably, he did not publish his medical findings in such reputable Anglophone journals as the *Journal of the American Medical Association*, *Plastic and Reconstructive Surgery*, or the *British Medical Journal*.

Prior to developing Surgifoam, Franklyn experimented with commercial plastic foams. However, he found these to be lacking due to "poor cell structure" and the strong chemicals that were used to preserve them. Openly detailing his unusual research journey to his readers, the self-claimed "beauty-parlor" surgeon asserted that, during World War II, he discovered the ideal implant properties of a synthetic foam not yet available on the US market. Franklyn, who claimed to be a chief plastic surgeon for the US Armed Forces, recalled examining a "captured German fighter plane" on public display in Canada during World War II, while on a work trip to observe plastic surgery being carried out by the Royal Canadian Air Force in Toronto at the time.[23]

The photograph in figure 5.3 was taken at a location likely to be the Toronto Royal Canadian Air Force Staff College; labeled "RCAF Day June 1948," it shows a captured German bomber, the Messerschmitt Me 262, marked "AM52," on display.[24] Perhaps this is the type of bomber, or possibly even the same plane, to which Franklyn referred? Although the photograph is dated later than Franklyn's recollection, it shows that captured bombers were on display in the area at the time. Franklyn also noted that "the staff sergeant in charge of the plane explained that because the Germans had no access to rubber during the war, they had made an imitation foam rubber where in America we had made an imitation solid rubber."[25] Again, the sergeant's comments match with the history of polyurethane foams, which were not available on the US market until after World War II. The seats in German cockpits were padded with this synthetic foam rubber, and "a piece of the upholstering had been cut away to show the imitation foam rubber beneath," which, Franklyn observed, "looked like a revolutionary type of plastic sponge to me."[26]

Franklyn's decision to reveal the apparent German military source of his first foam implant experiments is striking. It neither problematizes nor disguises the military-industrial origins of the material, nor its wartime context as a relic from an enemy bomber plane. The story's inclusion reflects North American public interest and curiosity about wartime advancements in materials and technologies developed in Germany, and a desire to interact with these artifacts. Once shunned as unreliable, plastic materials that had "stood the test of war" were increasingly perceived by postwar American consumers as dependable, hygienic, and functional.[27] Together, these factors could explain why Franklyn's story, which he referred to in each of his publications, gained traction in selling Surgifoam to women in the postwar United States.

5.3 A Messerschmitt Me 262, probably at the Toronto Royal Canadian Air Force staff college, on Royal Canadian Air Force Day, June 1948.

Remarking that "the trick was to get a piece," Franklyn was desperate to obtain the plane's plastic foam, which he felt was unrivaled by any other foams on the US market at the time. First, he explained to the sergeant that he was a medical doctor doing research and would like to have a sample. However, this did not get him very far: "[The sergeant] replied, reasonably enough, that if he gave a piece to every souvenir hunter who asked him there wouldn't be any plane left." Franklyn persevered: "Finally by going through proper channels, I was able to obtain about a square foot."[28] Bombers in Canada, which had once toured across the country in public displays and were protected from "souvenir hunters," were later sold off in parts and even scrapped in the 1950s.[29] When Franklyn finally got his hands on a prized piece of the bomber's upholstery, he "treated it as if it were the crown jewels."[30] However, the "treasure" itself was marked by the mechanics of war, marred with grease and oil stains.

Indeed, there wasn't enough clean material for Franklyn to do much experimenting. The limited amounts available would not suffice for the breast contouring he had intended. However, small, clean slivers could be sterilized and used instead on a small number of patients to help elevate facial scars. This worked with "stunning success" and demonstrated that the salvaged bomber foam "would be tolerated in the tissues of the human body."[31] Franklyn enthused, "The results were amazing. The body did not react to it in any way and it helped enormously in filling out missing tissues."[32] Franklyn's bomber anecdote, which could of course be exaggerated, is invaluable for providing a rare primary source that materially links developments in breast implants to wartime developments in plastics. Furthermore, Franklyn's story, perhaps influenced by the sensationalism with which he was prone to promoting his services, does correlate with sources on captured World War II enemy planes. As will be shown, it also aligns with primary sources related to the development and transfer of plastic foams.

Franklyn correctly identified the soft foam he found in the German plane as originating from IG Farben, even detailing the damage to the factory caused by Allied bombing.[33] He noted, "Later on in the war the Allied Forces completely bombed out the I. G. Farben factory in Germany which manufactured the soft foam material," causing disruptions in the supply of this material for the rest of the war and for several years afterward.[34] It is likely that he is referring to the polyurethane foam that originated from Otto Bayer's lab at the Bayer Leverkusen plant, which was part of the IG Farben conglomerate during World War II, discussed in detail in chapter 3. How did Franklyn have access to this military intelligence on foam? Did he read about it in *German Plastics Practice*, which made the findings of the US plastics experts' Quartermaster reports available to a wider public audience?[35]

Inspired by "this plastic foam 'cushion,'" Franklyn soon became aware that "there was no plastic material in the [United States] that could match it for lightness and resilience."[36] Again, this aligns with the history of plastics, as Otto Bayer developed a unique type of plastic foam that was not yet available in the United States at the close of World War II. Franklyn wrote to multiple plastics factories, perhaps even using the index provided in *German Plastics Practice*, but "almost every letter came back." However, he finally heard back from the IG Farben conglomerate: "With the war over, there was a possibility the Farben Company might export the imitation foam rubber."[37] Interestingly, while waiting for this to happen, Franklyn claimed he ordered an "American product" that had come on the market; however, there were no manufacturers of the actual material itself. Instead, Franklyn needed to order the material in its raw form, mix the chemicals together, and bake it himself. Again, the material fell short of Franklyn's augmentation mammaplasty plans, as it would not rise more than two and a half inches. He tried stitching it together with nylon, but this was not an ideal solution. Were these raw materials, available on the US market so quickly after the close of World War II, inspired by the polyurethane findings in *German Plastics Practice*?

After some research and time had passed, Franklyn claimed he discovered that German plastics companies had finally entered into a licensing agreement with a number of chemical manufacturers in the United States to distribute "various types of German plastics," including the plastic foam he was interested in.[38] He later enthused, "Finally I was able to import through an American company the raw material that made up the imitation foam rubber I first had seen on that German plane."[39] Franklyn ordered the material, experimented with it, and, in his 1959 publication *On Developing Bosom Beauty*, claimed to have "refined it for surgical purposes into a polyester plastic foam called 'Surgifoam.'"[40] Here, Franklyn gives his readers the impression that the material he is implanting in the breast is a new type of material that he has explicitly developed for use in the human body, free from the strong preservation chemicals he had previously complained about. He also later noted, "I located a small fabricator who was very interested in the internal application of the plastic foam."[41]

In his general-audience publications, Franklyn did not share the name of the manufacturer or supplier he was using, nor where and how this foam was produced, leaving many important facts of Surgifoam unanswered. However, a medical journal provides some more detailed information. In 1957, an advertisement for Surgifoam accompanied an article by Franklyn in *General Practice: The Medical Journal of the West* about his augmentation mammaplasty technique. It detailed that "Surgifoam is the specific destination of a gas extended polyurethane foam plastic used for surgical implantation" and "technically is a thermoset cellular plastic material obtained through polyester reaction."[42] Standard, unmolded blocks measuring 6 by 3.5 by 10 inches could be purchased for eight dollars each or sixty dollars per dozen from California Surgical Solutions Co. Ltd.

By coining the name Surgifoam, Franklyn was able to distract from what type of plastic foam he was actually using. Surgifoam could essentially be used as a mask for any type of plastic foam, and certainly throughout his publications, Franklyn interchangeably referred to it as polyester foam (1959), polyurethane foam (1957, 1960, 1961), or plastic foam (1976).[43] That said, did Franklyn at some point source his plastic foams from the polyurethane foam specialist company Mobay, formed as a merger between Bayer and Monsanto in the United States in 1954? This would link Franklyn's practice even more closely to the plastic foams that originated from Otto Bayer's lab in his bomber Surgifoam origin story. Franklyn's public-facing comments on his sourcing of foam can help us trace its most probable provenance and entanglement in international military-industrial plastics networks.

Franklyn Versus the AMA

Franklyn's promotion of his Surgifoam invention and Breastplasty operation in the medical and popular press attracted the attention of respected plastic surgeons, such as Herbert Conway and Jerome Pierce Webster, as well as the AMA, all of whom hold materials he published in their papers and archives.[44] Unlike these figures, Franklyn did not share his papers with an archive. The AMA's Bureau of Investigation papers indicate Franklyn's wide appeal and material impact: He is the only named bust augmentation surgeon for whom they have dedicated folders. Whenever he published an article in a popular magazine, there was a spike in national and international correspondence with the AMA asking for further information on his practice and credentials, as well as first-person reports of fraud and malpractice.[45] These documents provide a useful timeline of Franklyn's press activity and the AMA's mounting concerns, as well as its public warning against his services.

An August 1953 issue of *Pageant* magazine featured an article that triggered the AMA's public condemnation of Franklyn's breast augmentation surgery. The AMA still holds this issue in its archives, its cover wrapped in a white sleeve emblazoned with red text: "In This Issue—First Time in Pictures: THE OPERATION THAT REMOLDS FLAT-CHESTED WOMEN" (figure 5.4). A paper band conceals the bustline of a white, brunette female model dressed in a white swimsuit; revealing her bustline requires a *Pageant* purchase and removal of the wrapper in an act of undress and consumption. The cover text repeats, "WATCH THE OPERATION THAT REMOLDS FLAT-CHESTED WOMEN," further emphasizing the voyeuristic nature of this illustrated photo series that promises readers entry into the operating theater.

Inside, an article titled "Breastplasty" introduced this eponymous "new type of operation" developed by Franklyn.[46] The piece, by journalist Henry Lee, began by claiming that over four million young women in the United States suffered from "micromastia (immature breasts)."[47] In Franklyn's original medical journal article on breastplasty, he reasoned that since "the fully developed breast has become a standard of female beauty . . . spinsterhood and unhappy marriages" could be "traced"

5.4 Cover of *Pageant*, August 25, 1953.

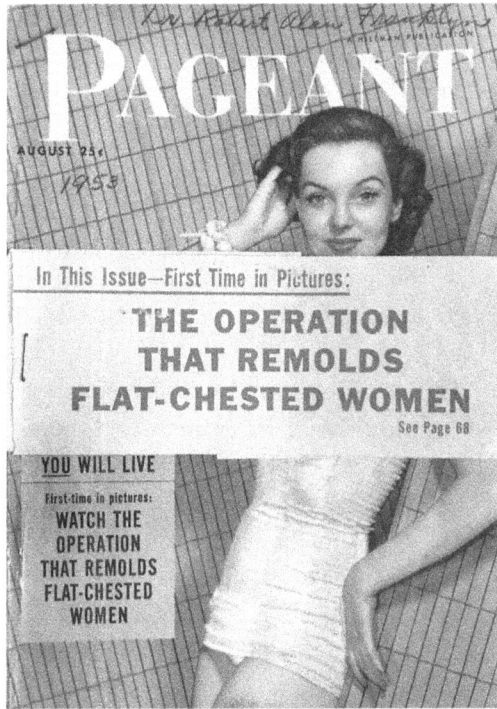

to this condition.[48] As with the "Counterfeit Bride" discussed in the previous chapter, contemporary commentary such as Franklyn's and Lee's used pathologizing language to explain how women's supposed failures to perform heteronormative gender roles were linked to their breasts. In these accounts, falsies were ridiculed and not seen as an answer. Lee proposed that Franklyn had the solution to all these problems. His procedure was reported to take only ten to twenty-five minutes and did not even require a hospital setting.[49] Perhaps hoping to challenge and undermine the surveilled, formal structure of the hospital and the costs it involved, Franklyn argued that Breastplasty could be quickly performed in a more informal office setting: "I want people to regard this kind of surgery with the same attitude as going into the beauty parlor or visiting the dentist."[50] He stressed that breast augmentation was to be viewed as a simple, painless procedure, nothing more than an addition to a well-established health and beauty regime. As discussed in the previous chapter, the patients featured in Franklyn's case studies were almost exclusively unmarked cisgender white women, and while he boldly claimed, "The operation is within the budget of the average woman," the price tag of up to $1,200 made his services inaccessible to most.[51]

Franklyn quelled any concerns readers might have about foreign body reactions, reassuring them that his choice of material promised subcutaneous acceptance. The implants, concealed within flesh, melded into the body. Lee assured readers, "As a matter of fact, because of [the implant's] porous construction, the foamed plastics tend to become a part of the body. Though they do not in any way affect the natural processes, they act as a framework for new tissues and blood cells which grow in and around them."[52]

Franklyn's foam implant was covered with a protective "nylon membrane," similar to how Maidenform sheathed its "Masquerade" foam falsies in satin.[53] His design harnessed foam's ability to be molded into "a closed pore system" that was further protected by a nylon covering, setting it apart from the sponge rubber materials that preceded it. Franklyn claimed that the nylon sheath ensured the implant retained its flexibility and proper contour. He also later experimented with a smooth silicone coating, which he claimed could also act as a protective layer to prevent tissue from entering the implant.[54] Franklyn argued that the great danger of other sponge rubber implants lay in their porous structure: If this was not covered in a fibrous synthetic material such as nylon or a smooth silicone coating, body tissues would infiltrate the air holes, causing the material to shrink and harden. The breasts, inseparable from the implants themselves, would become painful and lumpy, requiring surgical removal.

Franklyn also claimed that his "poly-plastic" invention, Surgifoam, was superior to previous synthetic foam implants, such as Ivalon, made from polyvinyl sponge.[55] He asserted that unlike its predecessors, his invention could be "sterilized at high temperature and pressure," minimizing the risk of infection. Previous "obsolete vinyl plastics," such as Ivalon, he argued, were sterilized by boiling and would crumble under pressure, resulting in a loss of shape.[56]

Surgifoam, Franklyn asserted, offered many advantageous properties as an implant: it was inert and unlikely to set off a foreign body reaction; lightweight, flexible, and compressible, allowing a surgeon to easily "shape [it] artistically"; unabsorbable by the body (unlike human fat or other plastics); nonallergenic; invisible on X-rays, so that patients did not feel embarrassed if they chose not to disclose their operation; and nonaging, "so that breasts will always remain youthful."[57]

The AMA took issue with Franklyn's work for a number of reasons. First, Franklyn did not cite the source for his statistics on "micromastia." Second, he claimed he was essentially the only surgeon able to perform this important surgery. Third, he failed to produce or cite exhaustive animal experiments with the materials before their implantation into the human body. Fourth, the AMA claimed Franklyn advertised his services in the classified section of phone directories.[58] Fifth, Franklyn refused to disclose to the AMA the technical specifics of the foam composition he was using, responding that "the substances used by me are not on the market—having [been] ordered specifically from the supplier for test purposes."[59] Lastly, it also called into question his credentials as a cosmetic surgeon. The AMA found no evidence that he completed a surgical residency, only indications that he completed residencies in anesthesiology and dermatology. It also stated that he was neither a member of any specialty group nor was he formally connected to any surgical subspecialties.

Shortly after the *Pageant* article was published, the AMA retaliated with a press release and an article in its journal titled "The Business of Bolstering Bosoms," which critiqued Franklyn's blatant advertising of his services.[60] The association's article denounced Franklyn's widely publicized surgical procedure and cautioned against his services, claiming that the material he vaguely described in the article as "a soft plastic foam substance—actually a nylon-like resilient material impregnated with penicillin" was likely to be Ivalon, a material known to be problematic as an implant.[61]

Auto-Design, Body Shops, and Bumpers

Cosmetic surgery was a particularly popular topic in celebrity gossip magazines like *Confidential*. Journalists linked cosmetic body modifications to the entertainment industry and Hollywood, glamour, and consumer goods such as fashion accessories. An article titled "What Are American Women Made Of?" depicted a wide variety of products, including eyelashes and peach-colored foam falsies, and compared them to spare parts of a car that could be easily switched out (figure I.4).[62] In 1954, the journalist Audrey Minor wrote in *Confidential* that Hollywood was now subject to "a new science of fantastic fakery . . . built-in or built-up bosoms."[63] She clarified that these alterations were subcutaneous and permanent: "These false fronts are not to be confused with the rubber kind you can purchase in a dime store and stuff into strapless dresses . . . the new curves are nature's own, remodeled, stretched or trimmed on a hospital operating table by modern science's wonder boys, plastic surgeons."[64]

In keeping with *Pageant*'s framing of Franklyn's Breastplasty as an operation that "remolds" women, Minor commented on how male plastic surgeons were able to permanently "remodel" the body's contours and curves. She also compared a private cosmetic surgery hospital to a "factory," where the body could be redesigned and overhauled—"the old chassis can be remodeled."

The chassis, a load-bearing structure that serves as the framework for technology—particularly cars—was also used by other journalists as a metaphor to describe women's bodies during this period. Vicki Starr, a Latinx transsexual female topless dancer who rose to fame in 1960s San Francisco, kept a newspaper clipping that praised her as beautiful and blessed with a "handsome chassis."[65] Using terminology that references automobiles and industry—similar to McLuhan's *The Mechanical Bride* and other contemporaries—Minor paints an image of a secretive space "behind closed doors" in the heart of Hollywood.

During what was known as its golden age, Hollywood and the studio system of celebrity was referred to as the star factory. Under the racialized studio system, artists were owned exclusively by studios, which dictated their image, both on-screen and off.[66] A famous example of this is Margarita Carmen Cansino, whose name was anglicized by Columbia Pictures studio head Harry Cohn to Rita Hayworth. Cohn demanded that Hayworth, who was of Spanish Romani and Irish ancestry, remove an inch of her hairline by electrolysis, lose weight, and dye her hair red. Scholar Clara E. Rodriguez notes that, after this transformation, Hayworth was no longer "relegated" to playing "senoritas" but was instead cast in leading roles as "an ethereal all-American girl" and "love goddess."[67] Hayworth's ethnicity was erased to conform to a homogeneous American ideal of whiteness that was promoted by the entertainment industry during this period.

The studio system's control extended to the public persona and perceived private life of contracted stars. This included regulating whom celebrities could associate with, whom they could marry, and their bodily autonomy, including their reproductive agency. These restrictions on actors' off-screen lives reflected the largely conservative values depicted on-screen and the lasting impact of the 1930 Motion Picture Production Code. Later known as the Hays Code, these regulations were established to create a moral framework of censorship for the US film industry. The code, which remained in place until the late 1960s, prohibited on-screen depictions of what was deemed by its creators as immoral, including nudity, homosexuality, and interracial relationships.

White American bombshell actress Mamie Van Doren, who claims to have coined the term "bullet bra," believed the Hays Code and the undergarment were linked.[68] Van Doren reflected that the omnipresent Sweater Girl look in Hollywood, popularized by actress Lana Turner in Hays Code–era movie *They Won't Forget* (1937), and the conical bullet bra this look required, was a way of circumnavigating strict censorship rules around nudity and sexuality, particularly in relation to breasts and cleavage. This look accentuated the bustline without revealing any skin.

Hollywood had a major impact on shaping feminine beauty ideals. Franklyn commented, "In Hollywood particularly, the plastic surgery work on breasts must be of the highest type, for Hollywood sets the standards of beauty for the entire country."[69] Indeed, prospective patients would often reference the busts of famous Hollywood bombshells when describing their request. Minor reported that, behind the scenes, "movie stars with drooping fronts can be wheeled into surgery for an uplift. Actresses too bountifully supplied can be whittled down to a more sedate size. Starlets born flat-chested register for the big build-up."[70]

In her description, abject unsightly female bodies are "wheeled" into a medical space that is not unlike a mechanical body shop. She compares women to car bodies in need of repair and abundant, flabby raw materials on an assembly line to be shaped and shipped out in accordance with Hollywood's dominant on-screen beauty ideals of heteronormative, cisgender white women. The cultural comparison of women's bodies with car body shops was also referenced in foundationwear marketing. Gay Deceivers, a company specializing in falsies, famously advertised using the slogan "We fix flats" in 1934, and Frederick's of Hollywood used the same slogan too.[71] These ways of viewing and describing women's bodies were also part of everyday interactions. Franklyn recalled a patient coming to see him about augmentation mammaplasty and asking, "You fix flats?," to which he jokingly replied, "This isn't a garage or a gas station."[72]

Franklyn, perhaps seeking to counteract this kind of press that equated cosmetic surgery with a standardized Fordist factory approach, asserted, "When our surgery is completed, our patients do not look as if they've been machine-tooled on an assembly line. Each one will look like an individual—but an individual whose appearance is greatly improved."[73] Unlike the premade silicone breast implants that emerged in the 1960s, foam implants were rarely prefabricated in standardized shapes or sizes but had to be individually handcrafted by the surgeon, usually with scissors from slabs of plastic. Franklyn's comments articulate tensions between standardization and customization, not entirely dissimilar from the language of car customization during the "Style Wars," a competition between US car manufacturers during the postwar period to make their once-streamlined cars increasingly fantastic and adorned with ornamentation of "wild extremities," including exaggerated tails, pointed noses, fins, bumpers, and pillars.[74] For example, the 1952 Oldsmobile "Rocket" Ninety-Eight had a rocket missile as a hood ornament.[75] Car bodies were adorned with incremental visual cues referencing fighter planes and missile rockets—capitalist symbols of postwar US prosperity and consumer choice.

In 1946, Cadillac, famed for its exaggerated tailfin designs, added an aggressive, artillery shell–inspired pointed front bumper to a new version of the 1942 Sixty-Two series to protect the car's body by denting any other cars that came near it. In the 1950s, these "conical bullet-shaped projections" became known as Dagmar bumpers, or simply Dagmars, named after the eponymous, white, female blonde TV personality.[76] Virginia Ruth Egnor, known professionally as the "seemingly Swedish" Dagmar, was celebrated for both her curves and her penchant for pairing a bullet bra with a tight-fitting sweater or a figure-hugging, low-cut décolletage. She appeared on her own show, *Dagmar's Canteen*, on NBC from 1951 to 1952. Dagmar was reduced to her bustline and transformed into a protruding part of a car; Dagmars were eventually placed higher on the front grille and became even more pronounced. The fluid curves of bombshell women were so desirable that advertisers capitalized on the explosive power of their sexuality. A 1957 Hudson Hornet V8 advertisement described the car model as "trim," "a beauty," and "a bombshell."[77] That same year, Cadillac's Dagmar bumpers gained black rubber tips, making the comparison between car body and female body more prominent, before being toned down for the 1958 model and entirely removed from the 1959 model.[78]

In addition to bumpers, postwar US car manufacturers offered consumers a variety of fashionable colors to choose from. Henry Ford's infamous 1909 quip that "any customer can have a car painted any color that he wants so long as it is black" had long been challenged when General Motors established the Art and Color Section in 1927.[79] The company used Duco varnishes, developed in collaboration with DuPont, and was the first to establish an entire department dedicated to styling.[80] In the 1950s,

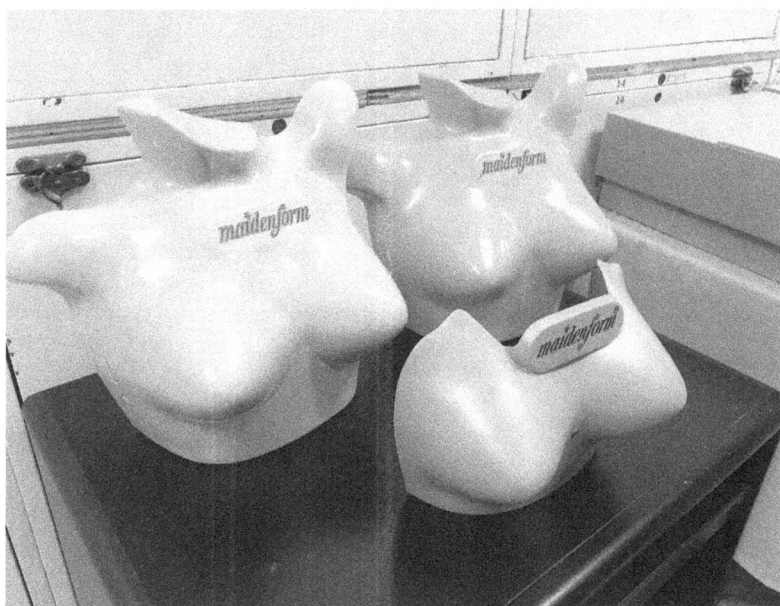

5.5 Maidenform sculptured bust forms in peach (1953), light blue (1957), and turquoise (1966). Division of Home and Community Life, National Museum of American History, Smithsonian Institution, Washington, DC.

pastel tones were a popular choice, perhaps most famously exemplified by the Chrysler La Comtesse and Dodge La Femme, both explicitly marketed to female consumers. The latter came with a plastic purse, complete with gendered plastic products, including makeup accessories and vinyl plastic rainwear ensemble—raincoat, boots, and umbrella—all coordinated to match the car's pink color palette.

In 1952, Maidenform launched a new bust form design for displaying its brassieres that was not dissimilar from popular car lacquers at the time. The first form, available in "flesh pink color," assumed a white consumer and was "sturdy, lightweight and adaptable to both window and interior displays." The Smithsonian's National Museum of American History holds three such forms, each designed to fit size 32B: one in peach (1953), one in light blue (1957), and one in turquoise (1966) (figure 5.5). Promoted by Maidenform as "a new bustform by famous sculptor available to dealers," it promised a "modern, decorative display."[81] Their streamlined curves recall the conical, pointed "Dagmar" bumpers.

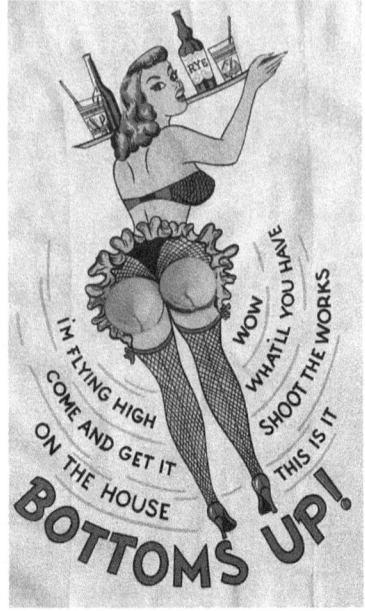

5.6 "Busty!" novelty tumbler, ca. 1955. Photograph by author. Collection of Sonya Abrego.

5.7 Detail of novelty apron, ca. 1955. Photograph by author. Private collection.

Beyond mannequins and the ideals they embodied and reinforced, a proliferation of mid-century plastics also gave shape to an industry of novelty, risqué "flesh pink" products that objectified white women, such as a "Busty" tumbler (figure 5.6), an apron with foam derriere pads (figure 5.7), a Raquel Welch "life-size" pillow, and inflatable "life-like lady's legs."[82]

Foundationwear designers, as well as plastic and cosmetic surgeons, equated the softness and pliability of foam with racialized and gendered concepts of American social mobility. Women's requirement to be pliable and malleable in society, the home, and the workplace—particularly in Hollywood, in the entertainment and sex industries—was likened to the softness of foam. In this paradigm, women's worth was dependent on how well male directors and agents felt they filled out their sweaters. Franklyn described "bosom measurement games . . . a popular pastime at private Hollywood parties," in which "young starlets—and sometimes established stars—will line up along a wall" to be measured by male judges.[83] Minor commented, "[A young starlet] knew that without something to fill tight sweaters she never would make the cinematic grade, that the girls who get places in Hollywood must look soft, appealing and well-upholstered."[84] By extension, women must be inviting to the touch, friendly-looking, and satisfactorily padded. Object-centric advertisements during this period compared women's bodies to items in domestic interiors, while simultaneously anthropomorphizing foam-upholstered furniture into women's bodies—an example being Eero Saarinen's Womb Chair.[85]

In postwar America, as troops returned from the front, women were actively encouraged to leave the workplace and return to the home. In his book *Design for Business* (1947), US design consultant Joshua G. Lippincott suggested that designers look to the latest trends in womenswear for inspiration on what would sell in the domestic environment. Lippincott commented that "a woman likes to be in harmony with her environment" and recommended a "parallelism between a woman's clothes and the furnishing of her home."[86] For some women, the parallels existed not only between the furnishings of their homes and the fashionable outer layers that encased their bodies or cushioned them as they sat, but also in the materials increasingly padding their bodies—both on the surface and sometimes within.

Parallel developments in industrial plastics and surgery were praised in newspaper articles such as Weldon Wallace's "Spare Parts for the Human Machine."[87] A 1953 *Ebony* article, similarly titled "Spare Parts for Humans," claimed, "In much the manner as manufacturers have developed spare parts for automobiles, surgeons and researchers have gone to great lengths to make available to the ill and disabled dozens of replacements of worn-out or destroyed parts of the body."[88] Articles like these often noted that implanting foreign materials in the body was nothing new, as ivory had been used for centuries as a bone replacement and sheep intestines for stitching. However, this was changing with the latest R & D: "The modern development of industrial plastics has placed a variety of potentially useful new materials at the disposal of surgeons."[89] Plastics were now being applied to medical applications thanks to their properties of being "lighter, less irritating, more flexible and easier to shape than many older materials." Wallace, like many of his contemporaries, was impressed by the ability of plastics to shift between the industrial, medical, and domestic realms. He remarked that "a sponge plastic useful to housewives in the kitchen is being used by surgeons to plug up cavities in damaged bones." The opportunities that postwar plastics offered for the body appeared endless.

Conway and Foam Sourcing on the East Coast

Articles on plastic foam as a material for bust augmentation, in both the popular and medical press, generally did not disclose how these foams were sourced. The most notable exception was Franklyn. Once he sourced the foam, Franklyn developed his own, allegedly medical-grade material, which he aptly named Surgifoam, etymologically positioning it in the medical realm, separate from military and industry. Similarly, in a 1961 PRS journal article on "augmentation mammaplasty," a group of plastic surgeons, including Milton Edgerton, recommended sourcing Scott Paper Company's Scottfoam (polyurethane) and Etheron (polyether).[90] The Scott Paper Company, an established firm associated with hygiene and the body, was likely familiar to many readers. Etheron was sold to medical institutions as a type of surgical plastic foam. In both these cases, as well as in Franklyn's, the female body was padded with foam materials presented as specifically intended for the body rather than for military or industrial use.

At this time, surgeons were experimenting with materials in the body that had originally been intended for industrial and other noncorporeal applications. Plastic surgeon Herbert Conway's archival papers on breast augmentation provide a useful insight into the inner workings of the industrial-medical complex. Conway originally sourced plastic foams for surgery from the Robbins Instruments Company. This specialist medical engineering firm provided him with tools for his surgery, such as derma-planing equipment. They also provided Conway with diisocyanate polyether, sold as Etheron, in sizes of five by five by five inches and six by six by three inches.[91] Etheron came with preparation and sterilization notes written by Elsa K. La Roe, a New York City–based plastic surgeon and author of *The Breast Beautiful* (1947), a guide to cosmetic surgery of the breast.

Robbins and La Roe promoted Etheron's properties as ideal for breast implants. Their promotional copy is similar to that of companies such as Mobay, which targeted the furniture market and stressed the supple pliability of their foams. Robbins promised that Etheron "contains millions of uniformly distributed inter-connected air cells and is not affected by chemical changes, ... is completely non-ageing, ... does not harden or shrink, ... is well tolerated by human tissues."[92] Furthermore, the material was durable, "highly elastic . . . with excellent cushioning effects," and could be "stretched and sutured without the fear of breaking." Etheron, Robbins assured, "is internally and externally non-toxic and stable. Will not decay or mold through fungi, bacteria or other biological agents. Odorless and tasteless." The material's alleged neutrality made it ideal for implantation. In these guide notes, Robbins appeared unconcerned about the long-term effects or the agency of the human body as an actant on foam. The rhetoric surrounding the materiality of Etheron and other plastic foams portrays them as easily manipulable materials that perform just as the surgeon or engineer demands.

In 1959, Conway pushed Noel Robbins for the exact chemical composition of Etheron. Robbins replied that the makers of Etheron "would not care to give [him] the formula."[93] This could suggest that the Robbins Instruments Company knew little about the material they were selling or were unwilling to share its chemical composition. As with previous examples of foam provenance, such as Surgifoam, there seems to have been reluctance to name or share detailed information on the plastics being cosmetically implanted. Judging by his papers, it is likely that Conway wanted to know the formula to include it in a future medical journal publication and to better investigate potential carcinogenic threats.[94]

In September 1959, Conway contacted Monsanto to request samples of Etheron foam in the same dimensions as the Robbins foam, presumably to compare the materials. Monsanto forwarded his inquiry to Mobay (their joint venture with Bayer that specialized in polyurethane foam), which replied, "We are not familiar with the name Etheron, but believe that it is a trade name for urethane foam manufactured by one of the companies on the attached list."[95] Mobay also offered to test a sample of Etheron to confirm whether it was a urethane foam. They concluded that Conway's sample was a polyether urethane foam and that "a similar type of foam product can be obtained from any of the companies on the attached list."[96] The list spanned twenty-seven US-based companies, predominantly located on the East Coast, and two Canadian companies (figure 5.8). It implies that no differentiation was made between industrial suppliers and "medical suppliers" of foam, and that they were essentially the same material marketed under different names.

Mobay was not the only producer of urethane foam with whom Conway was in touch. In July 1960, Conway contacted the General Tire and Rubber Company in Akron, Ohio, requesting foam samples for experiments in corporeal applications—first in animal studies and later in the

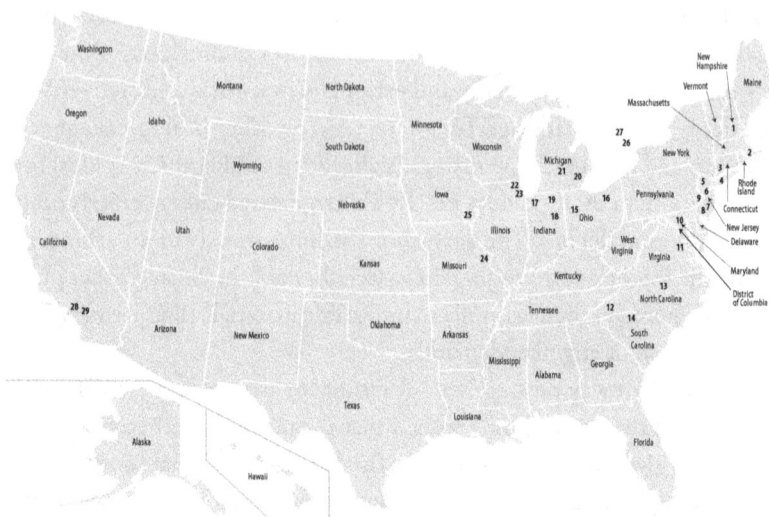

5.8 Map of twenty-nine polyurethane foam suppliers recommended by Mobay to Herbert Conway, 1960. Map by author. Source: Wikicommons.

Map Key
1. Davidson Rubber Company
   Dover, NH
2. Firestone Rubber & Latex Products
   Div. of Firestone Tire & Rubber Co.
   Fall River, MA
3. B.F. Goodrich Sponge Products Div.
   Shelton, CT
4. Curtiss-Wright Corporation
   Curon Division
   New York, NY
5. Hewitt-Robins Incorporated
   Franklin, NJ
6. Jersey City Foam Products
   Jersey City, NJ
7. Nopco Chemical Company
   North Arlington, NJ
8. Paramount Foam Industries
   Lodi, NJ
9. Scott Paper Company
   Foam Division
   Chester, PA
10. Wm. T. Burnett and Co., Inc.
    Baltimore, MD
11. E.R. Carpenter Company
    Richmond, VA
12. Dayco-Southern Corporation
    Division of The Dayco Corporation
    Waynesville, NC
13. Phillips-Foscue Corporation
    High Point, NC
14. Poly Products Company
    Subsidiary of Faultless Rubber Co.
    Spartansburg, SC
15. Dayton Tire & Rubber Company
    Division of The Dayco Corporation
    Dayton, OH
16. Goodyear Tire and Rubber Company
    Akron, OH
17. American Rubber Products Co.
    La Porte, IN
18. General Tire and Rubber Co.
    Industrial Products Division
    Marion, IN
19. U.S. Rubber Company
    Mishawaka, IN
20. Plastomer Corporation
    Detroit, MI
21. Reynolds Chemical Products Co.
    Whitmore Lake, MI
22. Midwest Foam Products,
    North Chicago, IL
23. U.S. Rubber Co.
    Chicago, IL
24. F. Burkart Mfg. Company
    Div. of Textron American, Inc.
    St. Louis, MO
25. Dryden Rubber Division
    Keokuk, IA
26. Hardifoam Products Ltd.
    Toronto, Canada
27. Robinson Foams Limited
    Woodbridge, Canada
28. American Latex Products Corp.
    Hawthorne, CA
29. Urethane Corporation of California
    Compton, CA

female breast. General Tire obliged and sent him two samples of Polyfoam P-17 Grade, supplied by its Industrial Products Division in Marion, Indiana. Responding to Conway's inquiries, company representatives noted they had no information on the carcinogenic properties of Polyfoam. Their reply includes a June 1960 article from *Air Engineering Magazine* titled "Cancer from Polyethylene, Polyurethane, Polysilicone Production Processes?," and claims that their specific foam did not have the same carcinogenic ingredients as those discussed in the article. Conway replied enthusiastically, noting that "as far as I know, no one has yet tried [General Tire Polyfoam] in the body."[97]

In May 1962, Conway and George Dietz coauthored "Augmentation Mammaplasty" in *Surgery, Gynecology and Obstetrics*. Presumably, this is the article Conway was researching when he contacted Robbins, Mobay, and General Tire from 1959 to 1960. The piece mentioned Polyfoam sponge, which was described as "furnished in blocks of 6 by 6 by 5 inches," and the specific chemical process to create the "polyetherurethane [*sic*] . . .

reaction product."[98] It appears that Conway was able to find the formula he had originally requested from Robbins and made this information available to readers of the medical journal. Material specifics were very important. In 1957, a journalist reporting on an annual American Society of Plastic and Reconstructive Surgery meeting of five hundred American surgeon members said that thirty of them performed breast augmentation surgery using plastic foams. However, some had abandoned the procedure due to disappointing results and had to remove the polyvinyl in a second operation.[99]

After a year of animal studies with polyether urethane, Conway found no malignancies. Conway and Dietz praised Polyfoam's malleability: "The material is carved easily into the desired shape with a knife or heavy scissors."[100] Foam's ability to be shaped into a conical prosthetic was a bonus. It could hold its shape better than the fat and tissue transplants that were shaped conically in earlier breast augmentation surgery.[101] Conway and Dietz described their process as follows: "The sponge is cut roughly to the shape of a truncated, eccentric cone. Recently, attention has been directed to the contouring of the prostheses to the shape of a virginal breast."[102] They instructed that Polyfoam should be shaped by scissors into a conical shape, pushing the breasts outward and upward in a manner similar to that of padded bras and pointed falsies available in the early 1960s United States, an aesthetic also followed by Franklyn.

Understandings of what constituted a desirable, racialized, and pathologized "virginal breast" fluctuated over time and were affected by (interdependent) changes in technology and fashion.[103] These male medical practitioners, as well as the outlier La Roe, pathologized the racialized female breast, and what they presented as the "healthy" or desirable "maiden-formed" (youthful and prelactation) breast was clearly shaped by technologies and fashions of the time. This is particularly significant, as two respected AMA surgeons in a reputable medical journal recommended the use of foam to sculpt conical prostheses for implantation within a professional medical context that adhered to the prevailing contemporary notions of fashion. Unlike most of Franklyn's articles, which addressed the general public, Conway and Dietz's article is likely to have been read by medical professionals both nationally and internationally. Although Conway and Dietz, like many of their peers, claimed that "an ideal prosthetic for augmentation mammaplasty . . . is not yet available," they effectively legitimized, endorsed, and recommended the material by concluding, "It was not necessary to remove any of the 30 [plastic foam] implants in this series of consecutive cases."[104]

Untangling International Foam Networks

Experimentation with and application of industrial materials in the surgical cosmetic contouring of women's bodies did not occur only on a national scale. Franklyn, who published in German and Italian medical publications, also boasted of sharing his plastics R & D on women's bodies with Russian businessmen, claiming they were particularly interested in his nylon-covered Surgifoam invention for mammaplasty.[105] Conway's papers similarly reveal an international network of surgeons with an interest in plastic foam breast augmentation who sought his recommendations.

Cosmetic surgeons working on breast augmentation participated in international research networks. In November 1958, Francisco Duran Acosta, medical director of Clinica de Cirugía Plástica Insurgentes in Mexico City, came to New York, and Conway showed him the Etheron sponge he recommended as the ideal material for bust augmentation. Acosta had been using the material since Conway's recommendation, but in a 1962 letter to Conway, he reported that his patients were experiencing foreign body reactions, as well as hardness and a decrease in volume. In addition to asking for Conway's advice on research and developments in implantable materials, he sought out pastoral support: "In both cases [hardness and loss of volume] the patients get angry."[106] Conway assured Acosta: "We have had no complications with Etheron synthetic sponge material for augmentation mammaplasty which have necessitated the removal of the sponge."[107] He stressed that "it is very important that the cavity be dissected widely so that the sponge is not compressed too much after insertion."[108] Here, Conway emphasized the importance of maximizing foam volume by minimizing flesh. Acosta was encouraged to remove as much tissue as possible to allow the foam to sit more comfortably in the cavity, without too much pressure and compression from the body.

Medical doctors, such as Conway and Edgerton, also received national and international inquiries about the sourcing of the materials they were using in breast augmentation surgery. In September 1962, Alexa Klossowski, a Berlin-based surgeon, wrote to Conway in response to his article "Augmentation Mammaplasty." She noted that this method was unknown in Germany and requested more information, along with a sample of "polyetherurethane" [sic] material or details of the suppliers.[109] In response to her request, Conway recommended obtaining polyether urethane from the General Tire and Rubber Company. By doing so, Conway further supported powerful American plastics companies by recommending their materials to surgeons working abroad. Conway informed Klossowski that General Tire's materials were "not expensive."[110]

Conway's article effectively argues that materials made for the automobile and furnishing industries were at low risk of foreign body reaction if they are correctly sterilized prior to implantation. His correspondence related to his May 1962 article such as with Klossowski, shows that he was recommending materials he had only recently been introduced to in a list of commercial manufacturers of urethane foam in the United States, provided by Mobay's district sales manager. The companies listed were working with Mobay Chemical Company supplies. This serves as another example of how US surgeons' choice of materials for aesthetically shaping women's bodies was influenced by nonmedical actors within the burgeoning plastics industry, such as sales managers seeking to widen the application of their products to the medical field.

Before the 1976 Medical Device Regulation Act, there was no formal testing protocol that new plastic materials had to pass before being implanted in the human body. Many of these materials have been grandfathered in and remain in circulation. This is significant because materials originally designed in the lab for military and industrial applications later found their way into the body, a site for which they were not explicitly designed. As will be discussed in chapter 7, Conway also turned his attention to a nonporous material for breast implants: silicones.

In the postwar United States, cosmetic and plastic surgeons used plastic foam materials to aesthetically shape feminine bodies. These materials were subverted from their original use by surgeons working both within and outside traditional medical structures. The relationship between the shaping of women's bodies, weaponry, and car parts not only functioned on a symbolic level but became increasingly material. These plastic foam materials, once expedited by war, were transferred to medical applications and used to permanently achieve a more exaggerated, streamlined, and fashionable conical ideal of bustline during a period when women's bodies were frequently compared to weaponry and automobiles. This reveals a complex medical and military-industrial assemblage of postwar US aesthetic body-contouring practices.

SILICONE / FLUID

6 SILICONES ON THE SURFACE / MILITARY- INDUSTRIAL R & D, POSTWAR CONVERSION, AND THE BODY

What are silicones? In 1952, Dow Corning employed an anthropomorphic approach to answering this question: Silicones are women. Seeking to expand its postwar commercial markets, Dow Corning cut the technical jargon and invited readers to informally meet the silicones. This approach resonated with an audience beyond scientists, chemists, and engineers, notably grabbing the attention of whitestream journalists. A newspaper headline summarized their approach: "Question: 'what's more perfect than any lady?' . . . Dow Corning has the answer: a silicone."[1]

In its promotional booklet *What's a Silicone?* Dow Corning introduced these materials as a glamorous obedient type of ageless, hard-working, malleable superwomen to an assumed white heterosexual male chemist or engineer. Readers could easily shape these silicones to suit their needs: "Take your pick . . . we make them in the form of fluids and oils, greases and compounds, resins, varnishes, and rubber."[2] It is important to note that while Dow Corning's Advertising Department compared silicones to women, the booklet made no mention of the gendered medical or cosmetic applications silicone would become increasingly associated with from the mid-1950s onward. So, how did silicone, a substance once hailed as an American industrial and military miracle, end up as a material used in body-contouring practices?

In the postwar United States, Dow Corning actively sought new markets for silicones. Formed as a wartime merger between Dow Chemical and Corning Glass Works, the company lost its sole client at the close of the war, when the US Armed Forces canceled all its silicone contracts. At this point, silicones were a mysterious scientific military development of synthetic substances, with "their wartime uses largely concealed."[3] However, by 1947, *Fortune* named silicone a "major new material to emerge from the war" and "a cornerstone of a new industry."[4] Commercial production was still in its early stages and remained limited, with prices reflecting the newness of the material. At an average of seven dollars per pound, the price of silicone was "among the highest in synthetic materials."[5] As with most new materials, an increase in demand could lead to a rise in production volume, a lowering of prices, and a subsequent increase in sales and applications. Silicones were no longer the reserve of US government military applications and needed to be reinvented for the postwar commercial market. It is important to address silicone's provenance to better understand the systems of corporate, social, cultural, and political power embedded within it from its inception, as these also shaped some of its later corporeal applications.

In this chapter, I focus on silicone's military-industrial origins and its immediate postwar introduction to the consumer market. Here, I drill into the more intricate and complex aspects of the silicone story that have been previously overlooked. Where did silicone come from? Who developed it? What industrial applications did it have? How was it used by the US Armed Forces? How did chemical company advertising department staff and sales directors pursue new commercial markets in the postwar United States? What role did race, gender, sexuality, and the body play within silicone's postwar promotion and applications? What impact did silicone's materiality have on decisions to associate it with, as well as use it on and with, women's bodies in the postwar United States?

In the first half of the chapter, I trace silicone's early industrial and military development. In the second half, I focus on silicone's postwar reinvention and applications. I closely examine some of the key actors involved in its postwar conversion, specifically addressing the role of racialized gender within this, offering new perspectives on the function of women's bodies in the promotion and domestication of Dow Corning's silicones. Silicone's provenance, its R & D, and its changing meanings and applications are examined using a variety of archival primary sources, including oral histories with chemists, chemistry engineering textbooks and journals, promotional material from chemical companies, media coverage, and advertisements. Silicone is explored as it traverses the domestic, medical, and cosmetic markets, moving from the surface of skin to inside the body, culminating in the establishment of the Dow Corning Center for Aid to Medical Research in 1959.

Shaping Silicones

As a material, silicone is a complex and chemically inert synthetic organic compound. Its molecular forms vary, and it can be either a liquid, a solid, a resin, or a man-made rubber. It can also be made into a sponge. The basic element from which silicones are formed is silicon, which is also fundamental in nature. Silicon, in the compound silicon dioxide (also known as silica), is one of the most abundant materials in the world and a major component of sand and many types of rocks.[6] Silica found an early application in the manufacture of glass around 3500 BCE.[7] One of the first known reflections on the nature of silica as a material was Roman scholar and natural philosopher Pliny the Elder's *Naturalis Historia*, written in 77 CE. He suggested that silica was a distinctive form of ice that resulted from extreme freezing.[8] In this chapter, silica—running through the sands of time—became instrumental in creating a material that could lubricate wartime technology, making it run faster and more efficiently.

The history of silicone is challenging to trace in its entirety, but it is important to provide some background to gain a better understanding of its military-industrial origins. Dow Corning largely succeeded in establishing itself as the "world's leading silicone company"—or, as it marketed itself, "First in Silicones."[9] This corporate victor story is often repeated in histories of silicone. However, silicone's story presents a complex trajectory of international development and global power politics.

Silicone's development began with Frederic Stanley Kipping, a UK-based British chemist who had studied in Germany and carried out the foundational work on the development of silicon polymers from the early to mid-twentieth century.[10] The established written narrative of Kipping's early work is one of "accidental discovery."[11] It claims he had unintentionally chanced upon a substance that would later "revolutionize both industry and medicine," describing the results of many of his experiments as "a series of uninviting glues."[12] However, Dow Corning chemist Franklin Hyde complicated this narrative.[13] He noted that the German chemist Friedrich Woehler had coined the term "silicone" in 1863, which appears to be omitted from the aforementioned texts, perhaps because they focus on Anglophone sources and actors, creating a clear linear Anglophone narrative.[14] Additionally, Hyde cited important contributions from French, German, and Russian chemists.

The work of German silicone chemist and scholar Walter Noll offers a more international perspective than most of his American peers, who tended to present silicone as an all-American wonder material.[15] For instance, in Germany, some silicone research was initiated at the start of World War II. Unlike the war's driving force for silicone advancements in the United States, in Germany, the war and its control of research impeded developments.[16] A journalist writing for *Fortune* in 1947 noted that, in the United States, "The war pushed silicone development faster and further than it could ever have been expected to go under normal conditions."[17] The versatility of silicone products in armaments, especially in aircraft manufacture, resulted in an accelerated effort toward their development in the United States during World War II.

British silicone chemist Stanley Fordham offers a competing narrative to Noll's. He asserts that "the true history of silicones" began between 1931 and 1940, when the US company Corning Glass Works employed its first organic chemist, Franklin Hyde.[18] Perhaps what Fordham means is that this is when the *commercial* history of silicone began. CGW was concerned that new synthetic transparent materials, such as Pollopas, might emerge as competitors to glass and hired Hyde to investigate whether organic silicon chemistry had commercial potential for securing their share of the flourishing polymer market.[19] Before World War II, Hyde had abstracted and translated the journal *Glastechnische Berichte* (glass technical reports) from German into English for *Chemical Abstracts* and *Ceramics Abstracts*. Hyde's challenge to the Anglophone narrative of silicone's history and development likely provided him with a solid understanding of international developments and networks in glass chemistry and production, as well as the culture of sharing research.[20]

In his new position at CGW, Hyde contributed his knowledge of polymers, with the abundance of silica at his disposal, to further developments in silicon chemistry, opening a new era for polymeric development.[21] CGW was looking for adhesives, binders, and lubricants to go with its new fiberglass materials. One of Hyde's earliest tasks, around 1931, was to develop a material that could hold together the company's 18 by 26-inch glass panels that were featured in a "fifteen by forty-foot architectural window over the . . . swinging doors at the RCA Building New York," the skyscraper centerpiece of the Rockefeller Center.[22] Still welcoming visitors today, the imposing Art Deco triptych by architectural sculptor Lee Lawrie bears the biblical passage "Wisdom and knowledge shall be the stability of thy times" and features one of the earliest public applications of silicones. Some of silicone's first industrial uses were driven by design's demand for innovation in architectural materials.

Hyde revisited Kipping's work when researching experiments in silicon chemistry and soon discovered that his "uninviting glues" had unique properties with great potential for industrial products, including architectural materials.[23] As with other scientific advancements like nylon, US actors, including journalists, were eager to repeatedly present silicone as evidence of American exceptionalism and scientific innovation, noting, "The silicones were carried out of Dr. Kipping's laboratory into commercial production by the practical genius of U.S. chemists and U.S. chemical engineers," and, "This last stage [of silicone development] was accomplished in the U.S., which again demonstrated its engineering talent for turning the basic science of Europe to practical account."[24]

Much of the Anglophone historiography of silicone has sustained this narrative, focusing on Hyde and Dow Corning; however, Noll offers a competing account. He suggests that in the late 1930s, General Electric independently began research on silicones.[25] The narrative provided by Noll (who does not appear to be closely affiliated with either company) aligns with chemists' oral histories held at the Science History Institute.[26] The 1947 *Fortune* article similarly noted, "Most of the intensive work [on silicones], however, did not come until the late thirties, when developmental lines became quite confused."[27] In 1937, a race to publish and patent silicone ensued. In 1940, Eugene Rochow, a GE chemist, made methyl silicone before Hyde had published or patented his work on silicones. There is no archival evidence to suggest that, at this time—only two years before the outbreak of World War II—silicone was being considered for aviation.

By the early 1940s, the collective research efforts pointed to the potential for increasing industrial applications for silicone materials. Silicone proved to have rubberlike properties, prompting further research during wartime, as rubber distribution shortages intensified. Synthetic alternatives to rubber were highly sought after in a military context, as US government agencies and chemical companies hoped they could contribute to greater wartime material independence.[28] The race to register patents accelerated.

In October 1941, a few months before the United States joined the war, Rochow was issued a patent.[29] Hyde recalls that, following the Japanese attack on Pearl Harbor on December 7, 1941, "secrecy orders were placed on all patents and publications . . . and so they were not available until the war ended in 1945."[30] At this point, GE tried to challenge a number of CGW patents but concluded that it would be most cost-effective for both companies to cross-license their patents until the 1950s and then work separately, rather than incur mounting legal fees.[31] CGW and GE's patent race is an integral part of the history of silicone, as it contests and complicates the narrative of silicone's origins and the victor story perpetuated by CGW, which later became Dow Corning. It also demonstrates the complex nature of the origins of these new materials and the difficulty of assigning a specific material to one person, one company, or even one country.[32] What this history reveals is the complexity of plastics history and the need for it to be revised, disentangled, and complicated to further demonstrate how the histories of chemistry and materials cannot be easily attributed to one specific "victor." Instead, plastics are produced via a complex web of individual actors, such as the many individuals involved in silicone's cosmetic application in the shaping of the body.

Though CGW's attempts to collaborate with GE may have been unsuccessful, its other major business efforts were not. Unlike GE, CGW actively sought to communicate and collaborate with companies it believed could provide the expertise, materials, and equipment it lacked, essentially buying them up and merging with them to eliminate potential competition. As a result, it further strengthened its profile over that of its competitor GE.

"I Want It Tomorrow": Wartime Applications of Silicone

GE and CGW were eager to expand military contracts for silicones. One of the earliest wartime uses of silicone was a silicone rubber gasket in US Navy searchlights, developed by GE.[33] Toward the end of the war, silicone gaskets, known for their resiliency at high temperatures, were featured in aircraft cylinder heads and diesel engines. Dow's Shailer Bass, a member of the US Army Air Forces Ignition Committee, theorized that a silicone compound might solve the problem of airplane ignitions breaking at high altitudes.[34] This was a key moment in the history of silicones, marked by the collaboration of several key figures in Big Science. Hyde recalled that CGW was "anxious to sell [Hyman G.] Rickover [navy engineer and then assistant chief of the electrical section of the Bureau of Engineering in Washington, DC] something. They wanted to sell electrical tape, and Rickover wanted more power out of small motors."[35]

The US Navy wanted Hyde's 990A resin for use in insulated electric motors with glass tape. At the request of Rickover, who hoped to source material solutions to improve the immediate performance of existing US Armed Forces equipment, Hyde collaborated with Mellon Fellowship chemists Earl Warrick and Rob McGregor to produce the first polydimethylsiloxane (a type of silicone) fluid in 1940, along with some early examples of silicone rubber.[36] Hyman Rickover, promoted to admiral in 1953, was a highly influential figure in World War II and Cold War US Armed Forces history and later led the Manhattan Project. The limited quantities produced by CGW were insufficient to meet the armed forces' requirements. Since silicone synthesis was associated with organic chemistry rather than glassmaking, CGW sought the assistance and established expertise of the Dow Chemical Company for further development.[37]

The formation of the Dow Corning joint venture proceeded quickly due to Rickover's interest and the promise of military contracts it held.[38] Rickover, later dubbed "Father of the Nuclear Navy," famously barked, "I want it [silicone fluid] tomorrow."[39] Dow Chemical's William R. Collings and his experienced staff "swung into silicone research, product development, pilot production and sales."[40] The Dow Corning joint venture, with the added approval of Rickover, was a highly advantageous and strategic corporate move, merging the expertise of two skilled industrial companies and ensuring the company's success for the next forty years. The establishment of Dow Corning was rooted in warfare and profited from military-industrial power relations.

Wartime applications of silicone prioritized its capacity to be used as a technology enhancer. In response to military demand, Dow Corning developed a silicone grease sealant for aircraft that prevented engine failure at high altitudes. This product was named DC-4 Dielectric Compound, described as "a smooth, colorless grease with the appearance and 'feel' of white Vaseline, but with the remarkable quality of not hardening, cracking, or flowing at any temperature from −40° to over 400° Fahrenheit."[41]

The DC-4 was in high demand and "proved its usefulness early in World War II by bringing the first flight of Thunderbolt [a fighter aircraft] safely over the ocean from Brazil to Africa." The product prevented "the ignition systems of high-altitude bombers and carrier-based planes from failing over the oceans and 'The Hump'; over Berlin and Tokyo."[42] At a relatively low cost, it enabled US aircraft to fly longer distances at the higher altitudes required to reach Japan after the Pearl Harbor attack. Military demand for DC-4 as well as 990A resin required full pilot-plant production efforts. This benefited from the wartime priority allocations for materials needed for new plant construction and the use of magnesium, both provided by the military, giving Dow Corning yet another competitive advantage.[43] Collings commented, "Our best boosters were from the Bureau of Ships of the Navy. They helped us to secure many things, [including] electrical equipment."[44] Silicones boosted the US military, and, in turn, the military boosted Dow Corning.

Expertise in established synthetics production was crucial to the successful development of silicone. Another advantage of the merger was the appointment of Collings as chief executive of Dow Corning. Formerly the manager of Dow's Cellulose Products Division, Collings had already led the successful development of a synthetic product. Dow Corning combined CGW's prominence in silicone research with Dow's established experience in the chemical business, synthetic materials, and commercialization, areas in which CGW was lacking.[45] GE chemist Charles Reed reflected decades later, "It is very interesting to note that the relative market positions of the two companies that finally emerged in the early 1950s remained unchanged until today. Dow is around twice the size of GE."[46] Dow Corning profited from the unique collaboration and sharing of expertise during wartime urgency, as well as from the fact that its silicones were designated as essential.

In 1943, the merger was expedited and encouraged by Under Secretary of War Robert P. Patterson. CGW and the Dow Chemical Company joined forces to formally establish the Dow Corning Corporation, focusing exclusively on the silicones needed for the war effort. It should be noted that it was rare for a company to be encouraged by a US government official to form a joint venture due to antitrust legislation; however, silicone's importance to the war effort outweighed this concern, and it appears this exceptional permission was never revoked, unlike other petrochemical mergers that took place after World War II.[47] Although the research origins of CGW's silicones "expanded concurrently" with those of GE, it is

Dow Corning that emerged as the "victor" on the other side, benefiting from both the merger and US Armed Forces contracts.

Sales of these silicones and other military applications enabled Dow Corning to report a profit in its first year of operation. Thus, in 1944, Dow Corning announced the first commercial production of silicones in the United States. As noted in *Fortune*, "In 1944 [silicone's] wartime uses were largely concealed, while peacetime uses were still in that dream-world of postwar products that has yet to materialize."[48] By the close of World War II, silicones were used as a sealant wherever the US Armed Forces employed electrical transformers. Dow Corning was therefore able to enter new markets after World War II, once production facilities no longer needed by the US Armed Forces, which had sanctioned their construction, became available. Whitestream journalists envisaged a prosperous postwar US utopia of better living through military industrial plastics, one where, "from the magic synthetic silicones, a new stream of products will soon be flowing into your home and into your work."[49] Dow Corning Corporation soon became one of the largest silicone manufacturers in the world, claiming in its advertising, "Dow Corning in Midland, Michigan, the world's first and largest silicone plant."[50]

Postwar Technology Transfer

At the close of the war, military contracts were canceled and Dow Corning lost its sole wartime customer. Immediately afterward, the company packaged its DC-4 compound in bright yellow tubes for retail sale, hoping that the many ex-GIs familiar with the product would purchase it, along with other items now available on the domestic market.[51] However, silicones and the production facilities that had been established largely remained without a market, leading to significant investment on the part of manufacturers in investigating alternative uses.[52]

In 1945, *Life* magazine introduced silicones with a report on "Bouncing Putty."[53] An anonymous journalist wrote, "Made of coal, sand, and oil . . . new silicone has freak properties," apparently including a viscous ability to slow time: "In effect the substance is a sort of slow-motion liquid. But for all of its delightful characteristics it has no known use." Silicones and the unique properties they offered urgently needed to find new commercial uses. Dow Corning continued to reorient itself to postwar market development, searching for applications in which silicone's unique properties would justify the high initial prices, which ranged from five dollars to six dollars per pound.[54]

Dow Corning's R & D on nondefense applications of silicone eventually created more than five thousand products.[55] A key discovery was made when silicone was tested on animals and found to be chemically inert, indicating its potential for corporeal applications. Silicone was subsequently marketed as safe for human digestion. Dow Corning's silicone defoamer products were used in wine, fed to cows in agriculture to remedy bloating, and used as a nonstick coating ("Pan Glaze") for bakers to maximize and accelerate production and prevent unnecessary wastage from sticking and burning.[56] In its postwar commercial applications, silicone was now moving closer to the body.

In addition to investigating new applications in-house, marketing was another major area of investment for Dow Corning. During World War II, silicone's research and development was kept under strict secrecy orders. With these now lifted, Dow Corning needed to ensure that actors in science and industry knew about its products and understood their properties to come up with more potential commercial applications. Dow Corning President Eugene Sullivan invested in public relations and communication in the hope that this could stimulate interest and orders. However, his early attempts at announcing silicone appear to have been unsuccessful. Unlike DuPont's announcement of nylon stockings, Dow Corning did not have an instantly recognizable, timely product with sex appeal that captured the attention of the American media and public. As noted in *Fortune*, "[Silicones] made their first public appearance in mid-1944 with a burst of publicity and no little confusion."[57] But what are silicones? The public remained confused.

The commercial success of silicones demanded a new marketing strategy that would appeal to both industrial actors and the public. In 1945, Sullivan hired Lou Putnam and his assistant John Church to set up and manage an in-house advertising department. Having already tested a conventional approach to communication, Putnam soon employed a more creative method for selling silicones. Silicones were a relatively recent material, their malleable identity not yet fully formed, and their story had not yet taken shape in advertising.

Seeking to appeal to the American postwar imaginary, Dow Corning created the silicones in female form in 1952. This was Putnam's most extravagant piece of salesmanship to date: *What's a Silicone?* a comprehensive thirty-two-page, indexed guide mailed to prospective clients at universities and research labs, as well as to the general press. Postwar race and gender norms shaped Putnam's anthropomorphic vision of silicone's identity. Silicones were introduced as a devastatingly flawless new breed of superhuman female, aimed at seducing an assumed white, male, heterosexual readership largely composed of chemists and engineers, as well as a more general audience (figure 6.1). At the time, a male journalist writing for *The Philadelphia Inquirer* crowned this booklet "one of the best examples of [public relations] to reach this desk in many, many months."[58] In addition to mailing the booklet, Dow Corning advertised it for years in US publications, including *Electronics* (1953) and *Scientific American* (1956), encouraging readers to request a copy and highlighting its high circulation rate.[59] It is worth analyzing the copy and imagery of *What's a Silicone?* in some detail.

What's a Silicone? reproduces the racial and gender power inequities of the chemical company environment in which it was produced. In the booklet, the silicones promised material perfection without compromise: "All of the virtues you dream about but never find in one woman."[60] By comparing silicones to idealized women, Dow Corning assumes its readership is male and heterosexual, reflecting postwar demographics in science and technology. In the United States during the 1950s, men with degrees in engineering constituted approximately 10 to 15 percent of all male college graduates, whereas fewer than 0.2 percent of all female college graduates majored in engineering.[61] This data does not mention race or ethnicity. As scholars have noted, statistics on African Americans in STEM fields during this period are limited.[62] Data from 1940 estimate that out of the seventy thousand chemists in industrial research laboratories, three hundred were Black.[63] However, it is important to be wary of these numbers, since some of the individuals surveyed may have kept their racial and ethnic backgrounds hidden. An assumed white male reader was in a position of power when it came to learning about and understanding new material developments. The copy suggests that women and their bodies were available to have this power exerted on them. While *What's a Silicone?* does not explicitly mention race, all the people photographed in the booklet are white or light-skinned.

A closer look at the language and references used in *What's a Silicone?* further reveal an unmarked norm of hegemonic American whiteness. A nationalist tone of American exceptionalism and settler colonialism informs much of the company's creative writing on the silicones: "Silicones are as typically American as the tall tales told by our forefathers. Like Paul

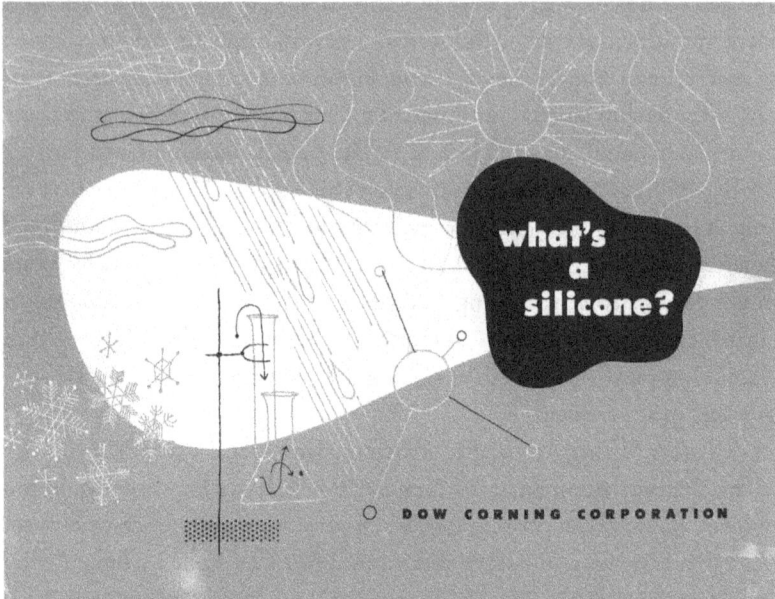

6.1 Introduction to the 1952 Dow Corning booklet, *What's a Silicone?* Trade Literature Collection, Smithsonian Libraries and Archives, Washington, DC.

Bunyan and Joe Magarac; like Pecos Bill and Davy Crockett, silicones were created to do the impossible."[64] A year later, in 1953, Dow Corning dedicated its latest booklet, *Tall Tales and Fabulous Facts: Dow Corning Silicone News—New Frontier Edition*, to these themes.[65] Clearly hoping to achieve maximum appeal among engineering departments dominated by white men, Putnam likened Dow Corning's research and development of silicone to the colonial expansion of the American frontier.[66] In doing so, he referenced the racialized American pioneer narrative of self-reliance and individual entrepreneurship amid colonial expansion. Unlike its previous booklet, in which the silicones were presented as a type of ideal woman, Dow Corning primarily anthropomorphized these silicones into mythologized white American "frontiersmen" and the aforementioned "folk heroes." It introduced these colonial characters, noting that "the giant strength, ingenuity, and native humor [our forefathers] put into these frontier heroes made their own overwhelming task of taming a wilderness seem more feasible." This dualistic silicone imaginary of colonization also related to a wider postwar nostalgia for the Old West and Westernwear fashion.[67]

Postwar heroic tales of American warfare played into Dow Corning's silicone imaginary. Promotional materials detailed numerous impressive wartime feats that the silicones had achieved in both industry and military applications: "They're compounds that keep radar from going blind on a foggy night" and "They're rubber that won't melt on aircraft engine cylinders or freeze on switches that operate bomb bay doors at 100 degrees below zero."[68] Such imagery evoked technological military advancements and silicone's precision as a high-performing and reliable material during warfare.

Putnam and Church created a complex, racialized, and gendered identity for their products that balanced glamour and sex appeal with propriety. It was important to stress that while the silicones were hard-working in war, their gender presentation was unquestionably sleek and spectacularly feminine: "The Silicones are not drab and dowdy. Quite the contrary. They have all of the sophistication and polish of that we associate with a Powers model."[69] This carefully constructed postwar vision of the silicones as photographic models had come a long way from earlier press descriptions of silicones as specialized industrial materials: "not yet highly photogenic, and perhaps not destined to be."[70] Likening the silicones to a "Powers model," Dow Corning referenced the first modeling agency, founded in New York in 1923 by the white actor John Robert Powers. Scholar Elspeth Brown notes that the strength of Powers's marketing lay in reinventing working-class women's identity into a commodifiable but managed type of sexuality, while still "containing its threat through whiteness, the respectability of middle-class propriety, and the distance of glamour itself."[71] Dow Corning's portrayal of an ideal silicone woman is in keeping with the commodified, managed sexuality and its unmarked norm of whiteness that Brown describes in the creation of the Powers model. The silicones could be easily shaped to fulfill the desires of their male creators while maintaining the same respectability and "sophistication" as the Powers model. This presented a nonthreatening, carefully managed vision of racialized womanhood and restrained sexuality, based on heteronormative white middle-class norms.

A political postwar return to established American gender norms and heteropatriarchal control of the racialized gendered body informed Dow Corning's vision of the silicones. They completed arduous chores without complaint: "Silicones will work all day long over a hot stove, year after year, and like it. Heat doesn't bother them a bit." They retained composure: "The Silicones . . . [are] always the same . . . they don't lose their shape all of a sudden when it gets a little hot."[72] In the context of American postwar shifts in middle-class gender norms, Dow Corning presents the silicones to a predominately male readership as the embodiment of domestic stability: Once formed, they will stay flawlessly in shape, unaffected by changes in climate or environment.

The silicones are new and improved: presented as nonthreatening, nonleaking entities, free of contagion or disorder, that can be controlled by humans. They are bombshells put to good use. Plastics scholar Heather Davis proposes that "the value of plasticity—that is, endless manipulability—fulfills the dream of mastery over the material world."[73] In *What's a Silicone?* powerful male actors are interchangeably able to shape materials and women to meet their needs. The implication in *What's a Silicone?* is that the silicones' visceral materiality is a desirable one: unlike their human counterparts, they are sterile, controllable, and presented as a product of a racialized, gendered binary of science and logic.

Dow Corning's silicones are presented as forever young. They can shield off decay, abjection, and the "bodily betrayal" that is aging.[74] In whitestream heteronormative Western discourse, women's aging bodies, unlike those of "silver foxes," are often regarded as abject and unsightly.[75] Putnam and Church reinforced these dominant ideals in the booklet's introduction: "Unlike any woman that ever lived, Silicones don't grow old. They are untouched by the passage of time. Their lifespan is at least ten times that of comparable organic materials."[76] Throughout the Dow Corning text, women and silicone materials are referred to interchangeably.

All these physical forms are materials for making. Agency is given to the implied white male reader, who can choose from a range of viscous, sticky, and pliable industrial materials. Putnam's decision to introduce the silicones as obedient, hardworking, sexy domestic "superwomen" who maintain their shape is highly emblematic of its time. These booklets make no mention of medical or cosmetic applications. However, this text goes beyond language and representation—silicones were not only used in the shaping of feminine beauty ideals but also materially shaped women's bodies.

Borrowed from Industry

Reflecting on silicone's advancement in the immediate postwar period, author and *Vogue* journalist Simona Morini framed silicone and its conversion to the cosmetic and corporeal realm as "borrowed from industry": "Either injected as a liquid or inserted through an incision, as a sponge, diaphanous sliver, or a carved rubbery chunk, silicone is the sensational discovery physicians have borrowed from industry after having experimented for centuries—often unsuccessfully—with precious metals, alloys and in the last century, with plastics to replace obliterated tissue in the human body."[77]

Morini's framing of silicone is in keeping with an established marketing narrative of science and industry benefiting the everyday, perhaps most famously encapsulated by DuPont's "Better Things for Better Living . . . Through Chemistry" slogan, which endured from 1935 to 1982. Dow Corning's public relations advisers similarly promoted silicone as being borrowed from industry and military applications to improve civilian life. Silas Braley, a key actor at Dow Corning who later directed the Dow Corning Center for Aid to Medical Research, reflected on the immediate postwar period and silicone's conversion: "[Despite US government–canceled contracts] the information gathered during [World War II] indicated that there *should* be civilian uses of these new [silicone] materials, and the silicones began to be used in new applications: for furniture polish, for high temperature paints, for high temperature rubbery insulation, for waterproofing and for mold release compounds—to name a few."[78]

Dow Corning's executives and promotional materials regularly stressed the company's military contracts to emphasize its credibility.

Braley's recollections described large-scale American corporate democratic military-industrial-civilian material adaptations, later known as spin-offs, which were used to justify and distract the taxpayer from costly federally funded R & D projects.[79]

American companies capitalized on military endorsement in their conversion back to peacetime materials production. Design historian Cynthia Lee Henthorn argues that the "techno-corporate order" felt that the "test of war" served as a powerful promotional tool.[80] The logic behind this was that if a material or product had stood the test of war, its postwar popularity was inevitable. Conversion/reconversion selling points thrived on the thinking that "the 'miracles' credited with winning the war had freed Asia and Europe and thus could easily emancipate the American housewife."[81]

By the mid-1950s, Dow Corning's efforts to find new domestic markets for silicones met with some success. In the postwar United States, journalists and advertising professionals alike introduced silicone as a miraculous, practical, and problem-solving substance "borrowed from industry." Silicone, once a secret industrial ingredient needed by the US Armed Forces, made its way into an array of consumer goods and was featured on the pages of various high-circulation publications, which included *Life*, *Fortune*, *Ebony*, and *Vogue*.

Silicones made their first appearance in *Vogue* in a feature titled "Changes for 1954": "Revolutionary hand-lotion containing silicones (which are chemically related to glass) used in industry as water repellents and lubricating agents. Thin, cologne-like, the lotion forms a sort of invisible glove to protect hands from water, dirt, and detergents. Things simply bounce off it! One application lasts through several hand washings; dirt comes off, lotion doesn't."[82]

Vogue's writer promoted silicones as providing instant protection. Inert like glass, silicone sat superficially on the skin's surface and was presented as a way of guarding hands from soap and dirt while cleaning. Introduced in relation to glass, silicone was no longer an unknown material and was positioned as a nonthreatening R & D innovation of wartime. It entered the domestic and corporeal realm via cosmetic and household products, predominantly marketed to white women, in products such as On-Hand protective lotion, Cara Nome silicone lotion, O-Cedar Dri-Glo polish, Syl-Flex fabric protector, and Wolco Glas spray cleaner.[83] Advertising copy and articles in whitestream magazines, as well as Dow Corning's spokespeople and promotional materials, frequently reassured the public that, in its domestic applications, silicone was an "invisible protector" that could be trusted; its properties were associated with cleanliness and a safety barrier from undesirable elements.[84]

The supposedly "inert" safety of protective plastics was inextricably bound to a wider racialized, gendered US postwar cultural war on germs and insects.[85] Chemical companies continued to use wartime rhetoric in their postwar transfer of military technologies, such as in the promotion of insecticides, on the home front, to great commercial success. Advertisements depicted white women in domestic settings as soldiers, armed with the latest chemical developments, such as insecticides—"super ammunition for the continued battle of the home front."[86] Endorsement of a product by the armed forces tended to endow it further with a variety of "hygienic benefits."[87] Design historians have shown how modernism was used as a tool to enforce cleanliness and racialized whiteness during the postwar period.[88] In the Cold War United States, the modern home—particularly the "rational" kitchen—and, by extension, the white middle-class housewife, were increasingly presented as a racialized and gendered marker of societal and technological progress. Although they have largely been excluded from design histories of the US postwar home, silicones were similarly shown in modern white domestic settings as an important, military-endorsed wartime discovery. For many white American middle-class consumers at the time, such an army affiliation signaled the "safety" of a product, often contributing to its domestication. In the case of silicones, this military-industrial affiliation further aided its civilian adoption as the material moved closer to the body.

By the late 1950s, silicones were advertised as key ingredients in beauty products. This 1959 advertisement for Re-Nutriv face cream by Estée Lauder in *Vogue* was aimed at a wealthier white readership and boasted "a goldmine of beauty" (figure 6.2). In this example, silicone was marketed as valuable, rare, protective, rejuvenating, and glamorous. Its precious droplets are captured in a golden jar, nestled alongside the essence of two of the ocean's now most endangered creatures: turtles and sharks. Re-Nutriv promised "to keep you looking younger, fresher, lovelier, than you ever dreamed possible," thanks to its "treasury of some of the world's costliest ingredients." In this instance, silicone did not just preserve the surface of everyday items and environments; it was also presented as an innovative protective material capable of the luxurious preservation of youth.

By the late 1950s, chemical and cosmetics companies increasingly promoted silicones as a tool for aesthetic transformation. In some racialized and gendered applications, this extended from skin care to self-fashioning

6.2 "What Makes a Cream Worth $115?," Estée Lauder Re-Nutriv advertisement, *Vogue*, October 1959, 52.

with makeup. From 1959 to 1961, the white-owned American cosmetics company Pond's ran a series of seventeen advertisements in *Life* for its "cosmetic-silicones makeup" Angel Face "powder and foundation in-one." Pond's use of silicones referenced scientific development: "new cosmetic discovery." The company claimed its application of silicones in makeup was exclusive, noting, "*Only* Angel Face has cosmetic-silicones for soft subtle shades that will never darken or discolor!"[89] The series exclaimed, "Change your skin tone to look lovely in any fashion color" (figure 6.3), giving "you different skin tones for different fashion colors!"[90] Referring to a color chart that listed "costume colors" against complexion shades, Pond's advertisement carried a message similar to that of nylon fashion hosiery charts during this period: namely, that the implied white woman consumer could "wear a different skin tone" in a range of light "nude" shades to better complement the color of her seasonal apparel: "Find your own skin tone on the chart and see how you can wear two, three or more depending on your costume color."[91]

In the Pond's Angel Face series, the violent realities of racial segregation and inequality are both glossed over and clearly enforced. In these advertisements, skin color simply becomes a matter of seasonal trends—a fashion to be switched out—thereby revealing the social privilege of the consumers to whom it was marketed. Cosmetics historian Kathy Peiss has noted that "white women . . . could confidently declare . . . that skin tone was a matter of fashion, that a dark complexion was one choice among many—as long as the boundary between black and white was secure."[92] A Pond's Angel Face advertisement ran in an issue of *Life* that featured a "before and after" mirrored image of the white actress Shirley MacLaine on the cover, dressed as a geisha and promoting her latest role in *My Geisha* (1962), a movie about an "American actress who so convincingly turns herself into a geisha that she fools her own husband."[93] The article detailed MacLaine's racialized "transformation" into a geisha with prosthetic rubber eyelids and contact lenses applied by renowned Hollywood studio makeup artist Frank Westmore. The positive reception MacLaine received for posing as a Japanese woman stands in stark contrast to the lived experiences of Japanese American women in the postwar United States, most of whom had been forcibly relocated to internment camps during World War II. Pond's Angel Face offered consumers with complexions from "fair" to "dark olive" a series of "subtle," transformative shades developed for a "natural" look of "perfection."[94] In the Pond's Angel Face advertising series, silicones are presented as the key ingredient in this

You . . . glamorous in green

You . . . stunning in cerise

Angel Face makes all the difference. On the left, it's Blushing Angel Face. On the right, it's Tawny Angel Face.

New *Angel Face* with cosmetic-silicones lets you change your skin tone so you can wear any fashion color

Now you can make fashion's most fabulous colors your most flattering colors! How? With the first fashion cosmetic—new Angel Face by Pond's—the *only* compact makeup with cosmetic-silicones.

Precious cosmetic-silicones let Angel Face change your skin tone—yet you'll never look made-up! Cosmetic-silicones make possible softer, subtler shades and prevent skin moisture from darkening or discoloring them, ever. And cosmetic-silicones actually capture light to give your complexion lovely radiance!

Now—look lovely in *any* color! Take this chart when you go to buy your new Angel Face shades. —►

POND'S COSTUME-COMPLEXION COORDINATOR				
COSTUME COLORS	FAIR SKIN	ROSY SKIN	OLIVE SKIN	DARK OLIVE
	POND'S ANGEL FACE SHADES			
REDS-PINKS	IVORY	NATURAL	NATURAL	TAWNY
ORANGES-YELLOWS	GOLDEN	GOLDEN	GOLDEN	BRONZE
GREENS-BLUES	NATURAL	IVORY	PINK	BLUSHING
BROWNS-BLACK	PINK	IVORY	BLUSHING	TAWNY
WHITE-NEUTRALS	NATURAL	TAWNY	BLUSHING	TAN OR DEEP TAN

New! The fabulous "Fashion Case" holding *the finest powder and foundation in-one*, $1.25 plus tax

6.3 Angel Face advertisement, *Life*, January 25, 1960, 37.

skin-tone-changing technology, while simultaneously assuring readers that racial boundaries would remain secure.

During this period, skincare and makeup products featuring silicones are less frequently mentioned in Black publications than in the whitestream press. Conversely, hair products featuring silicones are advertised in Black publications but rarely appear in whitestream magazines. African Americans' use of hair straighteners and bleaches was in decline in the late 1950s and early 1960s. The rising political and cultural interest in Pan-Africanism and Black nationalism found expression in self-fashioning.[95] Companies sought to address these changes in the market. In *Ebony* magazine, cosmetics companies specializing in African American markets, like the white-owned company Posner's, advertised hair products containing silicones that appeared to cater to both natural and curly, as well as straightened, styles. Posner's water-repellent "special silicone formulas" Sili Press pressing oil and Sili Curl curling wax were "especially prepared to resist perspiration and water . . . [guaranteeing] longer lasting hair styles."[96]

In 1961, the Baltimore-based African American company Apex similarly advertised a product called "Naturalizer," containing "silicones, your greatest protection against the effects of perspiration and humidity . . . against hair reverting," implying silicones could be used to keep hair styled in place.[97] Texas-based Nulox, later listed in *Ebony* by the American Health and Beauty Aids Institute as Black-owned, also advertised a silicone-based product: Sheen, a "hair and scalp conditioner-groomer," claiming it was "the ONE preparation with genuine BERGAMOT and SILICONES."[98] Nulox and other companies listed silicones as a key ingredient, along with bergamot, which appears more frequently in hair products marketed to African Americans.[99]

Gender is not always explicitly mentioned in the advertisements discussed. An exception is Apex's white Naturalizer tub image, which features an illustration of a woman (figure 6.4). In its advertisements, Posner referenced powerful women from African and Middle Eastern history, such as Cleopatra and the Queen of Sheba, as users of bergamot. Scholar Maxine Leeds Craig argues that Posner's references to Cleopatra were an attempt by the company to "Africanize" its hair-straightening products.[100] Natural hair became increasingly popular among African Americans throughout the 1960s. By the mid-1960s, "Black is beautiful" imagery proliferated in mainstream Black publications, as well as the visual cultures of the civil rights and emerging Black Power movements. *Ebony*'s June 1966 cover

6.4 Apex advertisement in *Ebony*, September 1961, 117.

celebrated the arrival of "The Natural Look" with a photograph of a Black woman wearing her hair "in a short natural" Afro style. It appears that Posner, in its ads, attempted to respond to these changes by combining "scientific developments" like silicones with historical and "natural" references: "Thanks to modern science, new miracle discoveries [are] added to this same age-old hair beautifier," alluding to the use of synthetic chemicals derived from petroleum and manufactured to imitate the chemical profile of bergamot oil and other essential oils.[101] While the pages of whitestream publications celebrated silicones and other postwar plastics, Posner and other brands advertising in African American publications weren't always as direct in referencing silicone as "borrowed from industry."

Black consumers were absent from the whitestream "world of tomorrow" postwar progress shaped by wartime R & D rhetoric.[102] African Americans had contributed to the war effort, but in reality, most were deliberately excluded from the GI Bill's postwar promises of progress and democracy, including racial disparities in access to federally funded housing, education, and higher-paying jobs.[103] Seeking to appeal to Black postwar consumers, haircare brands advertising in *Ebony* focused their promotional efforts on other synthetically manufactured ingredients, such as bergamot, which was presented as a "natural" ingredient with a longer history than silicones.

Dow Corning Center for Aid to Medical Research

The first step toward the use of silicone as an implant came in 1953, when Braley compounded Silastic S-9711 and its extrudable counterpart, Silastic-2000.[104] These were the first silicone rubbers created expressly for medical use. While Dow Corning's lengthy reference guides—*What's a Silicone?* and *Tall Tales and Fabulous Facts*—did not mention cosmetic or medical applications, they did encourage readers to contact the company's "staff of technically trained men," based across the United States, "for more information or for reference of your problem to our product development laboratories."[105]

As with other chemical and plastics companies, Dow Corning representatives encouraged clients to contact them with "problems" that could be solved by developing a special silicone formula. Silicone's properties could be shaped in numerous ways, and Dow Corning was still figuring out what properties would best suit an application. This approach allowed people to bring their own ideas and needs to Dow Corning, rather than relying solely on the company's in-house research chemists and engineers to do this work.

Soon after Braley's development, John Holter, an engineer whose baby had hydrocephalus, wrote Dow Corning to inquire about the material and went on to co-develop one of the first silicone medical devices: the silicone hydrocephalic shunt tube, first implanted in 1955. Silastic rubber's satisfactory biosafety record was fundamental to the Spitz-Holter valve's successful implantation and acceptance by the body. The result was groundbreaking, as implanted materials were typically expelled by the body within a short period of time. When medical practitioners discovered that silicones were not rejected, Dow Corning was soon inundated with requests for new medical designs. Mel Hunter, in his role as Dow Corning director of research, had been handling these requests. In 1955, following the success of Holter's design, correspondence markedly intensified. Dow Corning's star chemists, McGregor and Braley, were brought in to further develop Hunter's work.

The spread into new markets had already been noted by McGregor in 1954, when he reflected on the company's efforts to find new markets for silicone: "What had been started as a search for further knowledge proved to be the groundwork for technological advances that have proved helpful to industry, and in so doing have contributed to improving our standard of living."[106] In this rhetoric, what was good for industry was good for improving postwar standards of living and thereby the body.

In 1959, McGregor, as director, and Braley, as executive secretary, established the Dow Corning Center for Aid to Medical Research within the corporation's Research Department.[107] According to Braley, the center was set up to collate "all known information of the medical uses of silicones, and to answer (to the best of its ability) any questions that came to Dow Corning."[108] McGregor and Braley were organic chemists without medical qualifications. This may not have been unusual at the time, as engineers were often called on to design prostheses for surgeons.[109] Before the 1976 Medical Device Regulation Act, federal legislation in the United States regarding the implantation of medical devices was limited.

Actors such as Dow Corning staff members Warrick, McGregor, and Braley, along with journalists like Morini, presented a heroic corporate narrative, portraying Dow Corning as a philanthropic entrepreneur. Warrick wrote that the company "felt an obligation to [make] an important contribution to medical science."[110] McGregor and Braley decided to publish a newsletter to distribute among medical practitioners worldwide, as they were keen to promote the medical use of silicones on an international scale. They also met with physicians and surgeons to discuss how silicones could be used in medical applications. Dow Corning strove to provide samples to interested medical practitioners but was not equipped to offer finished products to the medical field on a large scale. Although commercial manufacturers had access to Silastic rubber, they were not interested in producing small-scale, elaborate items with low profit margins for Dow Corning to sell to the medical industry, nor were they capable of fabricating these in a sterile environment. As a result, Dow Corning discouraged medical research into devices and materials that it was unable to provide on a large scale or that did not generate sufficient demand. Arguably, Dow Corning served as gatekeepers in this role.

According to Warrick, the first products developed were blocks of silicone rubber and sponges of various hardness (surgeons carved the required shapes from the blocks for human implantation), scleral bucklers (used for detached retina procedures), and a small range of silicone rubber tubing (for blood pumps and the Spitz-Holter valve).[111] These items were all produced at Dow Corning's Midland, Michigan, plant in the silicone rubber product engineering laboratory. Because Dow Corning manufactured only a limited range of small-scale medical products, it lacked the resources to sell to and service the wide-ranging medical field, prompting it to form partnerships with well-established distributors like Becton, Dickinson.[112] Braley, who later became director of the DCCAMR, reflected on McGregor's observations from the 1950s on silicone's transfer from military to commercial use: "Not only did these technological advances [in silicones] contribute to our standard of living, but they would contribute also to our actual physical well-being by their implantation into the human body."[113] Industrial materials developed to realize architectural designs and accelerate and improve the technologies of World War II were now being marketed as materials with which to build, maintain, support, and shape the human body.

Silicone was marketed as a material that promised enhanced performance. It was used in wartime to improve and remedy faulty equipment, as well as to make it move faster and more efficiently. At the end of World War II, Dow Corning had lost all military contracts and needed to forge new identities and pursue new commercial markets for silicones. Race and gender played a key role in the domestication of silicones on the postwar US consumer market. Dow Corning researched, developed, manufactured, and marketed variations of its materials for industrial, domestic, cosmetic, and medical use. Industrial chemists, chemical company managers, sales executives, in-house advertising department staff, and journalists in whitestream media celebrated silicone as a "wonder product" of American industry that materialized from World War II. As well as being promoted as safe, digestible, and inert, silicone also became a desirable and precious material. The postwar reinvention and commercial launch of silicones demonstrate how a range of actors racialized and gendered their promotion and application. In these examples, the postwar American reimagining of silicones was largely shaped by the unequal, dominant, white, heteropatriarchal power structures that had created them. Silicone was increasingly anthropomorphized and cosmetically associated with racialized and gendered bodies. By the mid-1950s, silicones no longer just rested on the skin's surface—they were being implanted within it.

7 SILICONES BENEATH THE SURFACE / FLUID OTHERING AND JAPAN

It is difficult to pinpoint when, where, and by whom the first liquid silicone breast-enlargement injection was administered. Established US-centric scholarship on cosmetic surgery indicates that this took place around the end of World War II, and largely locates the origins of subcutaneous silicone shots in Japan.[1] It claims that vats of industrial silicone fluid disappeared from Yokohama Harbor, stolen by "the Japanese," and found their way into local women's breasts. John Byrne's heavily cited 1996 writing exemplifies this sensationalist narrative: "In the aftermath of World War II, transformer coolant made of silicone was suddenly disappearing from the docks of Yokohama Harbor in Japan. The silicone fluid was used by cosmeticians to enlarge the small breasts of Asian prostitutes who knew that a more Western appearance would enhance their appeal to American servicemen."[2]

Much of the foundational and some recent scholarship on cosmetic surgery reinforces a similar unsupported and racially charged rhetoric of the origins of silicone shots for bust augmentation in Japan, ignoring the development of silicone's medical applications in Japan. This literature fails to use primary sources from this period or reference Japanese medical literature.[3] This scholarship repeats the racialized rhetoric established by actors in the US cosmetic surgery and plastics industries in the 1950s and 1960s, such as that of the Dow Corning executive Silas Braley, who told *Science News* in 1968 that the practice of silicone injections for bust augmentation originated in Japan, with "up to a pint of silicone injected into each breast."[4] Braley was not alone in his claims, as they were repeated by medical professionals, chemical company staff, and journalists.[5]

Unlike the established scholarship on cosmetic surgery and silicones, I draw heavily on a wide range of historical and archival materials to present an alternative history of silicone bust injections. This chapter pays attention to key issues that have been sidelined in the past, such as powerful international postwar networks between chemical companies and surgeons, Japanese medical research on silicone, and how this shaped postwar American racialized gender ideals. Using archival materials, such as the papers of Herbert Conway—a plastic surgeon based in Manhattan who corresponded and collaborated with the Dow Corning Center for Aid to Medical Research and researched and published on silicones—as well as FDA and AMA papers, I seek to challenge the American-centric narrative of the development of silicone, which presents Japanese research as subpar when archival sources reveal that it was actually similar to what was happening in the United States.

Germany, Italy, the Czech Republic, Mexico, Singapore, South Africa, and many other geographical contexts are also featured in the archives. Much of the foundational scholarship presents a US-Anglophone-centric story, largely focusing on Dow Corning. It is important to note that Dow Corning faced competition from Japanese companies and brands, such as Shin-Etsu. In this chapter, I tackle the complex international networks that silicone encompasses, focusing on Japanese-US relations and representations within these frameworks. I analyze archival material to argue that medical research structures in Japan, contrary to established scholarship, were similar to those in the United States and can be understood as competitive rather than "unsophisticated"; many of these companies are still in operation today. Archival documents, medical journals, newspaper articles, and visual culture are used throughout to show how silicone's usage in breast augmentation was frequently othered as Japanese in postwar US culture.

Fluid Othering: Challenging Bust Injection Myths

"Even though silicone was created in the United States, its use in breast injections did not originate here. That application began in Japan shortly after World War II, when American forces still occupied the country."[6]

One thing is clear in mythologized and racialized Anglophone narratives such as these: Silicone is definitely an all-American wonder product; however, it was "the Japanese" who stole it from the military and industry and misused it, illicitly experimenting with female bodies before it was deemed "safe."[7] There is an urgent need to challenge and address the racial dimensions of this origin myth that has been and continues to be perpetuated in the established scholarship on cosmetic surgery. It neglects to cite or interrogate the primary sources, thus perpetuating postwar US power structures that uphold misogynist and racist ideology, failing to challenge and expose these ideologies. In this chapter, I present archival material to problematize, challenge, and complicate this established narrative of what I term "fluid" othering. I use this term to describe the way US actors, including chemical company and medical professionals, as well as journalists, conveniently racialized and othered the origins of silicone injections as something "un-American," associated with sex work and Japan.

The connotation of these stories is that "the Japanese" were "stealing" from the occupying US forces and "pumping" tanks of military-industrial materials into the bodies of Japanese female sex workers. They suggest that local Japanese women needed to modify their bodies to meet the white American busty ideal to sexually cater to the American servicemen. Japanese studies scholar Laura Miller has discussed the impact of the Allied occupation on the shift in erogenous zones in Japanese culture. Before World War II, the ideal female form traditionally centered on the kimono, which accentuated the nape, as well as the hips and ankles.[8] The influx of American culture during the postwar Allied occupation—including the arrival of the curvaceous Hollywood ideal and tight-fitting, Western-style clothing—began to change this focus.

There are gaps and unexplained discrepancies in the established historical narrative of silicone breast injections. Miller's scholarship is also critical of this origin story, as well as US-based scholars who have not investigated it further.[9] My own research shows that there are a significant number of Anglophone scholars who continue to repeat this myth.[10] This is puzzling, as some also cite Harry Kagan, an American osteopath, as experimenting with silicone injections a decade earlier, in 1946, using Dow Corning 200 fluid, a material intended for industrial and military applications.[11] While Kagan, who was not a board-certified plastic surgeon, appears to be one of the earliest known advocates of silicone shot treatment and links the practice to the United States, this fact is glossed over in the established scholarship. The dates align, since Dow Corning first announced the launch of silicones on the US commercial market in 1944.

Indeed, silicone was not the first foreign body material to be used in injections to increase breast size. Japanese doctors, like their American, German, and Austrian peers, were already familiar in the 1930s with the practice of injecting paraffin during cosmetic surgery. There are also a small number of equally unsubstantiated references to silicone shot treatment originating in 1940s Switzerland and Germany.[12] However, these alternative origin stories have not been referenced in the established scholarship as frequently. German physician Ludwig Lenz is cited by self-proclaimed "Hollywood beauty doctor" Robert Alan Franklyn as an early innovator. Franklyn even named his silicone subcutaneous injection technique "Cleopatra's Needle" after Lenz's further development of the technique in Egypt.[13] From 1925 to 1933, Lenz—who is said to have performed one of the first gender-affirming surgeries—was head of the gynecological department at famous sexologist Magnus Hirschfeld's Institut für Sexualwissenschaft in Berlin.[14] Given that paraffin had been used to augment the body, silicone would logically follow as the next material to be tested—not just in the United States and Japan but also internationally—as it offered similar viscous materiality with the added bonus of being perceived as chemically inert.

Because of complications with foam implants, silicone was presented as a safer alternative. Unlike foams that required surgical implantation, silicone shots could be administered directly into the body using a needle. They promised a new kind of inert materiality, which could range from viscous to rubberlike. Foams made of plastics had been prone to crumbling and becoming brittle or rigid, but silicone offered a soft, glistening substitute. In the postwar United States, Dow Corning promoted silicone's use in the food and agriculture industry, presenting it as a material that was safe for the body.

The unique properties silicones offered were now also praised for their ability to mimic flesh.[15] Plastic surgeon William St. Clair Symmers reflected, "The needs of ethical surgery have long encouraged the search for a bland material that can be placed in human tissues safely and easily, with the object of altering outward appearances while avoiding undesirable effects, immediate or later. Successive foreign substances have been described over the years as meeting these requirements: without exception each has eventually proved unsatisfactory, even dangerously so. One of the latest of these substances is silicone fluid (dimethylpolysiloxane), which in pure form or mixed with other compounds has been injected into the tissues to act as [a] prosthesis."[16] Silicone was the latest discovery in a long quest for materials that would be accepted by the body in plastic surgery. Its novel materiality held the promise of fluid turning into prosthetic; it could enter the body via a needle and then set internally to transform into an object.

Silicone was not the first industrial liquid substance to be injected into the body. In the late nineteenth century, Austrian doctor Robert Gersuny adopted the use of paraffin for breast augmentation, injecting the material in its soft form directly into the breast.[17] Patients and doctors soon discovered, however, that these injections resulted in just as many, if not more, complications than their internal application to any other body part, since paraffin had the tendency to migrate and form lumps.[18] These undesired side effects meant that by World War I, the practice of subcutaneous paraffin injections had "become largely abandoned."[19] Medical sources reveal a plethora of materials being injected into the body to achieve a fuller bust. "Paraffin waxes, beeswax, silicon wax, silicone fluid, shellac, shredded oiled-silk fabric, silk tangle, glazier's putty, spun glass, and epoxy resin" are all materials identified in specimens taken from individuals with foreign-body mastitis (i.e., inflamed breast tissue) between 1946 and 1968.[20]

Unlike paraffin, silicone fits into a wider history of US World War II military-industrial materials R & D. An emphasis on materiality offers different perspectives. Silicone had similar properties to paraffin, with the added advantage of presenting as chemically inert, so, as with previous material developments, it appeared to be an improved successor for cosmetic surgery. To date, scholars have not considered the interdisciplinary complexities of the silicone story nor investigated the international network it depended on, leaving many questions unconsidered and unanswered.

What if the coolant vats were not stolen by Japanese individuals, but were taken by US Armed Forces medical officers and servicemen? This scenario has not been considered in established scholarship. Former Harvard medical law history student Sharon Webb is the exception, describing in a paper how her father, Dr. Charles Webb, a US medical officer in Japan during the occupation, was acquainted with an army master sergeant who administered breast injections to local women for a fee.[21] The sergeant illicitly obtained syringes from the US military medical facilities and used what Dr. Webb identified as cooling fluid that was readily available on the base. According to him, this behavior was met with no official censure. He did not know the exact chemical composition of the cooling fluid; however, it is worth noting that Dow Corning exclusively supplied the US Armed Forces with silicone fluid and grease.[22] Even if the fluid were not silicone, Webb suggests that breast injections with military chemicals were familiar to some army personnel.

US-Japan Silicone Technology Transfer

As part of the process of unraveling silicone's arrival in Japan, it is important to note that Dow Corning, founded in the United States in 1943, presented itself as the world's first commercial supplier of silicone. US Admiral Hyman Rickover gave military orders for Dow Corning's production and supply of dimethylsiloxane, which was used throughout the US Armed Forces as an aviation engine lubricant. In 1945, silicone was exported from the United States to Japan and, as with all other materials, was controlled by the Allied occupation forces there. Access to silicone production in Japan was restricted by these forces and regulated by figures like General Douglas MacArthur, the Supreme Commander for the Allied Powers.

Japanese companies seeking to establish postwar industrial production of silicone needed to submit formal plans to the SCAP to receive permission to do so. In March 1950, Yosuke Suzuki, president of Kyoto-based Shimadzu Seisakusho Ltd., presented a "Plan for Execution of Industrialization Experiments" to the SCAP's Economic and Scientific Section, Industrial Division, for "Experimentation on Industrialization of Silicone Production."[23] A year prior, Suzuki had completed studies on the industrial-scale manufacture of silicone and set up a small-capacity pilot plant: "Having thus had full confidence in the production, we wish to start a trial industrialization of the same, in which we propose to turn out every month 600 kg of silicone in the first year beginning in March 1951."[24] Suzuki's application to the SCAP requested approval to scale up production and a grant to support Shimadzu Seisakusho Ltd. for "the industrialization experiment."[25]

Shimadzu, established in 1875, remains a major firm in advanced tech development. A manufacturer of precision measuring instruments and medical equipment, it was not the only company to begin silicone research shortly after the United States entered Japan. Shin-Etsu Chemical Co. Ltd., the largest chemical company in Japan, which originally was set up in Tokyo in 1929 as a specialist in nitrogen fertilizer, claims that, like Shimadzu, it began basic chemical research on silicones in 1949.[26] Dow Corning's Earl Warrick offers a competing narrative, claiming that Shin-Etsu was among a group of Japanese military and electrical companies that explored silicones during the war after discovering and studying them in downed American fighters and bombers.[27] This would date their research into silicones a little earlier, to the US air raids in Japan between 1942 and 1945.

These dates appear to correlate with those of the famous Japanese plastic surgeon Rin Sakurai, who later claimed to have begun experimenting with silicone augmentation in 1943 by injecting the material into his slender face and thighs as a filler.[28] Following seven years of successful trials in animals, he developed the Sakurai formula, which was said to combine Dow Corning 200 fluid with 1 percent vegetable or animal fats, including sesame oil.[29] From 1945 to 1963, Sakurai claimed to have used variations of this formula in 72,648 breast and body-contouring cases.[30] He said that 65 percent of his patients received treatment for the face, 20 percent for the breasts, and the remainder for other parts of the body.[31] However, how Sakurai sourced silicone fluid in Japan at the time remains unclear.

By 1953, Shin-Etsu had started commercial production of silicones and, in current publicity, claims to have been the first Japanese firm to venture into the silicone business.[32] Two years earlier, during his 1951 visit to Japan and Shin-Etsu's factory, Shailer Bass, a Dow Corning chemist, noted that Shin-Etsu's World War II studies of silicone in fallen US aircraft were so advanced that the company had already replicated and applied the direct process for producing silicones.[33] Warrick took a different view: He claimed that Shin-Etsu's success had been enabled by entering into a licensing agreement with General Electric in 1953 and with Dow Corning in 1957.[34]

As with Dow Corning in the United States, Shin-Etsu was also investigating silicone's potential as an implantable medical and cosmetic material.[35] Starting in 1951, Dr. Takeya Shirakabe of Osaka's Shirakabe Hospital and Yakuo Fujimoto, the hospital's chief of the technical service department, collaborated with the Shin-Etsu Chemical Company on the development and production of their own silicone formula for bust enhancement.[36] Shirakabe claimed that the Monsanto Chemical Company in the United States acted as a consultant on the project, further indicating an international network of involvement.[37] As part of this complex unfolding of the United States' attitude toward breast augmentation, Monsanto later distanced itself from this claim: "We are surprised to learn of Monsanto's identification with the development of a new plastic material used in [breast augmentation] surgery, designated as a modified silicone. The fact that it is a silicone leaves doubt as to our association therewith, since we do not manufacture this type of product. There is, however, the possibility that the Shinetsu [sic] Chemical Company . . . received from Monsanto certain raw materials of our manufacture which they have incorporated in a product."[38]

By November 1960, nine years after the development of its own silicone formula, Shirakabe Hospital declared that it had successfully used its secret Shin-Etsu silicone augmentation formula in more than two thousand patients for breast enlargement.[39] Shin-Etsu, however, was not the only company in Japan to produce its own formula of bust-enhancing injectable solution. There were several other branded materials available in Japan during the postwar period. They included paraffin-like materials (such as Organogen and Bioplex, which were mostly associated with "beauty parlor services") and silicones—Elicon (dimethylpolysiloxane), first produced by Koken Kogyo Co. Ltd., Zeflon, and Linsacrite zelm (comprising 98 percent liquid silicone and 2 percent snake oil).[40]

Herbert Conway in Japan and the DCCAMR

The Shin-Etsu/Shirakabe narrative is recorded in correspondence between Herbert Conway and Takeya Shirakabe. Conway—chairman of the American Board of Plastic Surgery and director of the Plastic Surgery Department at the New York Hospital-Cornell Medical Center—is an important figure in this complex story of the US-Japanese relationship with injectable silicones, as well as a key player in the development of plastic foam for cosmetic surgery. Rather than being linked to a tale of stolen silicone at Yokohama Harbor, Conway, an invited guest of the Japanese Society of Plastic and Reconstructive Surgery, in fact engaged in professional research exchange. Conway, an internationally established authority in his field with an active interest in injectable silicones, visited Shirakabe Hospital in September 1960—one of many stops on his tour in Japan. He was particularly interested in what local cosmetic surgeons demonstrated and described as "modified silicone." He recalled, "This was a thick, jelly-like, somewhat rubbery, clear plastic which was injected into subcutaneous tissues from a syringe through a needle of about 20 gauge."[41]

Conway continued, "I saw examples [of breast enlargement patients] in which the material had been injected two years earlier. It had become firm and had not caused any reaction in the tissue."[42] The body appeared to have neither rejected the material nor allowed it to migrate or disintegrate. At the time, this would have been considered a major success and a revolutionary breakthrough. Historically, materials that were unsuitable for the body tended to be rejected and expelled by the body within a short period, whereas the injected silicone Conway observed in Japan did not seem to cause any immediate adverse effects. He recognized the enormous potential of such a seemingly inert injectable material for augmentation but was frustrated to find that the surgeons refused to provide him with any further details about its composition: "They would not tell me the exact name of the material or its chemical make-up, nor where it could be obtained. These cosmetic surgeons actually told me that this was 'secret information.'"[43] Peppered throughout Conway's correspondence is a sense of annoyance at the Japanese surgeons' refusal to readily share the formula or provide samples during his visit. Local surgeons were protective of their formula and wanted Conway to pay for access to their knowledge and skills.

Conway returned to New York intent on reproducing the silicone injections he had seen in Japan without paying Japanese surgeons for access to their research. He recalled, "One of the surgeons in the crowd informed me later that [Dow Corning] had been consulted by the Japanese firm which supplied the material."[44] Shortly afterward, Conway contacted Dow Corning to further investigate. Many years of regular correspondence and exchange ensued between Conway and Dow Corning. This was not the first time Conway had approached the company. By 1959, he had already registered his interest in silicone. Rob McGregor and Silas Braley of the Dow Corning Center for Aid to Medical Research asked him for clarification about its intended use and sent him limited information. It was only after his visit to Japan that Conway became more specific in his queries and requests, which led the DCCAMR's staff to become more responsive to the details of discoveries he shared. McGregor asserted that the DCCAMR was inundated with correspondence, and although Conway experienced a delayed response to his initial inquiry, he was assured this was "occasioned by pressure of correspondence rather than by lack of interest."[45]

Conway's correspondence with the DCCAMR is particularly valuable, as it presents a rare perspective on the complex social, medical, and industrial networks and internal processes behind the use of silicones in body contouring. It also demonstrates a culture of plastics material sampling and documents the development of the implantation of polyurethane foam and injection of silicones in the US and internationally. Conway initially wrote to McGregor, director of the DCCAMR; however, he largely dealt with Braley, who was executive director at the time.

Braley was popular in the general press, where he was presented as a "youthful, sophisticated and sympathetic" figure, crowned by *Vogue* journalist Simona Morini as "silicone's guardian."[46] He was described to *Vogue's* predominantly white, affluent female readership as elegant, charming, protective, and patriarchal. The white American Braley was represented as the opposite of othered foreign "quacks." Morini described him as "a brilliant chemist who feels for silicone an almost anthropomorphic affection."[47] Braley was fascinated by the possibilities of military-industrial materials R & D, remarking, "The use of silicones in medical applications is a happy example of the medical and industrial fraternities working together to adapt an industrial synthetic material to the needs of the patient."[48] Medicine and organic chemistry were now working together to bring postwar plastics to the body.

Conway's papers reveal that Braley and McGregor shared Conway's interest in further investigating the silicone formula rumors from Japan.[49] They informed Conway that Dow Corning's Japanese distributor shared his frustration over the failure to obtain more precise information. The DCCAMR claimed it was further investigating "the possibility of developing an injectable fluid that will set up subcutaneously": however, its labs had not yet fully accomplished this. Conway was promised samples as soon as they became available and were "sufficiently far along to make it worth testing."[50] At this point, silicone in all its materialities—fluid, rubber, or foam sponge—was not yet subject to any formal FDA jurisdiction or testing. Plastics, an increasingly burgeoning technology in medicine, were difficult for the FDA to legislate according to established regulations, as they were neither a drug nor a device.

The DCCAMR was set up in 1959 to respond to and supply researchers' requests for small samples of "silicone fluid, resins, tubing, block, sponge, simple, molded parts, etc."[51] These samples were sent out free of charge. If any items were asked for in larger quantities than the laboratories could meet, Dow Corning would try to locate a manufacturer capable of fulfilling the request. Dow Corning was always looking to expand its production, so additional sourcing services were also freely provided to international researchers. Physicians and engineers worked with Dow Corning's fluid silicone samples in various countries, including Israel, Poland, and Germany.[52] Those who were unable to obtain samples would resort to other means, such as experimenting with industrial-grade silicone or, in Australia in the late 1960s, cutting open silicone implants to use the gel inside for injections.[53]

It should be noted, however, that Dow Corning not only distributed samples to those who requested them but also sent samples and a range of elaborate promotional booklets out in large-scale mailouts, hoping to attract doctors' interest—and their research funds. Since silicones had been developed "under a veil of secrecy" during the war, the end of the war left people in the United States and internationally with little knowledge about the material.[54] Silicone research also attracted the funding and support of federal grants. One such example was "The Present Status of Silicone Fluid in Soft Tissue Augmentation" (1967), which was supported in part by a US Public Health Services grant, thereby legitimizing silicone research.[55] As has been noted by STS scholars of the Cold War United States, securing federal and defense funding for R & D projects was immensely lucrative for labs and was certainly factored into decisions on what was being researched.[56] It is particularly notable that Braley is listed as a coauthor, despite having no medical qualifications. Furthermore, the article's introduction, which Braley reviewed, states that "silicone fluids have been used as medical prosthesis for at least a dozen years, particularly in Europe and the Orient," thereby effectively distancing Dow Corning and the United States from the practice of silicone injections.[57] Dow Corning worked hard to expand its markets, both domestically and internationally. The business, which began on March 1, 1943, with $15,000 per month in revenue, grew to $1.5 million per month in 1953.[58]

Dow Corning's free distribution of silicone samples and booklets was designed to encourage engineers, including those working with surgeons, to experiment with prototypes and come up with new ways to use silicone, effectively outsourcing the firm's R & D. This was common practice in the ever-expanding plastics industry since its inception. Sample makers were hired by chemical companies to research and prototype potential products and relay their findings directly back to the company.[59] Correspondence reveals that Dow Corning, like many other plastics and chemical companies of the pre- and postwar period, also employed salesmen who traveled the country with medical samples, literature, and demonstrations to further encourage product development.

At times, demand for orders outstripped production capacity. In a letter to Silicone Committee member Joseph Murray, a DCCAMR surgical products division supervisor, assured him, "We will not be supplying our salesmen with demonstration samples until we can fill all of your orders."[60] Salesmen with demonstration samples were put on hold as demand for Dow Corning medical products was already higher than Braley and his department had expected. In postwar America, plastics were increasingly entering all aspects of everyday life, no longer requisitioned by the US Armed Forces, and chemical companies were keen to exploit and explore the potential of their material production facilities for commercial everyday use. Hoping to promote their products and expand their postwar markets, Dow Corning and other Big Science companies including Dow Chemical, General Electric, and DuPont sent out plastics samples to engineers, universities, hospitals, and even to artists.[61]

Conway's correspondence with the DCCAMR reveals that Dow Corning was actively investigating the chemical composition of subcutaneous breast augmentation shots, known as "liquid flesh" or "fleshy injection" in Japan and sometimes referred to in the US press as the Sakurai formula.[62] US actors, including chemical company directors, doctors, and journalists, implied that Japanese silicone research for medical and cosmetic use did not meet the superior standards of Western medical practice—a racist perspective that continues to be perpetuated in contemporary, Western-centric, Anglophone histories of cosmetic surgery.[63] However, Conway's private correspondence offers a previously unconsidered, alternative narrative that challenges this view. It reveals US surgeons' and silicone manufacturers' frustration with Japanese manufacturers and their clients, such as Shirakabe Hospital, and Shin-Etsu's unwillingness to provide them with the formula.

Shirakabe and Fujimoto were only prepared to share their research with Conway in exchange for a fee. In the correspondence, Conway, McGregor, and Braley express outrage that Shirakabe and Fujimoto wanted to be paid for sharing their research. They refused to pay for access to Japanese medical knowledge. In turn, they engaged in correspondence, pooled their limited knowledge of Japanese advancements, and exchanged lay and medical articles (privately translated from languages such as Japanese, German, Polish, and Czech), as well as communication and leads, to map and chart silicone's development as a soft-tissue augmenter. Ultimately, Conway, McGregor, and Braley sought to replicate Shirakabe and Fujimoto's discovery. Dow Corning had an active interest in pursuing this research, as breast augmentation injections certainly promised a financial reward. Indeed, by 1959, American journalists reported that while silicone bust injections were circulating in select beauty salons in the United States, including in Boston, some American women who could afford it traveled to Japan in pursuit of treatment.[64]

In early January 1961, Conway received another letter from Braley. Unlike previous correspondence, this letter had a small plastic pocket stapled to it, containing a sample sliver of a new type of silicone developed at the DCCAMR's lab in Midland, Michigan. Braley wrote, "We have never been able to track down any firm information concerning the rumors we've had of a Japanese injectable silicone for facial and mammary augmentation. Your letters have told us much more than we have been able to discover ourselves."[65] Braley rewarded Conway's effort to share this information with the DCCAMR by including a sample designed to simulate the material he had described seeing in Japan only a few months earlier. Conway's overseas observations were becoming manifest in US materials R & D.

While still in an experimental phase, this US-based collaborative investigation into injectable silicones began to firm up. Braley sought out Conway's feedback to help further shape the sample's future materiality: "I am enclosing one of our early tries for you to examine. Our feeling is that this is too firm, and we are attempting to make it less so."[66] He continues, "From the chemical nature of this material, we'd expect it to be as implantable as our other medical grade silicones. It is fluid when injected and sets up to a non-fluid after implanting." This new type of silicone had the potential to be a game changer for Dow Corning. Inserted via injection rather than the invasive surgery required of foam breast implants, the material transformed from a viscous substance into a device: an agentic liquid that rapidly set, becoming a semisolid prosthetic object.

Conway was "thrilled" with the material. He enthused, "It looks to me as though it is just about the same as the material which I saw in Japan."[67] Wanting to know more, he wrote, "Will you be good enough to give us data regarding its chemical content, name, technique, etc.[?] We should have a fairly large amount, as I would plan to insert about one thousand grams in each two breasts, provided that I can find a suitable patient who will give written permission in advance for experimental use of this material for breast augmentation."[68]

As soon as the material became available, Conway indicated his keenness to test it and use it in breast augmentation. Despite Braley's cautionary comments that the silicone material was still too firm and that he and his colleagues would like to produce something softer, his concerns were ultimately ignored. There was a sense of excitement and urgency. Conway and his colleague Dicran Goulian Jr. wanted to immediately start using this material. As with all the materials that preceded it, as soon as silicone became available, there was a desire to experiment with it as a bust-enhancing implantable material. This is particularly noteworthy because Dow Corning's public-facing statements between 1959 and 1976 repeatedly distanced the company from the use of silicone injections for breast augmentation.[69]

In one such instance, Braley accused William St. Clair Symmers of bias for naming Dow Corning as the silicone supplier in his 1968 *British Medical Journal* article "Silicone Mastitis in 'Topless' Waitresses and Some Other Varieties of Foreign-Body Mastitis."[70] Braley wrote to Symmers on August 1, 1968, that "we at Dow Corning have never felt that [silicone injections for breast augmentation] was a proper use of the material and have consistently done all that we can to prevent it."[71] He continued, "Although we have done nothing in the area of mammary augmentation, we have done considerable work, under FDA protocol, in studying soft tissue augmentation other than in the breast." This contrasts starkly with what is revealed in Conway's correspondence and in publications such as Conway and Goulian's 1963 article, which clearly states that "the material under investigation is Silastic RTV S5392 which is manufactured by the Dow Corning Corporation," thus naming Dow Corning's support for his research in this area.[72] Braley also wrote to Symmers that "clandestine injectors" used silicone fluid adulterated with additives, and it was these additives that could cause reactions—not the (Dow Corning) silicone itself. He emphasized, "We have come to the almost inescapable conclusion that the reactions reported in patients are due to the unspecified additives and not to the silicone itself."[73] This corporate claim was later refuted by a group of medical doctors in Las Vegas.[74] Braley continued, "We have never condoned [the use of Dow Corning 360 Medical Fluid for breast injections] and have endeavored in every way possible to keep our material out of the hands of those doing breast injections." Dow Corning may not have publicly condoned Dow Corning 360 Medical Fluid for breast injections, but as this chapter shows, there was a culture of distributing silicone samples in various viscosities (e.g., Silastic RTV) and other material states.[75] Braley enclosed a bibliography on silicone fluids in soft tissues for Symmers; however, research on breast augmentation injections was neither mentioned nor included in it.[76] Dow Corning's public-facing research, newsletters, and annual reports from that time largely omit research conducted on breast enlargement.[77]

Contrary to claims that Dow Corning had no knowledge of breast injections before 1963, Conway's papers reveal that Dow Corning was corresponding with him on silicone injections for bust augmentation as early as 1959.[78] Conway's letters show that McGregor and Braley were already aware of silicone injections for bust augmentation and expressed interest in the practice even before Conway wrote to them. McGregor, then director of DCCAMR, responded on October 29, 1959, to Conway's request for injectable silicone materials for breast augmentation: "We would be very glad to supply you with experimental samples of such silicones [fluid when injected, solid after injection] as you might consider to be helpful," demonstrating Dow Corning's readiness to supply injectable silastic silicones for breast augmentation R & D.[79] McGregor advised Conway that the DCCAMR had also supplied silicone materials for experimentation in soft-tissue augmentation to other US-based doctors, but he does not specifically mention what parts of the body they are working on. Unfortunately, Dow Corning has no public archive or records of the DCCAMR correspondence, so it is not possible to fully trace and map the international medical actors with whom they corresponded and to whom they supplied samples.[80]

In October 1962, Conway and Goulian presented a report on their "Experience with an Injectable Silastic RTV as a Subcutaneous Prosthetic Material" at the annual meeting of the American Society of Plastic and Reconstructive Surgery and later published their findings in the PRS.[81] In this article, they make some of their research public. They outline the purpose of the research as sharing their findings on Dow Corning's Silastic RTV, an inert liquid silicone that can be transformed using a catalyst into a "jellylike solid with properties which offer distinct advantages over the implants of synthetic sponge."[82] As with previous materials, Silastic RTV is presented as a novel material improvement over its predecessor, plastic foams. Silicones were the latest in a series of materials to be promoted as a safe option for shaping the body. However, its materiality differed greatly from that of plastic foams, adding to its "novel" appeal.[83]

Conway and Goulian's PRS article stands out in that it cites and acknowledges the medical research being carried out in Japan while also drawing comparisons with it. Conway and Goulian noted that early in their investigation, they realized they needed a syringe of far greater mechanical advantage than those conventionally available on the US market. They sought the aid of the medical technology company Becton, Dickinson—a Dow Corning business partner that from 1952 worked as the company's medical silicones distributor, and which DCCAMR recommended for creating prototypes—to develop an instrument "not unlike the devices which have been described by the Japanese for use with similar silicone derivatives."[84] In this example, US surgeons openly looked to Japanese medical research to design a new surgical instrument. Figure 7.1 shows the original article's image of the syringe. Specially designed to administer large volumes of up to 600 mL of Silastic RTV S5392, the stainless-steel syringe, with its enormous and seemingly industrial proportions, penetrates the page diagonally at an acute angle, held by gloved hands. One hand holds up the syringe's barrel at a slant, and the other grips the plunger. This gesture is also illustrated during the procedure in figure 7.2. In both figures, the design looks like a weapon—more specifically, a rocket launcher. However, the 600-mL-capacity cannula is likely to have been designed in this way so that it would only need to be filled once, thereby reducing the risk of infection.

It is worth noting that the article omits Conway's visit to Japan, where he was most likely to have seen such an instrument in the procedures he observed. Instead of mentioning Conway's firsthand witnessing of procedures, it instead refers to "devices which have been described by the Japanese."[85] On the other hand, it mentions that "Dr. Takeya Shirakabe of Kyoto, Japan . . . kindly supplied a sample of the silicones which he has injected for augmentation of soft tissues."[86] This sample and the accompanying correspondence appear to be missing from Conway's papers housed at Weill Cornell Medicine. Perhaps Dow Corning and Conway finally paid Shirakabe's requested fee for the sample. However, the investigators were keen to assert Silastic RTV's superiority over the Japanese product: "The Dow Corning Research laboratories have found [Shirakabe's] samples to be substantially different from Silastic RTV. Whereas Silastic RTV sets into a semisolid mass upon mixing with the catalyst, the silicones injected by the Japanese (available in variable forms ranging from a thick fluid to a

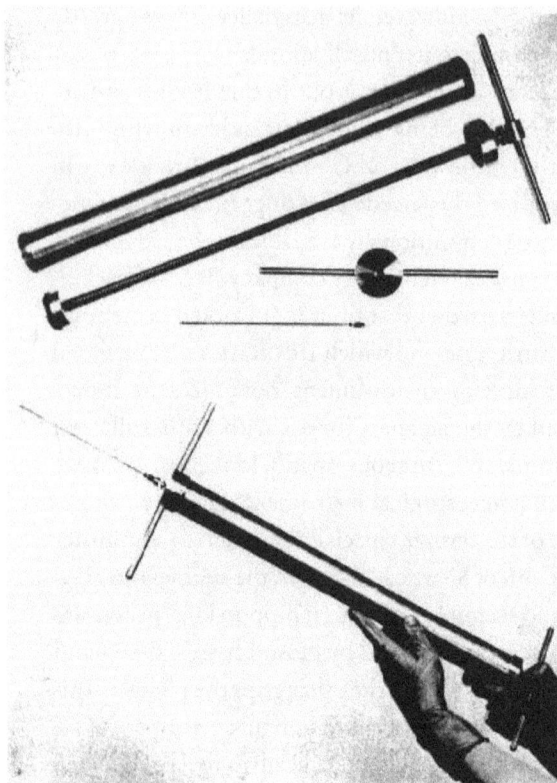

7.1 Photograph of syringe designed with Becton, Dickinson, in Conway and Goulian, "Experience with an Injectable Silastic RTV," 296.

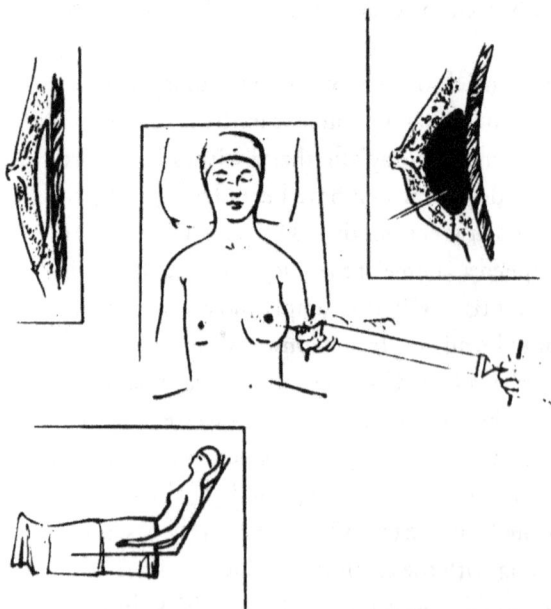

7.2 Mammary augmentation illustration in Conway and Goulian, "Experience with an Injectable Silastic RTV," 298.

puttylike mass) retain their original physical nature in situ. That material has been injected directly into the tissues, frequently in multiple stages, to accomplish the desired volumetric increment and surface contour."[87]

The Silastic of Conway and Goulian's investigation was unique in that it was liquid when injected and would soon afterward form into a semi-solid shape: essentially an injectable prosthetic. Once delivered via the high-gauge needle, Silastic would set in the body, and it was hoped that its semisolid consistency would prevent migration and the physical complications this could cause. Here, silicone's materiality shifted in the process of its application to the body: The material became an object, entering the body as a liquid and setting within it.

Silicone and Orientalism in *Women of the World*

As noted by medical scholars, the history of injectable materials in breast augmentation is international and dates to the late nineteenth century. However, throughout the 1960s, important actors in US whitestream media continued to attribute breast injections to Japan, often with an Orientalist tone. In American postwar popular culture, the "mad professor" was rarely a US doctor or scientist but often a "foreign" figure.[88] In keeping with this mad scientist as other, American and international press perpetuated the idea that silicone injections originated in postwar Japan with Panpan girls, sex workers, or entertainers responding to the tastes of "the Occupationaire."[89] *Women of the World* (1963), by Italian film directors Paolo Cavara, Gualtiero Jacopetti, and Franco Prosperi, presents the Allied occupation's influence on feminine ideals in Japan (figure 7.3). Known as a "mondo" (*world* in Italian) movie, *Women of the World* belongs to a genre of exploitation pseudodocumentaries, also known as shockumentaries, made in the early 1960s. Mondo movies were dedicated to sensationalizing the Other to a largely heteronormative, white, Western audience.

These montages of often staged scenes depicted sex workers, cosmetic surgery procedures, subcultures, and ethnocentric racist scenes of the "non-Western" world. The footage can also be understood as demonstrating the racialized "pornotropic gaze," which historian C. Riley Snorton—writing on the history of the violent and racist treatment of Black female patients in the United States by white male doctors—observes "informs,

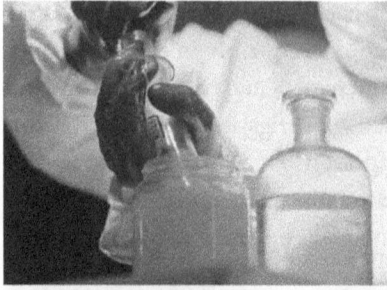

7.3 "In Japan a surgeon prepares to administer fluid breast augmentation." Dialogue and stills from the 1963 mondo film *Women of the World*, directed by Paolo Cavara, Gualtiero Jacopetti, and Franco Prosperi (Embassy Pictures).

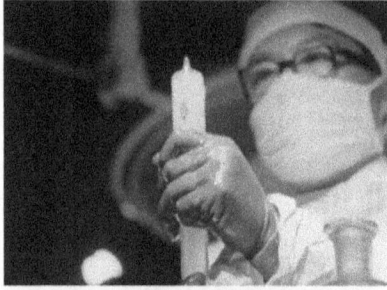

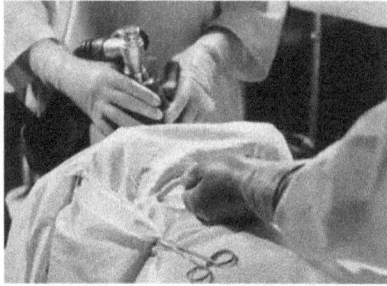

subtends, and frequently subsumes the medicoscientific [gaze]," showing medical spaces as "a libidinous site."[90] *Women of the World* provides a rare visual interpretation of the Allied occupation's impact on Japanese women's bodies. It shows how imported plastic materials and ideals gave shape to Western-inspired toy dolls with almond eyes and "busts à la [Jayne] Mansfield," before moving on to the material feminine body.[91]

Following an immediate cinematic cut after the depiction of these plastic dolls, the narrator introduces a sequence of cosmetic surgeries in Japan. The first sequence features Eurocentric eyelid reshaping surgery.[92] The second presents breast augmentation through injection via an oversized syringe. The surgeon fills what is described as a hydraulic pump with an unspecified fluid, which he uses to inflate the patient's breasts (figure 7.3). In this sequence, it is unclear whether this is a genuine depiction of breast augmentation or what exact material is being used.

The syringe does appear larger than what one might expect. It is noteworthy, however, that the famed American white topless dancer Carol Doda's description of the procedures she undertook in 1960s San Francisco correlates with the on-screen depiction: "Once a week for a year I went. [Silicone is] pumped in with a big needle, like a horse needle. At first it felt too firm. But it's like a breast now, it feels soft."[93] These images also appear to mirror those from a Japanese medical journal article published the same year that investigates Elicon injections.[94] In this 1965 study, the Sapporo-based physician Yasuo Mutou included trans feminine individuals.[95] As will be discussed in the following chapter, some trans feminine and gender-nonconforming people in the United States also shaped their bodies with silicone injections. But they were not explicitly identified in US medical reports on such procedures during this period. Mutou provides a rare medical citation from this era that shows a wider range of women undergoing silicone shot treatments.

The Japanese sequence in *Women of the World* had a lasting impact on its international viewers. In September 1967, the AMA received a letter seeking information on the practice of fluid injections to augment the bust. Having watched the film a few years prior, the inquirer wanted to start their own breast augmentation business:

> I have done some research on what it would take to remodel women's busts . . . I am thinking about the possibility of starting a business that specializes in remodeling women's busts but plastic surgery requires one to be a doctor. I have read about women in California (topless waitresses and dancers) having silicon [*sic*] injected into their busts but plastic surgery requires one to be a doctor. I saw a movie a few years ago titled "Women of the World" and it showed a doctor in Japan literally "pumping" up the bust of a woman with a hydraulic pump. Do you have any literature on this subject? Is there anyone in Japan or this country I could contact and base this process on? Or have a doctor come over and work for me in such a business? Is it against the law if someone other than a doctor engages in such a business?[96]

Seeking to imitate a film sequence that portrays silicone shots as a straightforward process, the AMA correspondent sees a lucrative opportunity. They envision bringing a Japanese doctor to the United States or, better yet, mastering the technique themselves without the inconvenience of obtaining medical qualifications. The AMA replies, clarifying, "The injection of a foreign substance into the human body to enlarge the female breast involves the practice of medicine, and one must be a licensed physician in order to engage in this activity. We are sorry, but we have very little additional information that we can provide you, and suggest that you consider the legal implications involved in setting up a business of this kind."[97]

The bust augmentation sequence in *Women of the World* mentions that European and American women undergo similar eye surgeries to the Japanese. Breast augmentation injections, however, are presented as something solely Japanese. Postwar Western representations of this procedure and the established scholarship around it are frequently Orientalist. They suggest that Japanese doctors, portrayed as incompetent in meeting the stringent and superior guidelines of Western medicine, injected syringes filled with unsterile, industrial-grade silicone, or silicone that was "adulterated" with local ingredients, such as peanut or sesame oil. There is little to no acknowledgment of the fact that European and American doctors also diluted silicone with other substances.[98]

The Hiroshima Maidens and Akiko Kojima

The role of the Japanese female body and the body politic within US and Japanese postwar relations have been explored by scholars from a range of disciplines studying the Hiroshima Maidens.[99] American journalist Norman Cousins published a series of editorials in the *Saturday Review* with the aim of raising funds for a group of twenty-five young Japanese women from Hiroshima. They had been scarred and injured by the August 6, 1945, dropping of the atomic bomb and became known in the United States as the "Hiroshima Maidens." Scholar Christina Klein notes that Cousins coordinated the voluntary efforts of individuals and institutions "who represented the forces of global military and economic integration, as well as of humanitarian internationalism."[100] Arthur J. Barsky, a high-profile plastic surgeon, headed the operations and convinced New York's Mount Sinai Hospital to contribute surgical facilities; a number of Quaker families supported by the American Friends Services Committee housed the women; the US Air Force provided the outbound flight from Japan; and Pan American Airways, the return flight after their operations. According to this narrative, US religious, media, medical, and military structures collaborated to "save" this group of young Japanese women by medically reconstructing their bodies, which had been affected by the atomic war.

During World War II, US propaganda represented Japanese people in racist and dehumanizing ways. However, at the end of conflict, Japan's global role shifted from enemy to Washington's most prized East Asian ally.[101] The US government promoted the occupation and reconstruction of Japan as a prototype of American self-sacrifice, presenting it as evidence of America's altruistic international ambitions and as a strategy to bolster American influence in Northeast Asia against Communism. Klein argues that in Cousins's editorials, the plastic surgery the Hiroshima Maidens received serves as "metaphor for the 'reconstruction' of postwar Japan."[102] As Serlin has discussed, the treatment of the Hiroshima Maidens by US surgeons also served to further privilege Western cultural and technological forms over those of Japan.[103]

Four years after the US Air Force flew the Hiroshima Maidens to the United States for plastic surgery, Akiko Kojima, Miss Japan at the time, was crowned Miss Universe 1960 in Long Beach, California. Japanese studies scholar Jan Bardsley frames the coronation of the first Asian victor of the Miss Universe beauty pageant in racialized and gendered postwar politics. Bardsley posits that the victory of the twenty-two-year-old, long-legged fashion model from Tokyo was celebrated by Americans and Japanese alike as "a major breakthrough for Japanese women's rights."[104] Akiko's unusual stature and long, straight limbs were "read as signs of postwar progress," alluding to changes in diet. At first, Kojima's win was celebrated; she returned to Japan, where she was inundated with promotional bookings. Beauty contests across Japan recorded a phenomenal increase in applications.

It was not long, however, until rumors began to circulate in the US and Japanese press, suggesting Kojima's victory was fixed as an American tactic to court approval in Japan.[105] Kojima was also accused by Japanese journalists, critics, and writers of having undergone breast injections. Additionally, they alleged that the American cosmetics company Max Factor Corporation, one of the central sponsors of Miss Universe, had orchestrated the competition to facilitate the launch of a new range of cosmetics in Asia.

Scholars argue that Japan was reimagined in the American consciousness, transitioning from destructive enemy to ally, through a sympathetic focus on Japanese women and children. This was facilitated by a range of charity projects, such as the Hiroshima Maidens and the Christian Children's Fund adoption scheme, as well as popular cultural representations such as the 1957 film *Sayonara*.[106] Japanese men are notably absent in imagery of Kojima and the Hiroshima Maidens; instead, US forces are depicted as masculine protectors and saviors.

Unlike the Hiroshima Maidens, Kojima was eroticized. Her body was praised by the Miss Universe judges and embraced in the United States, as her hourglass figure and long legs aligned with dominant white American beauty ideals. And unlike the Hiroshima Maidens, Kojima's exposed body was glorified. This victory was celebrated in America and Japan as a sign of progress for women's rights in Japan, although it was not without criticism.[107] When Kojima arrived in Japan, she was questioned about rumors that she wanted to marry an American and relocate to the United States, which she denied. The "authenticity" of her Japanese identity was scrutinized and threatened. Likewise, the "naturalness" of her body was also attacked; for instance, a surgeon sold his story, claiming he had created her thirty-seven-inch bustline using silicone shots.[108] Regardless of the veracity of this claim, the accusation of breast augmentation highlights how postwar gender ideals and Japan's close ties to the United States were played out in body modifications. Bardsley reflects that this drawing of attention to cosmetic bust enhancement is likely to have brought up uncomfortable memories of the sexual politics of the occupation, particularly to the practice of postwar sex workers in Japan undergoing silicone shots to appeal to American clientele.[109] These treatments were often damaging to the body and, in some cases, fatal.

American military technology and materials development had a devastating impact on Japan and its people. Silicone, commissioned by the US Navy and developed by US industry, was developed to safely carry bomber planes to Japan during air raids, including the Hiroshima and Nagasaki bombings. This imported technology was, in some cases, reapplied by qualified (and unqualified) medical practitioners in the reshaping of Japanese women's bodies. Regardless of whether Kojima underwent such a procedure, her appearance was both celebrated and reviled for resembling a tall white fashion model with an hourglass figure.

The men who made important decisions about both women's bodies and the materials implanted within them participated in a homosocial, heteropatriarchal culture that focused on constructing and consuming women's bodies. In keeping with dominant masculine, heteronormative work norms at the time, male actors such as surgeons and chemists, operating within these networks of power associated with plastics, engaged with and participated in strip culture, both in the United States and internationally. When Conway wrote to another white American male plastic surgeon due to visit Tokyo in 1962, he recommended visiting both the Queen Bee night club and the Papagayo Club, which "has some nicely undressed shows, and these seem to be the rage in Japan at the present time since their appetite for entertainment is thirty or forty years later in development than that of the Western World. A friend took me there to scrutinize the breasts of the performers for needle marks of injection of their fluid polythene for augmentation mammaplasty. There is a special show at the Papagayo Club for men only at 5pm daily."[110]

Besides being Orientalist, Conway's comment on nude shows being "the rage in Japan," which is behind "the Western World," is inaccurate, given that nude shows remained a popular form of entertainment across western Europe and the US West Coast. Clearly, Conway's research was not simply confined to hospital visits in Japan, as he also enjoyed and recommended visits to nude shows, allegedly to scrutinize performers' breasts for needle marks. At a dinner honoring the respected plastic surgeon Milton Edgerton, a member of the Dow Corning Silicone Committee, Edgerton and his colleagues were served "Augmented Breast of Chicken with Smithfield ham" at Baltimore's Elkridge Country Club.[111] Only augmentation mammaplasty is joked about in the menu selection, further demonstrating the inequality of gendered power structures within the US network of medical actors. As these examples show, key actors within the postwar US network of silicone's application, such as respected US surgeons, engaged in homosociality, objectifying women's bodies as sites of entertainment and sexual pleasure.

Other professionals in the silicone network also engaged in a culture of sexualizing and sometimes mocking women's bodies. Consider the *Chemical Peddler* (1920s–1980s), published by and distributed among members of the Salesmen's Association of the American Chemical Industry Inc. It regularly featured misogynist imagery of semiclad and often entirely nude, predominately white women in chemical company advertisements. This imagery embodied an ideology that positioned white women's bodies as objects of experimentation and pleasure, often crafted by chemists themselves. The imagery presents women's bodies as a site for improvement and consumption, shaped by developments in the chemical industry. A 1955 Monsanto advertisement that appeared in the *Chemical Peddler* depicts the transformative qualities of the company's "Rejuvenation x 516" (figure 7.4).

7.4 An advertisement for Monsanto's Rejuvenation x 516, *Chemical Peddler*, 1955. Science History Institute, Philadelphia.

A seemingly endless line of "abject" white women—characterized by aging features, sagging breasts, short hair, stout figures, long noses, and thin lips—is sent through a Monsanto Rejuvenation chemical showering device.

The suggested solution is to undergo a procedure of some kind. In the case of the women in figure 7.4, an instant treatment with Monsanto's "x 516" fluid can quickly fix their alleged predicament. Once they have passed through the Monsanto-fueled contraption and emerged from behind its curtains, these women are nude, youthful, and svelte. Their bodies are smooth and hairless, except for perfectly coiffed hair on their heads. Their faces are now altered, with long lashes and fuller lips. In the ad, the desirable, rejuvenated body is a white body, evoking the smoothness of classical sculpture with exposed, lifted breasts.[112]

This image, created to appeal to salesmen working in the chemical industry, embodies a postwar US imaginary where white heterosexual men are able to shape the bodies of women to align with dominant white American beauty ideals. It illustrates the racialized and gendered power structures at play within the postwar American chemical industry. It is telling in that it reveals the homosocial ideology that shaped these industries and serves as a representation of how these unequal structures affected the development of plastics and their medical application in the aesthetic shaping of feminine bodies, a profitable market pursued by companies like Dow Corning. However, as the next chapter will show, many silicone users applied this technology to resist, subvert, and challenge dominant white, heteronormative gender norms of the postwar United States.

8 QUEERING SILICONES / CAROL DODA AND THE COGS OF THE FDA

In the summer of 1964, Carol Doda, a twenty-three-year-old, petite, white, blonde cocktail waitress on the San Francisco go-go circuit, established her reputation as a dancer by donning one of Rudi Gernreich's "monokinis," a topless swimsuit that replaced the bikini she previously wore. Dressed only in this design and nothing else, Doda mounted the piano at the Condor Club, having opted against wearing the nipple tassels—also known as "pasties"—worn by burlesque performers in the preceding decades to avoid censorship. By removing the element of tease, she performed with her breasts exposed from start to finish. This was revolutionary. Shortly afterward, Doda secured her position as "queen of the San Francisco exotic dancers," as she was known on the go-go dancer circuit, by having a pint of silicone injected into each breast, which was an equally novel act.[1] Doda would perform five or six shows a night, seven nights a week, at the Condor Club, which was next door to Big Al's, where performers included the Japanese American Tosha McDonald (figure 8.1).[2]

Soon after, in October 1964, the *San Francisco Chronicle* crowned Carol Doda "the first topless dancing act of widespread note in America."[3] Articles featuring photo spreads of Doda appeared in major publications, including *Life* and *Playboy*.[4] Her status as an icon of 1960s Californian pop culture is immortalized in Tom Wolfe's writing and in the Monkees' feature film *Head*, in which she played Sally Silicone (figure 8.2), a "busty

8.1 Carol Doda and Tosha McDonald in *Playboy*, April 1965, 74–75.

blonde bombshell" draped in a mink coat, sitting beside an out-of-shot boxing ring.[5] A much-loved fixture in San Francisco's LGBTQ+ communities, Doda was regularly featured in the *Bay Area Reporter* and starred in a number of theatrical productions. These included *The Rise and Fall of the World as Seen from a Sexual Position* (1972), which featured the gender-queer performance collective the Cockettes; and *Geese*, "a psyche-sensual musical about homosexual love," billed alongside the renowned drag performer Charles Pierce.

Doda understood silicone injections as a form of body work and body capital. In the mid-1960s, *Jet* reported that Doda's bust was insured for somewhere between $1 million and $1.5 million and that she eventually earned the equivalent of around $5,300 a week in today's currency.[6] In a 1965 interview, shortly after receiving the silicone shots, Doda commented, "I believe in self-improvement. If you don't make yourself better you might just as well be dormant."[7] In total, Doda invested $12,000—the current equivalent of around $120,000—on silicone injections, enlarging her bust from a 34B to a 44DD.[8] As a result, Doda gained publicity as the proud bearer of what one journalist described as "the New Twin Peaks of San Francisco," in

8.2 Carol Doda as Sally Silicone in *Head*, directed by Bob Rafelson (Columbia Pictures, 1968).

reference to the city's well-known geographical landmark.[9] Another journalist commented that Doda "thrust silicone into the forefront of modern medical breakthroughs."[10] This chapter follows silicone as a technology of shaping women's bodies across different actors, networks, and communities in the United States, while progressively exploring agency and access.

1964: Silicone Committee

The year 1964 was pivotal in the history of silicone, for it was when Carol Doda first shimmied topless across the Condor's white baby grand and celebrated the liberatory potential of self-authorship through silicone shots: "Now a girl can be as large as her dreams."[11] Doda hit US headlines from the West to the East Coast with her technocultural, pro-silicone stance— "Science has invented all these new wonderful things, why shouldn't we use them?"[12] From the late 1950s to mid-1960s, between twenty thousand and fifty thousand individuals in the United States received silicone shots for breast augmentation, as well as injections administered to the buttocks, legs, and face. A precise number is difficult to estimate due to the sometimes underground and illegal nature of the practice; however, the FDA in Washington, DC, began to take notice.[13]

Implantable silicone materials proved challenging for the FDA to categorize in concrete legal terms, as it depended on an interpretation closely intertwined with the viscosity or firmness of the material. Was the silicone material a liquid (i.e., a drug) or an object (i.e., a prosthetic)? For instance, Silastic, a type of silicone rubber used in artificial heart valves and other implants, was considered a medical device by the regulatory agency. Therefore, it could not be governed by the FDA, as there was limited legislation during this period regulating implantable medical devices, which meant that it fell outside the FDA's jurisdiction.

Silicone's different fluid materialities could slip through cracks in the FDA's regulations. For example, although Silastic was fluid upon injection, once inside the body, it had the power to transform instantly into a prosthetic. This slippery visceral material, both fluid and object, was opening up novel approaches to shaping the body. Not until the 1976 US Medical Device Regulation Act was the FDA granted federal jurisdiction over the regulation of medical devices, including implants. Silicone fluid, however, could be interpreted as halfway between a drug and a device.

In 1964, under mounting pressure from the FDA and increasingly negative press coverage of complications and deaths linked to silicone injections, Dow Corning registered 360 Medical Fluid as a drug with the FDA. Simona Morini, *Vogue* contributor and author of *Body Sculpture: Plastic Surgery from Head to Toe* (1971), wrote that once silicone entered the "FDA complex machinery, it turned into a national headache."[14] Morini and some of her contemporaries felt that silicone was probably safe, and that the FDA was an inefficient, authoritarian, and bureaucratic institution that acted overly cautious. Now, jammed in the cogs of the FDA's "complex machinery," silicone was no longer fluid in the way it had previously been used as an engine lubricant, propelling military and industrial efficiency and advancement. Instead, it was ensnared in the system, unable to formally advance into medical use at its previous unchecked speed. New drugs could not be marketed to the public until they had been proven safe and valuable, and allegedly, new drug applications "may well remain on FDA officials' desks for years."[15]

Silicone was a controversial material, and both popular and medical opinion was divided. A number of baby boomer whitestream journalists were questioning the FDA's "puritanical" stance, arguing that these authority figures were unjustly forbidding something perceived as beneficial for their generation and their rights to autonomy over their bodies, sexuality, appearance, and choices.[16] Some medical actors within major academic medical institutions who worked with Dow Corning voiced frustration at the FDA's lengthy review processes. They saw silicone injectables as a miracle material for congenital conditions and dermatological treatments, believing that, if used cautiously, the material could be applied successfully to the body.[17] Others, such as vocal opponent Charles Vinnik—a Las Vegas physician who had seen widescale complications from silicone injections among individuals working in the local sex work and entertainment industries—were intent on having it banned.[18] Meanwhile, as will be discussed later in this chapter, coverage of silicone in trans feminine publications documents a nuanced approach that often centered community-building and harm reduction.

In 1964, in accordance with FDA protocol, Dow Corning set up an investigation committee, known as the Silicone Committee. The body was headed by Silas Braley, director of the Dow Corning Center for Aid to Medical Research, who was described in *Vogue* as "affable and easy-going but adamant when it comes to keeping track of every pint of the fluid, which is shipped only to an investigating committee of eight physicians— seven plastic surgeons and one dermatologist."[19] The committee notably included Dicran Goulian Jr., who—with Herbert Conway—had already been working with Dow Corning on silicone injectables for breast augmentation; Milton Edgerton, an established plastic surgeon based at Johns Hopkins and later a cofounder of the Johns Hopkins Gender Identity Clinic; Joseph E. Murray, a Harvard-based plastic surgeon; and Norman Orentreich, a high-end dermatologist and professor of dermatology at the New York University Medical Center.

Silicone was in high demand for medical applications. Since its founding in 1959, the DCCAMR had been inundated with requests for silicones. By 1962, Dow Corning claimed it was proving increasingly challenging to locate manufacturers equipped to efficiently supply medical-grade silicone items "properly," so it set up the Medical Products Division to meet the demand. In 1963, it expanded production and opened a medical products plant.[20] Braley claimed that between 1959 and 1973, Dow Corning corresponded with around thirty-five thousand physicians and medical researchers from around the world, each of whom was hoping that silicones could provide the solution to a medical problem.[21]

Since Dow Corning Medical Products Division's inception, the company claimed that it had grown to become "the world's largest manufacturer of implantable soft tissue substitutes." Dow Corning acted as gatekeepers to medical-grade silicone. Braley described the company's selection process: "If the solution was satisfactory and if the device seemed to satisfy a need, it was turned over to the marketing people for evaluation. If the economics of the device permitted, it was produced for the physician to use."[22] Braley's comments highlight the lack of formal testing rules before the Medical Device Regulation Act of 1976.

The official story of silicone access after 1964 was that only the eight members of the Silicone Committee had access to medical-grade silicone; this access was strictly monitored, and every pint of fluid was required to be traceable and accounted for.[23] The limited number of committee members with access to the fluid stands in stark contrast to the thirty-five thousand surgeons that the DCCAMR claims to have corresponded with. Doctors who requested the fluid were required to sign an affidavit, legally agreeing that silicone would not be injected into humans and detailing their proposed application of the fluid. However, there were many loopholes that were exploited; tracking medical-grade silicone fluid was particularly challenging due to its slippery nature and other factors. For example, samples often went "missing" or were "undelivered."[24]

Dow Corning appears to have had difficulty both collating data as well as monitoring and tracking samples. This also applies to its record-keeping of studies with silicone injections, which the FDA ruled was shoddy, as well as its documentation of who was given access to silicone in the first place. Noncommittee doctors circumvented the access restrictions by claiming to carry out research on nonhuman subjects, such as race horses, to ensure their continued ability to use silicones for illegal human applications.[25] In 1964, after silicone had been classified as a new drug, supplies of the product were seized en route to the osteopath and known silicone experimenter Harry Kagan, as well as other doctors.[26] This resulted in a federal grand jury indictment in 1967 against Dow Corning for shipping an unapproved drug via interstate commerce. The indictment named various executives, including Dow Corning President Shailer Bass. The company claimed it was unaware that doctors planned to use the fluid for breast injections. Finally, in 1971, Dow Corning pleaded no contest to the charges and paid a $5,000 fine.[27] Despite restrictions and legal battles, silicone continued to circulate in bodies and the media.

Injecting Silicone into Beauty Salon Culture

Silicone injections were offered as a quick beauty salon service by providers both within and outside established medical structures. But not all beauty salon services were the same, and access was shaped by social and economic inequities. Some providers advertised their services in newspapers' classified sections and the yellow pages, and profiles of others were featured in articles on the pages of glossy magazines including *Vogue, Harper's Bazaar*, and *Esquire*.[28] Contrary to much of the coverage surrounding Doda and her peers, silicone injections were not reserved exclusively for unlicensed providers serving sex workers primarily on the West Coast.

Silicone injections also circulated in more privileged spaces dedicated to beauty culture and body work. Silicone Committee member Norman Orentreich, a dermatologist who ran a cosmetic clinic in New York's affluent Upper East Side, served a clientele of rich and powerful celebrities and socialites, allegedly including Nancy Reagan.[29] He claimed to have first encountered silicone in 1954. A vocal supporter of silicone, Orentreich stated that silicone was "immediately needed" for cosmetic applications. Prior to his work with Dow Corning, according to Orentreich, he ordered industrial-grade silicone from General Electric and transformed it into purified, sterilized medical-grade fluid in his own laboratory.[30]

In theory, silicone injections offered the possibility of an instant beauty fix. Orentreich, who used silicone for over thirty-five years and claimed to have treated more than a hundred thousand patients, maintained his belief that silicone was harmless if injected correctly over prolonged periods at microdoses of less than one-fiftieth of a milligram, using his own specially designed syringes.[31] In Tokyo, plastic surgeon Rin Sakurai also offered silicone shots to counteract signs of aging and as a bust enhancer in a beauty salon setting. Similarly, Los Angeles–based, self-professed "Beauty Surgeon" Robert Alan Franklyn referred to silicone as "injection therapy [for] instant beauty," describing this treatment as "beauty surgery [without the] surgery." He assured prospective clients, "No incision is made, no skin is clipped away, no stitches are taken, no bandages are used, and there is no operating room or hospitalization."[32]

Orentreich believed that any issues with silicone related to the use of "impure silicone" and "improper technique." He attacked Japanese silicone formulas, such as Sakurai's, for their inclusion of additives intended to prevent silicone migration in the body.[33] Internal Silicone Committee communications reveal there was a concern about Orentreich's outspoken enthusiasm for silicone injections and his lack of transparency about the number of clients he was treating.[34] As the only dermatologist on the committee, his work was more directly linked to the commercial cosmetic realm. In the late 1970s, the FDA investigated Orentreich when he refused to limit his use of injectable silicone to congenital facial anomalies, as the amended protocol specified.[35]

In 1976, the Dow Corning new drug application review stated that Orentreich had treated several hundred patients with liquid silicone for cosmetic purposes but omitted these from the research records submitted to the FDA. He was eventually dropped from the investigation in 1978, but his clinic continued to be successful. Coverage of silicone shots administered in an exclusive dermatological setting, such as that of Orentreich's clinic, associated the material and its injection into the body with luxurious cosmetic properties, similar to those seen in promotional material for beauty products like Estée Lauder's exclusive Re-Nutriv face cream (see chapter 6).

While their medical backgrounds and qualifications differed, much-publicized medical figures associated with the beauty industry, such as Orentreich, Sakurai, and Franklyn, offered silicone injections to wealthy consumers and celebrities as a way of counteracting signs of aging and enhancing physical appearance. Coverage of their work arguably served to legitimize silicone in the eyes of the American public and prospective users as a reliable method of cosmetic body contouring. These associations could also explain why people were eager to pursue this treatment in less expensive and more accessible alternative economies.

West Coast US Strip Culture and the "Topless Craze"

The association of silicone with a glamorous and exclusive material administered by charismatic doctors was absent from US whitestream reports on West Coast sex workers. In *Mondo Topless* (1966), the American film director Russ Meyer, inspired by Carol Doda's notoriety, presented a midnight movie snapshot of San Francisco, considered the home of the topless phenomenon, its go-go dancing nightclub culture, and the dancers themselves. Actor John Furlong breathlessly narrates, over a montage of neon signs of San Francisco nightlife:

> Exploding from dusk to dawn with the way-out craze of the topless, fawned and nurtured by staid and stolid San Francisco and cut loose to rampage across the USA and even Europe. National publications such as *Life*, *Playboy*, and *Esquire* have documented the topless—the phrase and the craze that is changing the mood and the mores of people everywhere. *Mondo Topless* captures the basic quintessence of the movement, with movement: way-out, wild movement! There, go-go girls in and out of their environment will be revealed to you in scenes that can only be summarized as a swinging tribute to unrestrained female anatomy.[36]

Furlong's sensationalist text conveys the excitement of what became known as the "topless craze." By the mid-1960s, large scale social activism and subcultural youth movements in the United States challenged what the previous generation thought to be acceptable gender norms and sexual expression. Peggy Moffitt, a young white American woman, scandalously modeled California-based designer Rudi Gernreich's monokini topless on the June 4, 1964, *Women's Wear Daily* cover. The revolutionary design also featured in African American publications. *Ebony* magazine noted, "Back in 1945 B.B. (Before the Bikini), even the most daring of women would have considered the topless bathing suit a sartorial impossibility," and included an image of a Black model in Gernreich's design (figure 8.3).[37]

On June 22, 1964, Carol Doda was credited with being the first woman in the United States to perform in Gernreich's design, further challenging local and national censorship laws. Doda's legendary monokini performance in San Francisco, her subsequent prosecution and acquittal, and the press coverage these events generated had a domino effect on other striptease venues in the city and throughout the West Coast. By 1966, approximately fifty topless restaurants, nightclubs, and bars had opened in San Francisco. In Los Angeles, three hundred locations, ranging from beer joints to private clubs, followed suit. "In Seattle a topless dancer [shuttled] between two restaurants in a chauffeur-driven limousine," and in Las Vegas, "where *oo-la-la* Parisian revues" had flourished since 1958, the "Topless Watusi" hit venues.[38]

In the summer of 1964, this "topless craze" started when feminine dancers, especially go-go dancers, began performing topless throughout

8.3 Monokini featured in *Ebony*, November 1965, 215.

Topless wonder of 1964 left little to the imagination, as beachwear approached nudity. Rudi Gernreich creation made headlines across the country, but public officials hastened to ban it. Revealing bathing suit often made even models blush.

their routines, doing "swim-inspired" movements (as seen performed by Doda and McDonald in figure 8.1) such as the Watusi, the Monkey, the Jerk, and the Swim. The Watusi, the dance most closely associated with Doda's act, was named after the Tutsi, an ethnic group of the African Great Lakes region known for their dances. In 1962, the Orlons, a Black quartet from Philadelphia apparently named after a synthetic fiber, hit the charts with their eponymous Tutsi-inspired track, "The Wah-Watusi." Topless performances of dance crazes like the Watusi were often accompanied by live commentary on the history and cultural context to "redeem social importance"—a technical loophole in legal definitions of obscenity that was often used at the time to defend pornography and nudity.[39] No longer restricted by fans, nipple tassels, or bras required for striptease, these new dances conveyed a freedom of movement that coincided with changes in body-contouring practices, particularly breast enlargement.

A COUPLE OF OUR
CHEMICAL PEDDLERS

She doesn't mean V for Victory.
She gets off at two o'clock, that's all.

Another Saaci navel engagement.

"There's a party after the show but
you have to go dressed."

"I get all my clothes one size too
small."

BILL & VINCE

FINE CHEMICALS   INTERMEDIATES

BOFORS INDUSTRIES, INC.
1075 EDWARD STREET. LINDEN, N.J. 07036

8.4 "SAACI at Play" spread in the *Chemical Peddler,* published by the Salesmen's Association of the American Chemical Industry, 1972. Science History Institute, Philadelphia.

After the success of topless performances in restaurants and bars, topless-themed country clubs, resorts, shoeshine parlors, ice cream stands, and girl bands like the Ladybirds followed soon afterward.[40] Corporate clients like Smirnoff hired Doda and dancers like her for promotional functions.[41] Members of professional organizations such as the Salesmen's Association of the American Chemical Industry Inc. participated in homosocial corporate culture, visiting topless clubs and resorts together and sharing photographs of these excursions in their trade publication, the *Chemical Peddler* (figure 8.4).

The topless craze and its economies spanned a wide range of social groups and individuals. In San Francisco, topless culture permeated the city beyond infamous North Beach strip venues like the Condor. The Cellar, in upscale Nob Hill, was known as San Francisco's "gourmet topless restaurant" and "catered to local button-down-business-lunch trade rather than to tourists," with clientele that included members of the Junior

Chamber of Commerce.[42] North Beach's famous topless clubs had discriminatory hiring practices, as Black women weren't employed as topless dancers on the main strip but were instead hired elsewhere in the city as topless waitresses.[43] The Cellar's Black topless waitresses included Paulette Hefner, whom *Jet* reported was arrested in multiple topless raids, and Arline Marie Resnick, a law school student, who earned between $150 to $300 a week in tips for the lunch shift.[44]

Unlike trans feminine and queer publications, articles on the San Francisco scene in whitestream publications such as *Life* and *Playboy* did not mention trans feminine topless dancers such as Vicki Starr, a Puerto Rican transsexual woman, who performed at El Cid, and her friend and fellow performer Roxanne Lorraine Alegria, a transsexual woman of Mexican descent.[45] This omission of trans feminine individuals from the historiography continues to this day, with these dancers notably excluded from a recent documentary on Carol Doda.[46] Artist-scholar Gigi Otálvaro-Hormillosa provides an alternative counterhistory, stressing Alegria's importance as "a legend in the Bay Area trans community, not only as a topless performer, but also as an activist educator."[47] The majority of white female impersonators performed on stage at clubs like Finocchio's in North Beach, while "trans women of color survived and worked on the streets of the Tenderloin," a central San Francisco neighborhood.[48] Topless performers Starr and Alegria were connected to both of these areas and communities.

Topless venues became novelty must-sees for San Francisco tourists, with an audience described as consisting of "well-dressed couples, college groups, political heavyweights, and regular folks"; along with celebrities, including Frank Sinatra and Sammy Davis Jr.[49] A local reporter, reflecting on changing fashions, remarked, "Are we ready for girls in topless gowns? Heck, we may not even notice them."[50] A known figure, Doda was banned from some local establishments, including cafés, whose staff told her, "We don't allow topless here," Doda recalled. "But I wasn't *topless*. I mean I had a regular dress on, what I wear all the time."[51]

San Francisco's topless dancers were presented by some whitestream publications as embodying the sexual liberation the city had become known for, and their ongoing arrests generated much publicity. In 1964, a racist, anti–civil rights Republican campaign movie even included footage of a topless, light-skinned blonde woman dressed in Gernreich's monokini, alongside other white women, dancing to jazz trumpeter Dizzy Gillespie's music as part of a narrative of "moral decay" that framed increases in pornography, illegitimate births, and student activism as threats to America's social order.[52] For those opposed to changes in racialized gender norms, the message was clear: The body politic demanded discipline. By the mid-1960s, when US involvement in the Vietnam War increased, San Francisco's commercial sexual activity was subject to increased surveillance.[53] Military and civilian police, as well as public health officials, wanted to prevent the spread of sexually transmitted infections among troops. Consequently, the port city's cis and trans sex worker communities were subject to more frequent raids and arrests.

On one night in April 1965, twenty-seven arrests were made in North Beach, including Doda, other topless dancers, and club owners. The next day, Yvonne D'Angers, an Iranian performer known as "The Persian Lamb," was also arrested for wearing a topless dress during a topless fashion show at the Broadway, a venue across the street from the Condor.[54] In response, over one hundred protesters gathered outside a San Francisco police station, demanding the release of all the dancers, as well as Mario Savio, a free speech student activist already in prison.[55] Both Doda and D'Angers were quickly released on bail paid by the clubs they worked for.

In the 1960s, topless dancing and free speech were associated: Doda, who attended student protests, was known to have openly commented on the liberating feeling of not wearing a bra on stage.[56] In *Mondo Topless*, a dancer commented on the freedom of not wearing anything on top: "You don't need to bother with pasties anymore; sometimes they slip and then the glue shows, sometimes they fall off, and that's terrible [*laughs*]."[57] A judge ruled in favor of the North Beach topless cases, noting that performances were "in the nature of a theatrical production and thus were protected by Constitutional guarantees of free speech and assembly."[58]

In 1966, after returning to San Francisco from a less successful stint in Las Vegas, where the art of the tease still reigned supreme, Doda opted to further increase her bust with silicone injections to a size of forty-four inches as a way of ensuring sustained interest in her act.[59] She was now joined by Tura Satana, a Japanese American dancer who performed a topless nipple tassel-twirling act and would later star in Russ Meyer's cult movie *Faster, Pussycat! Kill! Kill!* (1965). Doda noted that she was unable to perform in this way: "[My breasts] don't move anymore . . . I just keep getting shots."[60]

Silicone was an important factor in topless West Coast entertainment culture. Big Al's, a topless entertainment venue and the Condor's competing next-door neighbor, boasted an in-house doctor who administered weekly, one-and-a-half-ounce liquid silicone shots to the breast.[61] This venue was also home to "Japanese Doll" Tosha McDonald, known as "The Glo Girl" (figure 8.1), at one point reportedly North Beach's only Asian dancer.[62] McDonald performed on stage among glowing neon lights in a topless and bottomless sequined swimsuit, with her buttocks exposed and a boa constrictor as part of her shows.[63] As the popularity of topless shows grew, Doda commented, "I don't believe topless is a fad. It's something that's going to stay—like burlesque."[64] Doda was subsequently proven right, as appearing on stage topless became the norm in Western strip culture. While other venues that opened during the "topless craze," such as Big Al's, have since closed down, the Condor Club and its iconic white hydraulic piano remain in operation to this day.

Fluid Complications

In theory, silicone shots gave individuals the agency to shape their bodies as they desired. In practice, this was often not the case, as many suffered complications and, in some cases, died. Reports and studies on silicone injections tended to focus on the West Coast and its hubs of topless entertainment. In 1965, the FDA, which publicly opposed silicone shots, estimated that seventy-five doctors offered this service in the Los Angeles area.[65] One surgeon claimed to treat twenty-five patients a week at a rate of $1,000 per treatment course.[66] In Beverly Hills, the treatment was known as "Cleopatra's Needle," a term the cosmetic surgeon Robert Alan Franklyn claimed to have introduced.[67]

In 1968, a medical journal published "Silicone Mastitis in 'Topless' Waitresses and Some Other Varieties of Foreign-Body Mastitis," a study of patients who had received silicone breast injections and experienced complications.[68] The article described the patients' medical conditions and complications resulting from the procedure, as well as their line of work. Some "worked in 'topless restaurants' [in the United States] and had been obliged to seek artificial means to maintain the excessively large bust that was necessary in that environment."[69] This included upkeep in the form of top-up injections, even when the individuals had already experienced granulomas. A teenage student who worked as a "get-together girl" in a topless country club had undertaken a two-year course of silicone injections to increase her bust from thirty-four inches to thirty-seven inches. Her first series of injections, spread across three weeks, entailed about 400 mL of DC-360 Medical, "an ethical proprietary brand of pure silicone fluid."[70] Spreading the treatment over multiple weeks, months, or even years was common, as it was believed that microdoses of silicone would be better tolerated by the body. It also enabled those seeking treatment to spread the cost over time.

Silicone body work, particularly in the form of breast injections, was often equated with sex work. As with the decision to wear synthetic foam padding or pursue foam implants, some individuals who underwent this procedure were responding and contributing to socioeconomic changes and pressures. However, silicone's fluid materiality perhaps appeared less invasive and more immediate. A journalist reasoned that women were likely to take the risk "because compared to breast implants, silicone is fast," as it did not require a hospitalization. Some individuals who injected silicone allegedly offered clients a faster turnaround: "[They could be] pumped up in the afternoon to become a perfect 36 and maybe get a job the next day."[71] Ruth Ponce, a light-skinned, recently divorced Las Vegas dancer with a young child, said she was not told she *had* to get injections but saw it as an economic investment. "I had to eat," she said. "I could get a job as a secretary for $90 a week, only you can't live in Vegas on $90 a week. You can live on $240 a week, and that's what I was promised for chorus work at the Sahara."[72] Silicone shots offered seemingly instant social mobility and financial security, but they demanded upkeep and could be hazardous to one's health.

Often more shots were required to maintain the desired volume as silicone tended to migrate and scar tissue would shrink. Adverse effects included granuloma formation; scarring, hardening, and discoloration of breasts; infections; gangrene, which required amputation; and, in some cases, death from silicone poisoning when silicone entered the bloodstream or lungs.[73] Silicone also complicated the detection of breast cancer.

Judy Mamou, who performed topless in North Beach at the same time as Carol Doda, recently shared that she received multiple injections from Vincent Spano, who was known as the famous silicone doctor of San Francisco and whom Mamou claims also injected Doda.[74] Mamou said she didn't have any issues with the silicone until she began breastfeeding, at which point she explained that the increased milk production led to circulation issues and ultimately gangrene in the breast. As a result, she had to have five surgeries to remove the silicone injections.

It was common practice for patients to be required to sign a statement acknowledging that the treatment was experimental in order to protect physicians and institutions from liability.[75] To mitigate further complications from silicone injections and remove the material from the body, a variety of medical interventions were performed. These included medical procedures such as mastectomies, attempts to remove the silicone injections via syringe, and replacing prior silicone shots with silicone implants.[76]

Health risks remained a federal concern. In August 1971, the AMA announced that complications from silicone injections to increase breast size "had reached such a volume that it was time to sound an alarm," and the FDA agreed.[77] The FDA supplied cases where autopsy reports showed that silicone had entered the bloodstream and traveled to the brain or lungs. One of these deaths was caused by an individual pretending to be a physician, who was later charged with murder by malpractice in Houston. An article in *Today's Health*, the AMA magazine, concluded "silicone became a national outlaw."[78] The AMA urged physicians to explain to inquirers that breast injections were illegal and dangerous, as the organization felt silicone was unsafe because of its propensity to migrate in the body. There was also concern that silicone could mask tumors because the material is opaque under X-ray imaging and could obscure the correct identification of pathology.

Silicone Committee correspondence in Murray's papers reveals that, from the early to mid-1970s, Dow Corning was increasingly worried about the number of deaths linked to silicone.[79] During this period, Dow Corning perhaps started to regret its self-proclaimed position as "First in Silicones." At a Silicone Committee meeting, Dow Corning staff warned that they had "a special risk which has been generated by the [DCCAMR] publicity."[80] It was feared that Dow Corning could be sued, whereas other companies might not be, due to their pervasive marketing to the public. To counteract the negative publicity generated by the 1971 interstate commerce indictment, deaths, and tales of illegal silicone rings, Braley and Bob Emmons, a Dow Corning public relations expert, toured the country to promote the benefits of silicone to reporters and editors. Emmons recalled, "Si [Braley] was an articulate, clever and witty individual. Si and I came up with a program to get the heat off of us. We went to New York and called on the ladies' magazines with press releases and other materials to get better publicity."[81] This tactic certainly seemed to work, as evidenced by Morini's effusive *Vogue* article.[82]

Journalists and medical practitioners reported that silicone injections were sought by women of different ages, from teenagers who described feeling inadequate to women complaining of flat chests and sagging breasts after childbirth. Most women treated by one Anaheim doctor were housewives and college girls, with even some anxious parents bringing their teenage children for treatment.[83] In Las Vegas, one doctor estimated that he had given some sixteen thousand silicone injections to two hundred women, about half of whom he claimed were housewives rather than showgirls.[84] Twelve thousand individuals in the area had reportedly received silicone breast injections between 1962 and 1976, and approximately 120 sought medical attention due to complications.[85]

At an average of $200 per silicone shot appointment, the cost of treatment was prohibitively expensive for many.[86] The cost of cosmetic surgery and bust augmentation in the United States made it inaccessible to most people of color in the United States during this period.[87] Furthermore, in *Ebony*, Michele Burgen argued, "Among many blacks (and to an extent, among Orientals), there is the feeling that altering the nose, lips or other facial features is a betrayal of one's racial heritage."[88] Moreover, as Burgen noted, there were not many Black plastic surgeons working at the time. Burgen, however, welcomed a new age, enthusing, "[Cosmetic surgery] just gives you a new way to look. It's nothing to be ashamed of. It's no different than losing weight, deciding you want to change the kinds of clothes you wear, changing your walk, or taking a new career. It's just shaping the house you live in—and the house you live in is your body."[89] Burgen's affirmative comments echo Doda's beliefs about the importance of being able to enact agency on the shaping of your own body. Technologies that were once unattainable for most people of color in the United States and had therefore received little attention in African American publications during the 1950s and 1960s were becoming comparatively more accessible by the late 1970s, leading to more coverage.[90]

In 1964, when the FDA turned its attention to silicone shots, a new type of silicone technology to permanently add curves was emerging. Inspired by the limitations and issues of foam implants, Frank Gerow and Thomas Cronin of Baylor University collaborated with the DCCAMR on the design of what was later known as the Cronin-Gerow silicone gel mammary implant. Surrounded in the hospital by new plastic bags for storing blood, which replaced sterilized bottles in the medical industry, Gerow envisioned a return to an implantable "falsie" style design—only this time, the object would be less structured: a silicone bag filled with fluid.[91] Profiting from the controversial coverage silicone shots were receiving, silicone implants would be publicized by the DCCAMR and surgeons alike as a safe and legitimate solution to shape the body. Since silicone implants could be prefabricated and mass-produced, they offered greater profits and greater control.[92] In some cases, patients whose breasts had been surgically removed after complications from silicone shots received Cronin-Gerow gel implants.[93]

Implants offered a very different solution to prospective clients, though not all could access or afford them. Las Vegas–based dancer Ruth Ponce was originally offered gel implants by a cosmetic surgeon, but she decided against them, citing economic reasons: "It would have meant six weeks out of work, along with a hospital bill, a bill from the surgeon and money for baby-sitters."[94] Unlike silicone injection treatment that required a comparatively smaller payment with each visit, implants demanded a higher upfront fee and longer recovery times. In addition to the costs associated with an in-patient procedure such as implantation of a silicone device, it was also an entirely different procedure altogether.

Prospective patients such as Ponce would be presented with a sizing device such as the Dow Corning Silastic mammary sizer demonstration kit (figure 8.5). The accompanying booklet for surgeons explained, "A seamless, thinwalled, contoured silicone sizer enveloping a translucent silicone gel which closely resembles the characteristics of the SILASTIC (R) Mammary Prosthesis Seamless Design."[95]

It appears the patient was presented not with an actual implant but instead with a sizer device designed exclusively as a demonstration piece to "assist the surgeon in office patient counselling."[96] The text in Dow Corning's booklet reinforces medical paternalism and gatekeeping, saying that the device "allows the surgeon to accurately decide in the office the appropriate size for patient requirements."[97] The sizer was also designed for "size demonstration or size determination in the operating room" to "provide assurance of correct size selection," again emphasizing the surgeon's agency in selecting the "correct" design for the patient's body. When the Cronin-Gerow implant launched, it was only available in three sizes—small, medium, and large. However, shortly after, Dow Corning developed another size to meet demand, which the company referred to as "larger than large."[98]

By defining implants as devices rather than drugs, Dow Corning marketers and designers ensured they avoided FDA regulation in the twelve years preceding the Medical Device Regulation Act of 1976. During this period, Dow Corning executives repeatedly distanced themselves from silicone breast injections with statements like "Dow Corning has never sold silicone fluids for such purposes and we have never promoted, recommended,

8.5 Dow Corning Silastic mammary sizer seamless design demonstration kit, 1970. Science History Institute, Philadelphia.

suggested or sponsored such usage. We do not consider it acceptable practice and we are prepared to cooperate in any way which would prohibit and eliminate such a practice."[99] In ongoing legal proceedings, the company stressed, "the medically acceptable means for enlarging the breast is to surgically implant a silicone device."[100] Dow Corning also continued to stress that, "for purposes of edification, the injection of liquid silicone originated in Japan and eventually found its way to this country."[101] Silicone shots were strategically framed as something "bad" and disapproved of by official medical structures, which enabled silicone implants to thrive as a "safe" alternative.

Fluid was now contained in a manufactured object; the implants required surgery and were therefore, in comparison with injections, more easily regulated and controlled within formal medical structures. Silicone implants appeared to offer a solution for cosmetic breast enhancement that could not easily be commercialized illegally. Silicone's viscous materiality was now confined to a sac, competing with the earlier free-flowing injection into the body, and this marked its new containment within established structures and the discriminatory medical industrial cisheteropatriarchal surveillance norms that this often served.

Silicone in Trans Feminine Communities

Trans feminine and gender-nonconforming people also had silicone injections. However, they are largely omitted from the articles on body contouring discussed in *Ebony*, *Harper's Bazaar*, and *Vogue*, as well as histories of silicone use in cosmetic surgery. US whitestream coverage on silicone shots from the mid-1960s focused on light-skinned cisgender women dancers, such as Doda and many others, often accompanying articles with full-body shots that showed off their prominent bustlines. In whitestream publications trans women—particularly trans women of color—were given minimal or no mention in this context. To gain acceptance in this press, white trans women needed to present as heterosexual, married homemakers who denounced homosexuality.[102] In these publications, trans women who made a living as exotic dancers and in other forms of sex work received negative publicity, if they received any at all. Outside of some trans feminine publications, positive coverage of Black trans women was almost exclusively found in African American publications including *Sepia*, *Ebony*, and *Jet*. Trans women were also rarely mentioned or explicitly identified in medical reports on silicone shots.[103]

Silicone injections largely circulated outside formal medical structures compared to implantable objects, which aided their ability to defy medical heteropatriarchal structures of surveillance. Oral histories from trans women attest to the prevalence of silicone injections within trans feminine communities in the late 1960s and beyond.[104] Dorian Corey, a Black trans woman and New York City–based house mother who was featured in the 1990 film *Paris Is Burning*, had silicone injections in the late 1960s. She later described her experience of visiting a "doctor in Yonkers who was doing silicone [injections]": "I should have gotten [a silicone implant] . . . but that called for one chunk of money, and with the injections, for thirty-five dollars you could get a double shot, and then when you had some more money you could go back and get some more. 'Til you were happy."[105]

Getting access to surgeries or hormones through official medical institutions outside major metropolitan areas remained nearly impossible for trans individuals until well into the 1970s.[106] Milton Edgerton, a member of Dow Corning's Silicone Committee who was based at Johns Hopkins Hospital at the time, specialized in the relationship between plastic surgery and psychology.[107] He was also one of the founding members of the Gender Identity Clinic, established at Johns Hopkins Hospital in 1966 by psychologist John Money. The institution claims to have performed the first "sex change" surgery in the United States. However, its opening followed decades of research involving trans and intersex people.[108] The facility was a central hub with in-house specialists, such as psychiatrists, plastic surgeons, pediatricians, and urologists, who alleviated the need for countless referrals. By the end of the 1970s, there were between fifteen and twenty major gender identity clinic centers in the United States.[109]

Most patients at the Gender Identity Clinic were already living in their chosen gender identities, and many had already undertaken cosmetic procedures such as breast augmentation. This was reflected in the admission forms for new patients:

Have you had any operations or procedures to help change your appearance?

Y / N Date
Breast surgery
Nose surgery
"Adam's apple" surgery
Silicone injections—place of injection

Are you presently receiving hormone pills or injections?

Pills Y/N
Injections Y/N.[110]

This form shows that cosmetic surgery was not uncommon among the clinic's prospective patients. "Silicone injections—place of injection" indicates that shots were applied in a variety of different areas of the body. Despite their prevalence, there was a lack of medical studies investigating the impact of these experimental technologies and their risks for trans feminine individuals. Some archival records suggest that mammaplasty was to be carried out at Johns Hopkins.[111] If that were the case, it is worth noting that Edgerton, a member of the Silicone Committee, would have had direct access to silicone and Dow Corning products at the time.

The mid-1960s to the late 1970s represent what Stryker refers to as the "Big Science" period of trans history, when changes in science and medicine affected how sex and gender presentation could be altered.[112] This period also notably coincides with increased professionalization and developments within cosmetic surgery, where the body was increasingly becoming a site that could be permanently altered. During this time, there was a rise in the number of academic medical institutions that offered gender-affirming procedures. Prior to these services becoming more available in the United States, trans people had to travel abroad to access them. However, trans people who sought out surgery and hormones at these academic medical institutions found that such programs often were more invested in "restabilizing the gender system, which seemed to be mutating around them in bizarre and threatening directions," than in "further exploding mandatory relationships between sexed embodiment, psychological gender identity, and social gender role."[113] Gatekeepers granted access to trans health care services only if individuals adhered to established binary concepts of gender and sexual attraction.

For most trans feminine individuals, gender identity clinic programs and the access they provided to gender-affirming care, including breast implants, remained out of reach. Most trans people had very different experiences and pursued DIY forms of transition.[114] This may have occurred due to limited access to gender-affirming care within official medical systems. Additionally, through these official structures, there was a fear of possibly being labeled as having a mental health diagnosis that could ultimately lead to harmful corrective therapies and institutionalization.[115] These barriers, as well as the desire for bodily autonomy and agency, are reflected in the political message of STAR (Street Transvestite Action Revolutionaries), an organization established by Sylvia Rivera and Marsha P. Johnson, both street queens, that provided shelter and support to Black and Brown street youth. Johnson and Rivera called for "the right to self-determination over the use of our bodies: the right to be gay, anytime, any place; the right to free physiological change and modification of sex on demand; the right to free dress and adornment . . . the end to all exploitative practices of doctors and psychiatrists who work in the field of transvestitism."[116]

Silicone injections were prevalent within some trans feminine communities. In 1975, Cook County State's Attorney Bernard Carey investigated what was reputed to be one of the biggest illegal silicone injection rings in the country.[117] Based in Chicago, the leader of the ring, Hal J. Ellison, had been injecting silicone from 1967 to 1975. On December 5, 1974, he injected Tammy White, a transsexual woman, and she died within an hour at Ellison's apartment.[118] The majority of Ellison's customers were queer, trans feminine, and/or sex workers, according to Assistant State's Attorney Nicholas

Lavarone, who carried out a six-month investigation in the Special Prosecution Unit.[119] In 1971, silicone injections had been completely outlawed by the FDA. Ellison, who had no formal medical qualifications or experience, had easy access to industrial-grade silicone via his industrial chemicals wholesale business, Silico Chemical Company. Apparently, he had the idea after his wife alerted him to the silicone work that she had done, and he was attracted to the potential profitability of such a business.[120] Prospective clients were screened by a beautician working at a North Side Chicago salon.

An administrative assistant at the silicone manufacturer Dalcorn Inc. testified at Ellison's trial, stating that the company had sold various quantities of silicone to him between 1970 and 1973. Ellison, like all Dalcorn's purchasers, was required to sign an affidavit that the silicone would not be used for human injection.[121] However, newspaper reports claimed that Ellison sourced his silicone from Dow Corning. Carey pushed for stricter consequences for the illegal and dangerous practice of unlicensed medicine, advocating for it to be treated as a felony, punishable by penalties such as prison sentences. Illegal silicone injectors such as Ellison profited from popular press coverage that presented silicone as a miraculous inert material in science and medicine.

According to the investigators, Ellison charged $200 per shot administered and averaged eight injections per customer. Ellison also injected silicone into other parts of the body according to his clients' requests. The material cost Ellison about ten cents per shot, and his profit was allegedly over $100,000 annually from his silicone injection business.[122] The court records and transcripts held at Cook County Circuit Court Archives offer rare descriptions of illegal silicone injections. On the evening of her death, White met with April Vaugine, a sixty-year-old transsexual woman originally from Los Angeles, who had been working as one of Ellison's go-betweens and had arranged silicone injections on his behalf since 1967. Vaugine, who had received approximately sixty total silicone injections from Ellison, described his usual procedure: "The person who was to receive the silicone injection would indicate the part of the body where silicone injection was desired. The person would then disrobe, [Ellison] would clean the skin with alcohol, mark off the area with an eyebrow pencil, and inject an anesthetic with a hypodermic needle."[123] Clients pointed to parts of the body they wanted enhanced, and Ellison would inject the fluid there. Ellison was known to use the same syringe for anesthetic and silicone, as well as for subsequent patients.[124]

White had seen another local individual for silicone injections before her appointment with Ellison, and he remarked that "whoever had done the prior silicone injection had done a poor job."[125] That night, White had multiple silicone shots to her bust and hips. She immediately felt nauseous and died within an hour of being injected by Ellison. The coroner's report noted "extremely large amounts of silicone in [White's] breast tissue," as well as vacuoles in the lungs caused by silicone particles.[126] Ellison was found guilty of manslaughter and sentenced to one to three years in prison. Due to the limited options available, trans feminine and gender-nonconforming people often had to rely on underground economies for hormones and cosmetic procedures for their survival.

Individuals within trans feminine communities actively circulated information on the dangers of silicone shots. A variety of publications increasingly warned of the complications and dangers linked to silicone injections and worked to fight misinformation.[127] In 1976, *Drag* published two articles detailing the complications associated with silicone injections: "Silicone Shots Busted Wide Open" and "Deaths Tied to Silicone Breast Shots."[128] Both warned of fatalities and noted that, for some individuals, complications were only beginning to emerge a decade after their initial treatments in the 1960s. *Drag* cited key structures and figures that opposed silicone injectables, such as the FDA and Charles Vinnik, both of whom were working to prevent the legalization of silicone injectables for body contouring. The publication noted that while silicone implants were legal, silicone injections, especially the kind that used industrial-grade silicone, were not. Here, *Drag* directly engaged with medical publications and the FDA silicone case to relay the most important facts on the dangers of silicone shots to their readers.

Trans feminine publications were vital resources for community-building, offering nuanced coverage of silicone injections. Readers were keen to connect with others and share their lived experiences, often writing in to do so. A personals listing stated, "Interested in silicone before and after."[129] Another publication described a theater production worker who had silicone shots done and moved across genders throughout the year, switching from looser-fitting menswear suits for professional wear to tighter-fitting womenswear for private and informal occasions.[130] Within these publications, Carol Doda is also referenced as having made silicone shots famous.[131] In the early 1970s, silicone busts even received humorous mentions, such as a drag ball advertisement that exclaimed, "More cleavage than you'd see at a silicone convention."[132]

Drag repeatedly wrote about silicone. Lee Greer Brewster, a white, self-identified drag queen, established the publication in 1971. Silicone's frequent mention in *Drag* is particularly noteworthy, as the publication was closely intertwined with organizing and activism within trans feminine communities. Brewster, who cofounded the Queens Liberation Front in 1969, was inspired by militant street activism led by street queens after Stonewall.[133] Brewster had been involved in the Mattachine Society, organizing a successful drag ball fundraiser that saved the organization from significant financial issues. Despite his assistance, members of the society's leadership felt that drag was "merely camp" and were, as Brewster remarked, "more than willing to sacrifice the drag in the interest of appeasing the straight."[134] The night he founded the QLF, he also established Lee's Mardi Gras Boutique, which started as a mail-order business advertised as "the sole establishment of its kind, owned by, operated for, and staffed by a transvestite."[135]

As an active community organizer, Brewster used *Drag* to call for action, collect funds, connect readers, promote events, educate on important issues ranging from historical representation to legislation, and circulate an updated list of medical professionals specializing in gender-affirming care. Brewster used the publication's profits and other commercial enterprises to fund QLF's activism, including a march on New York's state capital to legalize cross-dressing. At this march and others, Brewster was photographed alongside key figures in queer and trans feminine activism, including Sylvia Rivera and Marsha P. Johnson.[136] Brewster and the QLF ultimately were successful in ending New York ordinances against cross-dressing.[137]

Trans feminine publications actively circulated information on harm reduction alternatives to silicone shots. Linda Lee, a trans woman who wrote for *Drag* and later *Female Mimics International*, urged readers to avoid subcutaneous silicone breast enhancement altogether and instead consider using external silicone breast forms.[138]

Drag featured advertisements for a range of breast prosthetics. A 1973 issue included two advertisements listed by Jaru Inc. that read, "SILICONE BREAST—NEW DISCOVERY! Artificial breast so real, feels like nature itself. Fits any bra—32A—40D. Silicone. Lasts indefinitely. Prices start $79.00. Other forms from $30.00."[139] The advertisements did not include an image or mention any colors. Another advertisement featured the "Treasure Chest," a latex breast piece sold directly by Brewster's Mardi Gras Enterprises. This design was priced at thirty-five dollars, and while listed as available "only in size 36C," it did come in "light, medium, and dark tones." In 1975, an advertisement for the same piece featured an image of performer Toni Lee, a Jewel Box Revue dancer who billed herself as "The China Doll" modeling it (figure 8.6).[140] For an additional ten dollars, clients could have the design customized to "match your skin tone." The Treasure Chest was almost less than half the price of the silicone "artificial breasts" listed by Jaru Inc., which does not appear to be a queer or trans-owned business. It is notable that Jaru does not mention skin color. While mainstream breast forms marketed toward an unmarked assumed cis, heterosexual, white woman frequently did not mention skin color, products developed by and specifically marketed to trans feminine individuals often did.

Some readers may have preferred silicone breast forms, as they were made from the same material used in surgical enhancement. Implantable devices, such as the 1973 Surgitek Perras-Papillon Design Mammary Prosthesis (figure 8.7), were marketed and photographed to look like bras that could rest closely against the chest wall. These implants were described as "a soft seamless outer 'skin' of Surgitek medical-grade silicone specifically designed to follow the individual left and right natural anatomic contours of the human female breast with axillary prolongation," and promised to "simulate the natural weight and consistency of normal breast tissue."[141] The curved, "organic" teardrop form, shaped by silicone's material possibilities, significantly differs from plastic foam's conical shapes, popular in preceding decades, that were compared to "bullet," "torpedo," "rocket," or "missile" designs. When it comes to form, the Surgitek prostheses do not look dissimilar to the silicone bra inserts available today. In the ensuing years, trans feminine publications mentioned silicone breast prostheses more frequently as a way of shaping the body.

Throughout the 1970s and into the 1980s, the silicone breast form industry continued to grow and evolve. Some trans feminine publications advised that "the best technique [for creating and enhancing cleavage] is to use the special [silicone] gel-filled forms that are often worn by mastectomy patients," which could be purchased from "large corsetry departments or from surgical appliance supply houses."[142] Some designs required a special bra, while others could be worn with a regular bra. Silicone breast forms offered a more realistic feel than their more structured foam predecessors.

Silicone had replaced earlier popular materials. While "[latex false busts] that one sometimes sees advertised" might be "perfectly reasonable" for "photos or stage use," Linda Lee and others felt that "for streetwear it would not only be unrealistic," but also "extremely uncomfortable as

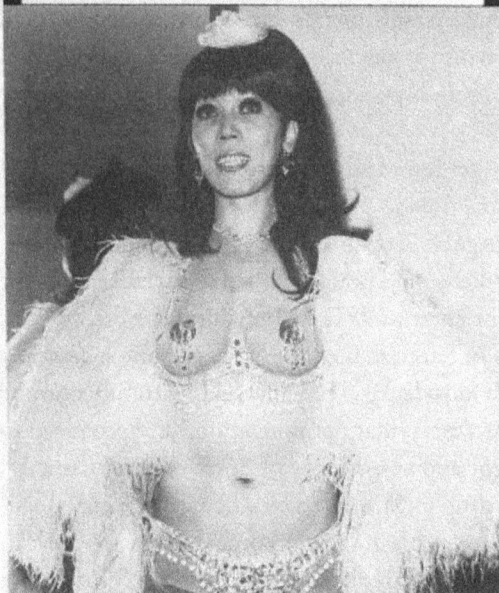

8.6 Treasure Chest advertisement featuring Toni Lee, *The Female Impersonator* 5, no. 8 (1975): 51.

8.7 Surgitek Perras-Papillon Design mammary prosthesis, 1973. Walter Spohn Collection, Division of Medicine and Science, National Museum of American History, Smithsonian Institution, Washington, DC.

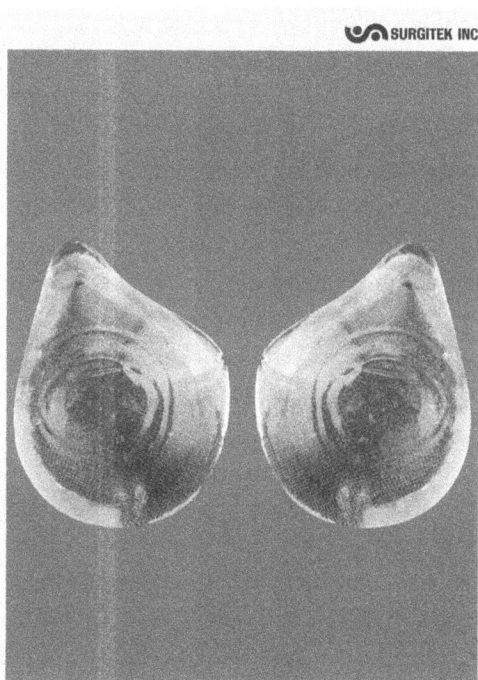

SURGITEK INC.

well."[143] Instead, individual silicone prostheses could be inserted inside the bra to enhance the bustline. Lee's account predates claims in fashion history that silicone-filled bras emerged in the 1990s, when Frederick's of Hollywood launched the first brassiere that incorporated pockets of silicone-filled gel pads.[144]

In the 1970s, surgical breast form companies began to market their products to trans feminine consumers. Transform is a line developed for these markets by Nearly Me silicone breast forms, which was founded in the 1970s by Ruth Handler, a breast cancer survivor and creator of the Barbie doll as well as former Mattel CEO.[145] These breast forms were made from liquid silicone enclosed within polyurethane and backed by rigid foam.[146] Another company that advertised breast forms to trans feminine markets is Mirage, which launched in 1976 after five years of research. The firm exclusively specialized in breast forms for "crossdressers who prefer not to deal with surgery or hormones."[147] The prostheses could be color-matched to the wearer's skin tone and attached to the body using medical adhesive rather than being worn in a bra. Its advertisements stated, "When Mirage is attached to your body you will be able to do almost anything you want with total confidence." Theoretically, this enabled the user to wear the prostheses without the need for support garments for extended periods, including during sleep, as advertised.

As contributors to trans feminine publications, including Linda Lee, noted, "[Breast forms] are temporary and for many of us, aren't enough. We want to actually develop more feminine bodies . . . usually the first step is . . . estrogen."[148] Throughout the 1970s and into the early 1980s, trans feminine publications consistently warned that Breastplasty (to use Franklyn's term) was still at an experimental stage and "strongly [advised] against it," remarking that hormones were widely used and "perhaps the most popular form of making the breasts large."[149]

Trans feminine publications served as a way for community members to share their varied experiences with taking hormones, detailing moments of frustration and disappointment with bust development, as well as satisfaction.[150] Sally Douglas offered detailed, practical advice on how to navigate medical gatekeeping to pursue treatment, including how to appear and what to say during an initial consultation.[151] She participated in community knowledge exchange by offering a practical guide to navigating hormone access at a time when information on and access to trans health care was scarce.[152] In a letter to Kim Christy, editor of *Female Mimics International*, Georgina reflected, "I'm longing for the day when hormone treatment means I don't have to use bra pads. The ones I have now are made of a silicone substance and so have the weight and movement of real breasts."[153] Some trans feminine individuals shared that while they had considered surgical enlargement using silicones, they ultimately decided against it. Sharon Davis, a Black transsexual woman, noted, "After all the planning and insecurity I've experienced about being flat-chested, I somehow realized that having B-size busts doesn't complete a woman. . . . Therefore, after giving the matter considerable thought, I cancelled the breast implant surgery and began to live with myself, depending on the slow progress of the hormonal shots."[154] Lee decided against subcutaneous silicone breast enlargement, citing distrust of the medical profession. She commented, "Plastic surgeons have a tendency to develop sort of a Pygmalion complex, and feel they know better than the patient what the patient wants or needs."[155]

In her 1987 "Posttranssexual Manifesto," Sandy Stone urged transsexual individuals to speak up on "the polyvocalities of lived experience."[156] She felt that trans people should embrace the vastness of gender identities and reject the singular understandings and definitions of sex and gender perpetuated within medical and feminist discourse, which effaced difference. This approach stood in contrast to the history of gender identity clinics and medical gatekeeping that required uniform compliance with the binaries of sex and gender as defined by medical providers, most of whom were predominately cis men. Stone encouraged readers to destabilize the narrative perpetuated by clinics, which invalidated their lived experiences and demanded they behave in ways that restabilized gender norms in order to gain access to their services. This included trying to pass as a cis person or concealing their transition from others by creating a new identity. By resisting these exclusionary definitions and structures, Stone stressed the importance of "reappropriating difference and reclaiming the power of the refigured and reinscribed body" and thus seizing the means of production.[157]

Following conical bullet bras, whirlpool stitching, falsies made from foam plastics for the bust, hips, and buttocks, and implants made from foam plastics, subcutaneous silicones were the latest technological advancement for femme body contouring. In keeping with the smooth, organic, round shapes offered by silicone-filled sacs, the fashionable breast shape also began to change, and bustlines looked rounder. The practice of free silicone injections exploited its viscous materiality, promising instant reshaping of the body. Challenging to regulate and control, silicone—a material whose research and development was originally expedited by wartime demand—slipped through established structures, enabling access for a range of socioeconomic groups.

Medical and industrial actors othered silicone injection as a way of legitimizing silicone breast implants. In legal testimony Dow Corning stated, "Doctors are not opposed to implants—precisely the converse applies—they endorse the use of implants. It is the injection of liquid silicone which they have questioned."[158] By containing silicone in a regulated and standardized object, breast implants ensured bodily transformations remained within formal medical spaces that were shaped by inequities of dominant power relations and binary understandings of sex and gender.

EPILOGUE / QUEERING PLASTIC FUTURES

A key theme that emerges in this book is the importance of equitable and sustainable access to technologies. It becomes clear that when access to and knowledge of these technologies changes, cultural imaginaries and visions of the future shift as well. Makers and users inscribe meaning into technologies that are shaped by the conditions within which they are created and experienced. But even as access increases, inequities often remain unchanged, which restricts the radical potential for new technologies of gender and the body.

Feminist, queer, trans, and crip thinking stresses the importance of bodily autonomy and seizing the means of production for liberation.[1] Bodily autonomy is under renewed and ongoing attack. I hope that by showing both hegemonic and counterhegemonic histories of military-industrial technologies in the form of plastics, this book has shown the vital importance of access to reclaiming these technologies. When users become makers, more expansive and radical futures are possible. What if a wide range of bombshells—instead of Big Science—had originally directed the materials research and development projects into technologies of the gendered body that they were using? Would safer materials and more equitable practices be available?

Even if access to technologies of gender is meted out equally, it doesn't change the systemic violence and oppression interwoven within social structures. The regulation of bodies within the framework of hegemonic gender and racial norms acts as a control over the legitimacy and safety of gender expression and sexuality. Technologies of gender as a form of self-fashioning and self-authorship have always existed both within and outside these norms, signaling tensions between standardization and customization.

In "The Android Goddess Declaration: After Man(ifestos)," micha cárdenas conceptualizes an android goddess from trans women of color's experiences and concerns. For cárdenas, an artist and scholar working today, reclaiming and subverting access to technology, including military-industrial materials, is a means of empowerment. Building on Haraway's original declaration, "I would rather be a cyborg than a goddess," cárdenas harnesses the liberatory potential of technology and declares that the android goddess "knows that she is made by the master's tools, yet she still seeks to resist the master."[2] Using metaphor as method in her action-centered practice, cárdenas repurposes technologies created within violent and oppressive systems, actively rewiring their accessibility as a means of collective resistance, survival, and community-building for trans women of color. For cárdenas, reclaiming and redistributing access to technology is an important means of reshaping the future.

Nylon

In the fall of 2021, a group of models dressed in red nylon swimwear gathered on New York City's Jacob Riis Beach, a historic LGBTQ+ site for sunbathing, gathering, activism, and cruising since the 1940s (figure E.1). This New York Fashion Week show celebrated a contemporary queer vision of the bombshell. The collection, designed by local artist and activist Tourmaline for Chromat—a queer-founded New York City–based bodywear brand—encompassed "swimwear for girls who don't tuck, trans femmes, non-binary and trans masc people who pack, intersex people, women, men and everyone embracing Collective Opulence Celebrating Kindred."[3]

In an article for *Them*, an online LGBTQ+ magazine, contributor Alex Jenny reflected, "I have searched far and wide to no avail for a gender-affirming line that accommodates girls like me—girls who don't always want to tuck, but still seek safety through various levels of coverage. To feel these garments, to run my fingers along them as they hung in Tourmaline's studio, was to feel a dream become, in the most profound way, a bikini."[4] For Jenny, Tourmaline's bikini design manifested her desire for visibility, her desire to be seen. Jordyn Harper, a model in the Chromat show, shared why this day was so important to her: "I'm untucked today. This is the first time I've ever been seen untucked. It's liberating. I feel

E.1 Chromat × Tourmaline ss22 campaign image, September 12, 2021. Photograph by Anastasia Garcia.

really connected to my body. I feel free. I couldn't ask for anything more. I get to live life and be me."[5] For Harper, Tourmaline's designs symbolize freedom and a way of feeling connected to her body. The eye-catching, vibrant red, lifeguard-inspired collection, crafted from nylon, is made with wider gussets and sizes from xs to 4x, catering to a greater range of wearers.

For Tourmaline, a community organizer for Black, trans, queer, gender-nonconforming, and disabled communities, the visual references to life-guarding were intentional and functioned on multiple levels.[6] In honor of Marsha P. Johnson's legacy, Tourmaline styled the models' hair with flowers like Johnson wore. As noted by Jenny, for Tourmaline, "creating a collection that centered the safety of trans girls who don't tuck is a personal project." Over the past sixteen years of beach visits, Tourmaline recalls feeling un-comfortable at times: "I can remember being in the water and being afraid to get out, not wanting people to see my body." Tourmaline hoped that her range of designs could empower wearers in the LGBTQ+ community: "This collection is about saving a life. It's going back to the basics of fashion and aesthetics: Our self-fashioning is a powerful act, and in our power, we have a greater capacity to create safety for one another."[7]

Fashioned from recycled nylon, Tourmaline's collection works toward destabilizing extractive tropes of plastics and challenges the inherent power structures embedded within them. Unlike the colonial and heteronormative rhetoric that informed plastics promotional material from the 1930s and into the postwar period, Tourmaline's design vision imagines queer futures for plastics. Chromat's online shop descriptions for her collaborative line reflect the brand's commitment to sustainable production methods. Designs are listed as "ethically made in Bulgaria with sustainable, regenerated nylon spun from recovered fishing nets."[8]

A tension exists between the need for ethical production and affordability, particularly among queer and trans fashion labels. Scholarship on visibly queer and trans fashion brands suggests that, at times, heightened ethical labor values can be merely performative.[9] The pricing for Chromat's nylon collaboration with Tourmaline ranges from $48 to $198, making it economically out of reach for many trans people, who are four times more likely to be low-income.[10] However, Chromat's approach to production mirrors that of other contemporary queer and trans brands, whose engagement "almost always, not surprisingly, supported queer and trans issues, threading their core values of queer and trans activism throughout all aspects of the business."[11]

Chromat's bombshell vision embodies queer kinship and sustainability. The company's creative team comprises scientists and models who are working toward an inclusive design vision for queer futures of plastics. Chromat's swimwear is made from "sustainable, regenerated nylon spun from fishing nets and post-consumer plastic bottles that have been recovered from the world's oceans." Its nylon mill collaborates with an international diving team to recover more than 160 tons of fishing nets from the ocean, transforming them into yarn combined with other nylon waste.[12] The nylon yarn Chromat uses is produced through a closed-loop process, which means it can be continuously recycled without affecting its quality.[13]

Tourmaline's acclaimed nylon designs celebrate and empower queer and trans individuals, queering a material that was originally launched to the public with live model demonstrations by a white "Test Tube Girl" for DuPont at the 1939 San Francisco and New York World's Fairs, who embodied a singular, unmarked norm of womanhood as exclusively white, cisgender, slim, young, able-bodied, and heterosexual. At that time, nylon did not yet have a legacy; but it does today, as an abundance of plastics in the environment has even resulted in nylon-eating bacteria that have evolved in response to nylon and its by-products.[14]

With recent visible queer and trans design efforts, such as Tourmaline's, reported in mainstream media, there is a reimagining of beauty ideals and a plastics futurity that is actively shaped by a wide range of intersectional identities. Tourmaline's recycled Collective Opulence Celebrating Kindred nylon bikini creations are emblematic of how responsible design practices can work collectively toward dismantling unequal power structures, both in relation to gender, sexuality, embodiment, identity, and representation, as well as materials themselves.

Bombshells Reimagined

Contemporary efforts to reclaim the bombshell are becoming more visible. Queer, nonbinary, gender-nonconforming, and trans feminine communities are embracing, redefining, and celebrating this term. Efforts include community events and organizing like Bombshell!, a weekly dance night in London, self-described as a "TV/TS/Drag/Non-Binary/Gender-Bending/Friends&Lovers club," and the New York–based drag queen Essa Noche's Bombshell Brunch.[15]

Bombshells also continue to be represented and reimagined in visual and material culture. *Bombshell*, a recent zine and photobook by photographer Ethan James Green, provides "an inclusive look at feminine sensuality in its most alluring form."[16] It features trans, cis, and nonbinary icons of the New York scene, such as cover girl actress Hari Neff, fashion editor and *Vogue* stylist Gabriella Karefa-Johnson, and *Interview* magazine fashion director Dara Allen.[17] Green reflects, "You can't do a bombshell project without being political. It's a word that's been reserved for a small group of people. But for me, bombshell is an energy, and it's a choice. I don't see why it should be limited to a single type of person. Everyone has their own version of it."[18] For Anetra, a drag queen from a Japanese, German, Filipino, and Puerto Rican background who starred in season 15 of *RuPaul's Drag Race* reality competition TV series, the bombshell is an identity. As she entered the show, she said, "Hey bitches, it's your girl, Anetra, the Sin City bombshell from Las Vegas, Nevada." As Green notes, "There are a variety of bombshell types. . . . It can get really campy and sometimes it's a bit more of the girl next door, if you want to call it that."[19] The plurality of bombshell identities is further evidenced by social media metrics that highlight the preference for #blackbombshells over the singular #blackbombshell. These examples continue to illustrate the many meanings of *bombshell* and how the word can also empower individuals with multiple intersecting and socially marginalized identities.

The bombshell continues to hold currency in Hollywood celebrity culture and the entertainment industries. In 2022, Kim Kardashian, once famed for her curvaceous hourglass body, gained headlines for wearing to the Met Gala a light nude-colored dress once worn by Marilyn Monroe, considered by many to be the embodiment of the white, American, bottle-blonde bombshell. Kardashian's internet-age media coverage demonstrates the longevity of the term "bombshell" as a sex symbol and its contemporary usage in popular culture.[20]

While the breasts were a focus of whitestream beauty standards in the postwar United States, much recent cultural discourse on the bombshell is focused on the buttocks. The Brazilian butt lift, a type of buttock augmentation surgery, is the fastest growing plastic surgery procedure in the United States, according to the American Society of Plastic Surgeons.[21] Scholars have described a long history of the racialization of the buttocks within Western discourse.[22] Kardashian, who has become known for her pronounced rear, has frequently been critiqued for appropriating Black culture and beauty practices.

The hyperfemininities associated with bombshells continue to break away from gender norms and have the power to subvert and reclaim gender binaries but arguably also conform to them.[23] The spectacular femininity that Kardashian is known for draws from queer and trans feminine practices, including makeup contouring and body modifications with padded foundations and prostheses. In 2022, *Interview* magazine featured Kardashian's exposed derriere on the cover of its September issue, titled "American Dream." Kardashian stood in front of the American flag, dressed in what is colloquially called a Canadian tuxedo, also known as double denim—denim jacket and jeans—and wearing a white cotton jockstrap, referencing queer and trans masculine aesthetics.

Imagery of Kardashian's buttocks is abundant online. In 2019, the Scandinavian artist duo Ida Jonsson and Simon Saarinen first created "The Bum," "an ongoing research project investigating media's fetishization of Kim Kardashian's behind."[24] In 2014, Kardashian was said to "break the internet" when her exposed buttocks appeared on the cover of *Paper* magazine.[25] The series, photographed by Jean Paul Goude, essentially replicated his 1976 controversial nude image of Black model

Carolina Beaumont balancing a glass of champagne on her buttocks. As part of "The Bum" project, over a two-year period, Jonsson and Saarinen collected thousands of paparazzi images of Kardashian's buttocks, which they used to create an "exact carbon 3D copy" of a one-to-one scalable online version of "The Bum," as well as a wearable piece made from silicone, created in collaboration with designer Beate Karlsson.[26] The controversial piece was worn in publicity stunts at New York Fashion Week and continues to be featured on the designer's Instagram page.[27] Perhaps breast prostheses makers working in earlier decades with plastics within trans feminine communities, such as Pudgy Roberts, would have capitalized on this cultural shift in beauty ideals by advertising custom-made forms for the hips and buttocks today, in addition to the chest.

Foam

Representation of trans feminine and gender-nonconforming people shaping their bodies with plastic foams is no longer solely seen in underground publications like *Female Mimics*. Today, *RuPaul's Drag Race*, an international franchise, has made the practice of using foam to shape the body more visible. Contestants are often shown backstage cutting padding from sheets of plastic foam. There is also a proliferation of YouTube tutorials on drag padding, with foams made from plastics. Unlike the femme foam padding guides circulated by queer and trans feminine community members in postwar American print publications that focused on chest pieces, these videos tend to focus on padding of the hips and buttocks. There are a variety of approaches. Some tutorials advise viewers on how to dress with ready-to-wear foam padding, purchased from businesses ranging from queer- and trans-owned companies to Amazon.

In a YouTube tutorial, drag sisters and self-described "big girls" Bucy and Sad Sally feature Planet Pepper, a gay-owned, New York–based business, that specializes in "booty" and hip pad undergarments in a range of skin tones for "drag queens and crossdressers." These custom-made items may be economically out of reach for many people, as their Astrobooty shorts and pants with custom pads range from around $100 to $200. However, Bucy complains that "kitchen pads" can take "an hour and a half to put in." The different materialities of plastics and their impact on the body also factor in Bucy's preference for foam. She notes that foam breast pads, such as those offered by Planet Pepper, are preferable to silicone breast plates, which can be very hot, especially on stage.[28]

Other videos offer a more DIY approach to crafting foam padding, even suggesting responsible sourcing of plastic foam from used furniture, such as couches.[29] In "How to Make Drag Queen Body Padding," famed vlogger Patrick Starrr collaborates with Body Pod creator and Drag Stop founder Victoria Trenta. In the tutorial, Trenta and Starrr instruct viewers to source the following common household items: plastic kitchen wrap, a marker, scissors, measuring tape, a turkey carver, nylon hosiery, and "high-density foam." Unlike the queer and trans feminine people who had experimented with making foam breasts in the 1960s, Trenta and Starrr don't need to mix chemical solutions to produce foam. Instead, they use a purchased, ready-made foam slab. The one depicted is light seafoam in color and purchased from the Joann fabric and craft chain, demonstrating how readily available these flexible foams are today. Plastic foams, Starrr and Trenta note, can also be sourced from furniture: "If you are on your grandmother's couch, steal it girl!"[30] Although these suggestions appear to be said in jest, they offer alternative, queer ways of reclaiming found and recycled materials for self-fashioning outside normative capitalist marketplaces. Notably, Trenta, who runs a business that sells foam pads, also shares online a free step-by-step guide on how to make the designs at home that she has perfected. Trenta offers people who can't afford her custom "body pod" designs a way to make their own. This can be understood as an example of community-building that operates outside traditional capitalist economies.

Sharon Waniz, a Black YouTuber based in Kenya, also offers her viewers a DIY way to shape the body. In a tutorial, she demonstrates how to add homemade foam pads to skirts and dresses in the lining so that, "when you take the skirt or the dress off . . . no one can notice that the skirt is really padded. No one can see the pads!"[31] Using a box cutter to craft pads from a plastic foam roll, she adds them to a pair of high-waisted spandex shorts. Waniz discusses adding pads to skirts and dresses, made from a range of materials, including nonstretchy natural materials such as African wax print fabrics, which are cotton-based, and she tests out a series of looks. She concludes, "As you can see how perfect I look. Like, no one can even tell 'cause [the padding] is actually balanced with my shape and doesn't look that exaggerated."[32] Referring to how she looks without her foam creations, she notes, "This is the 'before,' which is just OK, but now I want to look like an Instagram hourglass-shape model."

Admiring her padded body, Waniz remarks enthusiastically, "Look at your girl—I love what I am doing. This is what you can wear on occasions where you want to slay—padded dresses are pricey, sixty to seventy dollars. . . . Appreciate yourself. Don't get surgeries to look like an Instagram padded model."

While self-fashioning with foam padding has a positive impact for many wearers in terms of gender expression, foams made from plastics also negatively affect bodies—human, water, or "otherwise."[33] The US Environmental Protection Agency states that the manufacturing of flexible polyurethane foam is a major potential source of harmful air pollution.[34] Additionally, there are increasing amounts of polyurethane foams in landfills, and these foams emit noxious gases when combusted. Due to the flammable nature of polyurethane foam mattresses, manufacturers are legally required to add flame retardants. However, these chemicals contain endocrine disruptors that have adverse health effects, such as thyroid dysfunction, infertility, immunological issues, and cancer.[35] For example, polybrominated diphenyl ethers, a commonly used flame retardant, increase the risk of breast cancer and are associated with thyroid dysfunction.

While research exists on the potential hazards of polyurethane foam mattresses, I was unable to locate a study addressing the impact of polyurethane foam padding on the body.[36] Prolonged material exposure through wear is not totally dissimilar to that of sleeping, underscoring the importance of research in this area. Comparable concerns about the increased emission of toxic volatile organic compounds with body heat might also apply. CertiPUR-US is a certification program that works with mattress and upholstered furniture industries to create and enforce safer standards for the polyurethane foams it tests.[37] Since the organization is affiliated with industry and not an independent certification agency, there are divided opinions about the safety of polyurethane foam mattresses.[38] As with other materials used in fashion and design, identifying a truly sustainable material is challenging. The mass cultivation of "natural" materials, such as cotton, has also negatively affected the environment, disproportionally affecting populations in the Global South.[39] Plastics urgently need to be studied in greater detail to better understand their impact. Plastics consumption and waste disproportionately affect Black, Brown, Indigenous, and poor communities, the Global South, and "more-than-human" worlds.[40] It is important to explore the queer potential of plastics while also holding companies accountable for the damage they have caused.

Silicone

The American Society of Plastic Surgeons lists breast augmentation using silicone implants as the most popular elective procedure.[41] Across the United States, outside the structured medical system, the practice of silicone injections for shaping the body, also known as "silicone pumping," largely affects trans and cis women of color and low-income individuals, further reinforcing ongoing health inequities.[42] Although the FDA never approved free silicone injections for use in the breast, their use proliferated due to their lower cost and increased accessibility for underserved populations.[43] While there is limited research on free silicone injections in cis women, even fewer studies examine the impact of them on trans women, echoing continued research inequities.[44]

Breast augmentation remains popular among trans women.[45] It is difficult to estimate how common pumping is. A 2013 study of 234 trans women in San Francisco asked participants, "Have you ever injected substances other than hormones (e.g., silicone) to enhance your gender presentation?" Almost 17 percent of individuals reported ever injecting fillers, including silicone. The most commonly reported places to inject were the breasts (52.4 percent) and the face (51.4 percent). Notably, this community-based research was developed with trans leadership and utilized respondent-driven sampling methodology to ensure that participant recruitment matched the racial and ethnic diversity of trans women in San Francisco.[46] Almost 89 percent of the women interviewed reported living below the federal poverty line. Trans women continue to use silicone injections as an alternative to plastic surgery and other dermatological treatments within formal medical structures, citing comparatively fast results and low costs.[47] Additional studies measuring the prevalence of silicone injections among trans women have been carried out in Chicago and DC, as well as internationally in countries such as Brazil, Peru, the Netherlands, and Thailand.[48]

Some trans feminine community resources, often created by and for the community, continue to warn against the use of silicone shots.[49] In response to the number of community members injecting silicones, health care initiatives like the San Francisco–based TransLine and New York City's Callen-Lorde Community Health Center have created harm reduction resources for trans communities, as well as for health care workers. TransLine stresses the importance of safe injection practices.[50] Callen-Lorde includes a list of alternative ways to shape the body, including silicone prostheses.[51]

The recently closed, San Francisco–based St. James Infirmary, which was "a peer-based occupational health and safety clinic for sex workers of all genders," also provided harm reduction information on silicone and pumping parties. Its occupational health and safety handbook listed possible complications and cautioned users to limit injections to minimal amounts at one time.[52] The organization advocated for health equity, the anticriminalization of sex work, and trans rights, offering an intersectional model that challenged conventional health care structures that "divide patients and providers and foster unhealthy power dynamics."[53] Trans women, particularly Black trans women, sex workers, and those living in poverty, remain underserved within the larger, regulated health care system, and especially in relation to gender-affirming care. In an interview, Ruby Corado, a forty-nine-year-old Latinx trans woman and activist, reflected on her pumping experience: "When you go to one of these underground [silicone] providers, they're not asking you for a therapist letter. They're not asking you to be on hormones for a certain time. They're not asking you to live in your truth for a year or two. They're not asking you for a lot of things."[54] For some women, the providers "were seen as caregivers" who serve, and are sometimes part of, the community. As Corado notes, there are many different providers assisting with silicone injections, and harm reduction is essential.

⚛

By arguing for the importance of understanding materials as historical artifacts, I set out in this book to trace the changing meanings associated with emerging plastic materials, particularly in relation to their impact on the shaping of the "ideal," fashioned, and pathologized female body. The legacy of plastics, originally developed by chemical companies for industrial and military applications, continues to shape and affect the body today. Nylon blends and derivatives, including Lycra, are used in shapewear designed in a range of skin tones by companies such as Spanx and Skims, as well as trans masculine-owned companies like gc2b that produce binders. Polyurethane foam padding continues to shape the body within ubiquitous staple items, such as soft-shell T-shirt bras. Silicone pads marketed to women for temporarily padding out the silhouette are now sold in US drugstores, such as the Hollywood Fashion Secrets brand, whose products come in various skin tones and promise to "conceal, boost, and control" the body. Implants made from foam used to pad the body are no longer the norm. Silicone implants continue to be controversial; however, they remain approved by the FDA. Meanwhile, the practice of large-scale injectable silicone remains unsanctioned by the FDA. Nonetheless, the practice of body contouring using silicone injectables endures globally. Silicone continues to develop in its relationship to the body, particularly within the growing sex toy industry. Its unique materiality can both simulate the human body and shape queer futurities.

This project began many years ago when researching the conical bullet bra and how this signaled a major design change in relation to the body. I became interested in the weaponized language of the bullet bra, the bombshell, and the bikini. I wanted to explore how and why these comparisons were being made in popular culture, but also how military-industrial links to materials research and development coincided with the shaping of the bombshell via designs worn on the surface of the body, as well as plastics implanted within it. In whitestream media, this language was used to signal an unmarked norm of cisgender, heteronormative, able-bodied white womanhood, often with her sexuality harnessed, consumed, and put to "good use." This dominant vision of racialized gender was presented as dangerous if left to its own devices, and as something that needed to be regulated through white American heteropatriarchal social norms. Imagery of the bombshell could be consumed from a safe distance, but bombshell identities signaled a threat to the white American social order.

The bombshell, and the vivacious sexuality she signaled, became increasingly enmeshed with changes in plastics technology, which were further expedited by wartime demands. Her stockings were made of nylon, her bra and girdle padded with new plastic foams, and her curves eventually enhanced with silicones. The bombshell is a site of tension and contradictions, implementing the latest technological developments to both embody gender ideals and simultaneously offer the radical potential to challenge and subvert them. With *Atomic Bombshells*, I have sought to illuminate the relationship between plastic materials research and development and their impact on the shaping of a range of feminine bodies, symbolized by the bombshell. Throughout this book, I have endeavored to make complex power relations visible and show how a wide range of actors engaged in different processes of shaping the body with plastics, often reimagining and transforming materials from their original intended uses. These processes of self-fashioning using nylon, polyurethane foam, and silicone presented new material possibilities for shaping and experiencing the body on a personal and structural level.

ACKNOWLEDGMENTS

This project would not have been possible without the generous support of a number of individuals and institutions. I am immensely grateful for the opportunity the Techne UK Arts and Humanities Research Council Doctoral Training Partnership provided me with by funding the early stages of this research while I was based at the Royal College of Art. Here, I am thankful for my two PhD advisers, who provided instrumental support and guidance. Sarah Teasley encouraged me to think creatively with and through design and technology history across disciplines and geographies. Jane Pavitt enriched my research through her expertise in and enthusiasm for Cold War design. I also thank Alison Clarke and Joanne Entwistle for their input on early phases of this project.

Through the UK Arts and Humanities Research Council International Placement Scheme, I was able to conduct research at the Smithsonian National Museum of American History and later returned as a Smithsonian Institution Postdoctoral Fellow. Here, I am indebted to the ongoing kindness, enthusiasm, and support of Katherine Ott, my primary project adviser at the National Museum of American History, whose expertise in the history of medicine and material culture, as well as her sharp wit and immense generosity, continue to be invaluable to my research and time spent in the United States. There are many colleagues across departments, museums, and institutions at the Smithsonian whom I would also like to thank for their help and support, especially Margaret Vining, Margaret Weitekamp, Ellen Lupton, Nancy Davis, Abraham Thomas, Trina Brown, Franklin Robinson, Bart Hacker, David Haberstich, Elaine Nichols, Jia-Sun Tsang, Emily Orr, Caroline Fiertz, Alana Staiti, Anneleise Azua, and fellow plastics scholar Jessica Walthew.

At the Hagley Museum and Library, where I was a Henry Belin Du-Pont Short-Term and Dissertation Fellow, I am especially grateful to Roger

Horowitz for taking the time to support my research: His comments on my chapter drafts were particularly helpful and productive, as were seminar comments by Grace Lees Maffei and Elizabeth Higginbotham. I also thank Carol Lockman for her assistance and care, curator Debra Hughes for her expertise on nylon, and Yassin Abou El Fadil for his friendship, all of whom I was fortunate to meet during my Hagley fellowships. I thank the Science History Institute, where I was the Price-Doan 80/20 Postdoctoral Fellow 2020–22 and a Short-Term Doan Fellow. At the Institute, I am indebted also to the support of Andrew Mangravite, Ashley Augustyniak, Michelle DiMeo, Molly Sampson, Erin O'Connor, Charlotte Abney Salomon, Mia Jackson, and Christy Schneider, who ensured that my fellowship and postdoc were particularly productive and enjoyable. I thank the staff and students at Bard Graduate Center for hosting me as a Visiting Fellow and their engaging dialogue on design history. At the Philadelphia Museum of Art, Kristina Haugland's continued enthusiasm and generosity with her private collection of foundationwear is truly appreciated. This book has benefited greatly from conversations and work-in-progress sessions with Science History Institute fellows and friends Meg Piorko, Siobhan Angus, Gus Lester, Rebecca Kaplan, and Ruth Rand.

This project has grown through feedback and engagement with organizers and audiences at institutions including London College of Fashion; Goldsmiths, University of London; the Courtauld's Gender and Sexuality Group; Costume Society of America; Smithsonian Pride Alliance; Society for the History of Technology; The New School; New York University; International Council of Museums Committee for Conservation Textiles and Modern Materials and Contemporary Art Working Groups; Iowa State University; Museum Council of Greater Philadelphia; Business History Conference; Society for the Social Studies of Science; University College London; Wellcome Collection; University of Delaware; WHYY's The Pulse; Hagley Center for the History of Business, Technology, and Society; University of Gothenburg; Plastics Heritage Congress; and Techne Consortium.

I have published elements of research based on short sections of two chapters within this book in academic journals and benefited from feedback during the peer-review process. A more in-depth version of my content related to trans feminine padded foundationwear makers and wearers in the postwar United States appeared in the "Transgender Embodiment in Fashion and Beauty" special issue of Critical Studies in Fashion and Beauty (15, no. 1 [2024]: 49–74) as "Shaping Foundations."

Here, I am particularly grateful for the support of special issue editors Erique Zhang and Roberto Filippello and the anonymous reviewers, who gave me valuable feedback. *Fashion, Style, and Popular Culture*'s "Marginalized Identities" special issue (11, no. 1 [2024]: 45–64) included "Redefining Nude," my examination of the history of nude colors and their impact on nylon hosiery and other foundationwear from the 1930s to today. Here, I thank special issue editors Kelly Reddy-Best and Dyese Matthews and the anonymous reviewers for their thoughtful comments, which benefited my writing.

I have infinite gratitude for Ken Wissoker and Ryan Kendall at Duke University Press for believing in and supporting this book project from its early stages of development to its completion as a manuscript. I am also incredibly grateful for my anonymous reviewers' understanding of and commitment to the project. Their insightful and helpful comments and words of encouragement guided me as I worked through the book. At Duke University Press, I also thank Lisa Lawley, Nicholas Taylor, and Chris Robinson for their collective efforts during the editorial and production process. At the Fashion Institute of Technology, my colleague Nanja Andriananjason provided key support in preparing the images for this book.

Finally, I am forever thankful to my family and friends for their continued support and inspiration. Thank you Juliet Hill, Rie Takeda, Mirai Held, Kristina Heckeroth, Miriam Glaser, Mike Reichel, Johan Deurell, Josephine Rout, Paora Durie, Harman Bains, Sonya Abrego, Skye Nunke, and Jhenna Voorhis for the love and joy you have all given me. My glamorous Johanna, Big Lion, Hedi, Artur, Uncle Mike, and Rainer, every day, your glittering stars shine bright. Sidney and Susan, thank you for welcoming me into your loving family. I am beyond grateful for my incredible parents for being there for me and inspiring me never to stop being curious. Zeus, you taught me the importance of embracing life at its fullest. Gina, your creative passion for writing, steadfast support, and persistent encouragement have motivated me beyond words. M., M., S., and B., you are my world.

NOTES

Archives and Abbreviations

acc.	accession number
BMJ	*British Medical Journal*
DPA	DuPont Archives, Hagley Museum and Library, Wilmington, DE
DPAD	DuPont Advertising Department
DTA	Digital Transgender Archive, https://www.digitaltransgenderarchive.net/
GGIE	Golden Gate International Exposition
HCP	Herbert Conway Papers, Medical Center Archives, Weill Cornell Medicine, New York
JMP	Joseph Murray Papers, Harvard Center for the History of Medicine, Boston
JPW	Jerome Pierce Webster Papers, Columbia University Health Science Library, New York
JAMA	*Journal of the American Medical Association*
LOC	Library of Congress, Washington, DC
NMAH	Smithsonian National Museum of American History, Washington, DC
NMAH TLC	Smithsonian National Museum of American History, Trade Literature Collection, Washington, DC
NYT	*New York Times*
NYWF	1939 New York World's Fair
PRS	*Plastic and Reconstructive Surgery*
RAF	Robert Alan Franklyn Files, Series: 284 Franchises Frauds and Rackets, American Medical Association Archives, Chicago
SHI	Science History Institute, Philadelphia
WEGP	Walter E. Gloor Papers, Science History Institute, Philadelphia

Introduction

1 Franklyn, *Beauty Surgeon* (1961), 11.

2 As plastics expert Susan Lambert stresses, "There are literally thousands of different plastics, each with their own composition and

characteristics." I follow Lambert's lead in pluralizing plastics whenever possible to represent the plurality of the material. See Lambert, "Introduction," 1.

3   In *Work!*, historian and queer studies scholar Elspeth Brown uses the term "whitestream" to describe high circulation American media produced for a largely white audience; she also uses this term as a replacement for "mainstream" in order to mark the often-racialized bias of this word's usage.

4   Franklyn, *Beauty Surgeon* (1961), 13.

5   Franklyn, *Beauty Surgeon* (1961), 18.

6   For a selection of recent critical studies of plastics and their impact on bodies, human and otherwise, see Liboiron, *Pollution Is Colonialism*; Davis, *Plastic Matter*; Gabrys, Hawkins, and Michael, *Accumulation*; Cirino, *Thicker Than Water*.

7   For a selection, see Haraway, "Cyborg Manifesto"; Stone, "Empire Strikes Back"; Balsamo, *Technologies of the Gendered Body*; Seu, *Cyberfeminism Index*; Murphy, *Seizing the Means of Reproduction*; Murphy, *Sick Building Syndrome*; Kafer, *Feminist, Queer, Crip*; Hamraie, *Building Access*; Russell, *Glitch Feminism*; Liboiron, *Pollution Is Colonialism*; Davis, *Plastic Matter*; Angus, *Camera Geologica*.

8   Polyester refers to both a category of polymers and the common name for a specific textile fiber (aka polyethylene terephthalate). Lycra (aka spandex or elastane) is the trade name for fibers made from polyurethane.

9   Meikle, *American Plastic*; Handley, *Nylon*; Ndiaye, *Nylon and Bombs*; Hounshell and Smith, *Science and Corporate Strategy*.

10   Meikle, *American Plastic*, 1.

11   See Cowan, *Social History of American Technology*. US President Dwight D. Eisenhower is credited with coining the term "military-industrial complex" in his 1961 farewell address. For more on the historical development of the US Cold War military-industrial complex, see also Epstein, *Torpedo*; Leslie, *Cold War and American Science*; Pursell, *Military-Industrial Complex*; Ndiaye, *Nylon and Bombs*; Hitch and McKean, *Economics of Defense*; Koistinen, *Military-Industrial Complex*.

12   Cowan, *Social History of American Technology*; Pursell, *Military-Industrial Complex*; Leslie, *Cold War and American Science*.

13   Cowan, *Social History of American Technology*.

14   Henthorn, *From Submarines to Suburbs*.

15   Henthorn, *From Submarines to Suburbs*, 221.

16   Meikle, *American Plastic*, 2.

17   Serlin, *Replaceable You*; Haiken, "Modern Miracles."

18   Serlin, *Replaceable You*, 3.

19   Serlin, *Replaceable You*, 3.

20   See, for example, Haiken, *Venus Envy*, 7; Serlin, *Replaceable You*, 4; Gilman, *Making the Body Beautiful*.

21	Menon, *Refashioning Race*, 12. For more on self-actualization and cosmetic surgery, see Plemons, *Look of a Woman*.
22	Serlin, *Replaceable You*, 4; Haiken, "Modern Miracles," 184.
23	Serlin, *Replaceable You*; Hamraie, *Building Access*.
24	Farrell-Beck and Gau, *Uplift*, 124.
25	Hersch, "High Fashion"; de Monchaux, *Spacesuit*; Weitekamp, "Technology."
26	U.S. Food and Drug Administration, "A History of Medical Device Regulation and Oversight in the United States," FDA: U.S. Food and Drug Administration, August 21, 2023, https://www.fda.gov/medical-devices/overview-device-regulation/history-medical-device-regulation-oversight-united-states.
27	May, *Homeward Bound*, 106; Smith "Bombshell."
28	McLuhan, *Mechanical Bride*, 99.
29	Kakoudaki, "Pin-Up"; Buszek, *Pin-up Grrrls*.
30	For more on Las Vegas atomic viewing parties, see Titus, *Bombs in the Backyard*, 9; Knepp, *Las Vegas*; Denton and Morris, *Money and the Power*, 139–40. For Atomic Age culture in relation to the bikini and gender, see Xiang, "Bikinis and Other Atomic Incidents"; Taha, "Atomic Aesthetics"; Cole, "Bikini."
31	Voyles, "Anatomic Bombs."
32	"Rich Exotic Shake," *Vue*, May 1956, 126–30.
33	Bright, *Continental Defense*.
34	May, *Homeward Bound*. For more on wartime and postwar US gender roles, see Anderson, *Wartime Women*; Meyerowitz, "Beyond the Feminine Mystique."
35	See May, *Homeward Bound*, 92–93; Preciado, *Pornotopia*.
36	Crowley and Pavitt, *Cold War Modern*.
37	May, *Homeward Bound*; Smith, "Bombshell"; Preciado, *Pornotopia*.
38	May, *Homeward Bound*, describes the perceived threat of transgressive gender roles that became politically linked to Cold War Communist Eastern ideology. See also Loftin, "Unacceptable Mannerisms."
39	Franklyn, *Beauty Surgeon* (1961), 19.
40	Voyles, "Anatomic Bombs"; May, *Homeward Bound*, 106; Smith, "Bombshell"; Bailey, *From Front Porch to Back Seat*; Preciado, *Pornotopia*, 69–78; Sullivan, *Bombshells*.
41	See, for example, Vivian Henderson, "Stage Show at Yazoo Theater: The Bronze Manikins Will Be Presented Here Next Wednesday," *Yazoo City (MS) Herald*, December 23, 1938, 6; "Bronze Bombshell of Rhythm," *Alton (IL) Evening Telegraph*, March 20, 1939, 11; "Joyce Bryant's Best Kept Secrets," *Jet*, March 31, 1955, 58–61; "Vickie Henderson," *Ebony*, October 1960, 49–54; "Leslie, a Cool Bombshell," *Life*, June 23, 1967, 88–89; Tina Turner mentioned in "In Orbit," *California Eagle* (Los Angeles), March 26, 1964, 11; "Brown Bombshell" Eartha Kitt mentioned in "Broadway's Newest Darling," *Jet*, July 31, 1952, 56–57.

42 See "Oriental Bombshell," *Pittsburgh Post-Gazette*, April 23, 1964, 48; "Japanese Jewess to Entertain at Beth Israel's Donor Dinner," *Detroit Jewish News*, March 16, 1973, 34; Brazilian bombshell Carmen Miranda in "Beauty Arts," *Crenshaw-Mesa Southwest Wave* (Los Angeles), August 15, 1941, 13.

43 "Anatomic Bomb," *Life*, September 3, 1945, 53–54; Boyer, "United States"; Boyer, *By the Bomb's Early Light*.

44 "Vicky 'the Bombshell,'" *Female Mimics* 1, no. 11 (1968): 52–53, DTA.

45 See, for example, Meyerowitz, *How Sex Changed*; Stryker, *Transgender History*; Serlin, *Replaceable You*.

46 Snorton, *Black on Both Sides*, 140.

47 See Dorothy Kilgallen, "On the Bulletin Board," *Record-Argus* (Greenville, PA), January 8, 1953, 11; Aline Mosby, "Hollywood Keeps Up with Headlines," *Sandusky (OH) Register*, February 19, 1953, 7; Dorothy Kilgallen, "The Voice of Broadway," *The Mercury* (Pottstown, PA), January 6, 1953, 13.

48 Stryker, *Transgender History*, 66.

49 Serlin, "Christine Jorgensen and the Cold War Closet"; Serlin, *Replaceable You*; Skidmore, "Constructing the 'Good Transsexual'"; Meyerowitz, *How Sex Changed*, 52; Stryker, *Transgender History*, 104; Snorton, *Black on Both Sides*; Plemons, *Look of a Woman*, 8–9; Gill-Peterson, "DIY."

50 Gill-Peterson, "DIY."

51 Skidmore, "Constructing the 'Good Transsexual,'" 271.

52 Stryker, *Transgender History*, 66.

53 Gill-Peterson, "DIY."

54 For a selection, see Kafer, *Feminist, Queer, Crip*; Hamraie, *Building Access*; Haraway, "Cyborg Manifesto"; Stone, "Empire Strikes Back"; Stryker, *Transgender History*; Balsamo, *Technologies of the Gendered Body*; Serlin, *Replaceable You*; Cogdell, *Eugenic Design*; Carter, *Heart of Whiteness*; Strings, *Fearing the Black Body*; Gill-Peterson, *Histories of the Transgender Child*.

55 See Haraway, "Cyborg Manifesto," 151; Stryker, *Transgender History*, 30; Wark, "Introduction," 12.

56 For more on the concept of self-authorship see Menon, *Refashioning Race*.

57 Menon, *Refashioning Race*, 11.

58 Stryker, *Transgender History*, 105; Kafer, *Feminist, Queer, Crip*, 125; Balsamo, *Technologies of the Gendered Body*.

59 Haraway, "Cyborg Manifesto." Three aspects are crucial in understanding Cartesian dualism, also known as mind-body dualism. First, the mind and body are entirely separate. Second, the mind is privileged over the body, where the latter resembles a machine or object in which the self is located. Third, thought is performed by the mind, making it the producer of the self via cognitive rationalization. See Howson, *Body in Society*.

60 Serano, *Whipping Girl*; Gill-Peterson, *Short History of Trans Misogyny*; Stryker, *Transgender History*.

61 See, for example, Spillers, "Mama's Baby, Papa's Maybe"; Gill-Peterson, *Histories of the Transgender Child*; Schuller and Gill-Peterson, "Introduction"; Snorton, *Black on Both Sides*; Kafer, *Feminist, Queer, Crip*; Butler, *Who's Afraid of Gender?*; Halberstam, *Trans**.

62 Stryker, "(De)subjugated Knowledges," 7.

63 For scholarship that critiques universal definitions of woman, see Moraga and Anzaldúa, *This Bridge Called My Back*; Lorde, "Age, Race, Class and Sex"; Anzaldúa, *Borderlands*; Crenshaw, "Demarginalizing the Intersection of Race and Sex"; Crenshaw, "Beyond Racism and Misogyny"; Hill Collins and Bilge, *Intersectionality*.

64 Cowan and Rault, "Introduction." For more on the cyborg's complicated history, see Kafer, *Feminist, Queer, Crip*, 126–28.

65 Russell, *Glitch Feminism*; Seu, *Cyberfeminism Index*; Kafer, *Feminist, Queer, Crip*; Puar, "'I Would Rather Be a Cyborg'"; Balsamo, *Technologies of the Gendered Body*; Cowan and Rault, "Introduction"; Wark, "Blog-Post for Cyborgs"; cárdenas, "Android Goddess Declaration."

66 Haraway, "Cyborg Manifesto," 151.

67 Cowan and Rault, "Introduction," 6.

68 Williams, "Transgender," 232; see also Stryker, *Transgender History*; Rawson, "Introduction."

69 Stryker, *Transgender History*, 39.

70 Gill-Peterson, *Histories of the Transgender Child*, 8.

71 Stryker, *Transgender History*; Gill-Peterson, *Histories of the Transgender Child*.

72 Angell and Roberto, "Cataloging."

73 Gill-Peterson, *Histories of the Transgender Child*.

Chapter 1. Spinning Nylon

1 As is typical in the history of science, accounts vary as to who was "the first" team member to draw nylon. This is often difficult to pinpoint and disentangle, as many important actors collaborated in the lab over prolonged periods. See Joseph X. Labovsky, interviewed by John K. Smith, in Wilmington, Delaware, on July 24, 1996, SHI, oral history transcript 0148, 19; Hounshell and Smith identify Gerard J. Berchet in *Science and Corporate Strategy*, 259. Others focus on Julian W. Hill; see Meikle, *American Plastic*, 130–36; Handley, *Nylon*, 32–33. In his oral history, Howard E. Simmons Jr. claims that his former boss Donald Coffman was "the first to really spin nylon"; see Simmons, interviewed by James J. Bohning, in Wilmington, Delaware, on April 27, 1993, SHI, oral history transcript 0111, 19.

2 Handley, *Nylon*, 33.

3 Handley, *Nylon*, 35.

4   Handley, *Nylon*, 14.

5   Handley, *Nylon*, 14.

6   Meikle, *American Plastic*, 128.

7   For more on the history of Carothers and his work, see Handley, *Nylon*, 31–36; Meikle, *American Plastic*, 128–33; Hounshell and Smith, *Science and Corporate Strategy*, 223–74.

8   "Merchants of death" was a term first used in 1930s US investigative journalism to describe munitions companies who profited from war. For more, see *Forum* 92 (July 1934), 32–33. See also Hounshell and Smith, *Science and Corporate Strategy*, 317; Handley, *Nylon*, 34; Meikle, *American Plastic*, 134.

9   Handley, *Nylon*, 34.

10   Meikle, *American Plastic*, 133; Handley, *Nylon*, 34.

11   Meikle, *American Plastic*, 133; Handley, *Nylon*, 34.

12   For more on corporate showmanship, Batten, Barton, Durstine & Osborn, and DuPont, see Bird, *Better Living*.

13   Handley, *Nylon*, 38.

14   Handley, *Nylon*, 34.

15   "Nylon: Versatile Product of DuPont Chemistry," DuPont Nylon Division, 1941, NMAH TLC.

16   "History of Dupont Exhibits, 1935–1945," notes on 1939 GGIE display, DPA, acc. 1410, boxes 58 and 59.

17   "History of Dupont Exhibits."

18   For more on race and the NYWF, see Gregory, *Black Corona*, 44; Rydell, *World of Fairs*, 192.

19   Press release, November 22, 1940, DPA, acc. 410, box 44.

20   See Teague blueprints and promotional images at DuPont company exhibits, ca. 1939–41, DPA, acc. 77.242.

21   See "Lecturers Reports," NYWF Weekly Reports 1939, DPA, acc. 1410, series II, part 2, box 35.

22   W. M. A. Hart, "Report for Week of July 2–8, 1939", DPA, series II, part 2, box 35.

23   See also Meikle, *American Plastic*, 142.

24   John Boyko, "Lecturers' Reports, Shift A, July 22, 1939," NYWF Weekly Reports 1939, DPA, series II, part 2, box 35.

25   Hounshell and Smith, *Science and Corporate Strategy*; Meikle, *American Plastic*; Handley, *Nylon*; Ndiaye, *Nylon and Bombs*; Glickman, "'Make Lisle the Style.'"

26   "$10,000,000 Plant to Make Synthetic Yarn; Major Blow to Japan's Silk Trade Seen," NYT, October 21, 1938.

27   Lockwood, "Japanese Silk and the American Market," 34.

28   Lockwood, "Japanese Silk and the American Market," 34.

29   See Steen, *American Synthetic Organic Chemicals Industry*; O'Reagan, *Taking Nazi Technology*.

30   E. I. Du Pont de Nemours and Company, *Du Pont*, 92.

31   See, for example, "History of Nylon Scrapbook," DPA, acc. 1410, box 36, 2004.543.014; Joseph Labovsky Collection, SHI; NMAH DuPont Nylon Collection, NMAH.AC.0007.

32   Glickman, "'Make Lisle the Style.'"

33   For more on US/Japanese relations, see Davidann, "'Certain Presentiment of Fatal Danger.'"

34   See the Gallup public opinion poll discussed in Davidann, "'Certain Presentiment of Fatal Danger,'" 180.

35   Glickman, "'Make Lisle the Style,'" 579.

36   Glickman, "'Make Lisle the Style,'" 603.

37   Lockwood, "Japanese Silk and the American Market," 34.

38   Lockwood, "Japanese Silk and the American Market," 34.

39   Raymond Clapper, "Artificial Silkworm Developed by DuPont," *New York World Telegram*, January 17, 1939.

40   Cummins Speakman Jr., "Supervisor's Report June 18–24 1939, Thomas W. Witherspoon Report," NYWF Weekly Reports 1939, DPA, series II, part 2, box 35.

41   Arthur Simon, "Lecturers' Reports Shift B, July 22, 1939," NYWF Weekly Reports 1939, box 35; Leonard Waller, "Weekly Report September 14, 1940," NYWF Weekly Reports 1940, box 36, both DPA, series II, part 2. Emphasis added.

42   Brooks K. Johnson, "Lecturers Reports Shift B, July 22, 1939," NYWF Weekly Reports 1939, DPA, series II, part 2, box 35.

43   For more on Japanese beetle extermination, see Paul Sampson, "Report for Week of July 2–8 1939," NYWF Weekly Reports 1939, DPA, series II, part 2, box 35. For more on the economic complexities of US-Japanese relations in the run-up to Pearl Harbor, see Pak, "Complex International Alliances."

44   Russell, *War and Nature*, 110.

45   Bernice Dinwiddie, owner of the largest dress shop in Berkeley, California, quoted in Brooks K. Johnson, "Lecturers' Reports Shift B, July 22, 1939," NYWF Weekly Reports 1939, DPA, series II, part 2, box 35.

46   Irwin Heimer, "Lecturers Reports, Shift A," July 22, 1939, NYWF Weekly Reports 1939, DPA, series II, part 2, box 35.

47   N. M. Walling, "Lecturers Reports Shift A, Week Ending July 8, 1939," NYWF Weekly Reports 1939, DPA, series II, part 2, box 35.

48   N. M. Walling, "Lecturers' Reports Shift A, Week Ending July 8, 1939," NYWF Weekly Reports 1939, DPA, series II, part 2, box 35.

49   Charles M Hackett, DPAD, "Telegram to Watson Davis, Science Service, Washington, DC, New York, May 10, 1940," DPA, 1410, folder NYWF 1939–1940, box 44.

50   Charles M Hackett, DPAD, "Telegram to Watson Davis, Science Service, Washington, DC, New York, May 10, 1940," DPA, 1410, folder NYWF 1939–1940, box 44.

51   Handley, *Nylon*, 40.

52   See Leonard Waller, "Report September 7, 1940," and W. Uffelman, "Report of the Week September 1–September 7, 1940," NYWF Weekly Reports 1940, DPA, series II, part 2, box 36.

53   W. Uffelman, "Report of the Week September 1–September 7, 1940," NYWF Weekly Reports 1940, DPA, series II, part 2, box 36.

54   Dorothy McBride report cited in W. Uffelman, "Report of the Week September 1–September 7, 1940," NYWF Weekly Reports 1940, DPA, series II, part 2, box 36.

55   David, *Fashion Victims*.

56   DuPont, *Protecting the Public Health* (1951), 20, NMAH TLC.

57   DuPont, *Protecting the Public Health* (1951), 20, NMAH TLC.

58   DuPont, *Protecting the Public Health* (1951), 19–20, NMAH TLC.

59   Norman Walling, "Notes, September 14, 1940," NYWF Weekly Reports 1940, DPA, series II, part 2, box 36.

60   Handley, *Nylon*, 10–29; David, *Fashion Victims*, 194–201.

61   Named after its inventor, Count Hilaire de Chardonnet, a lab assistant to Louis Pasteur who was hired by the French government to find a solution to the silkworm saga of 1862, when an epidemic wiped out almost all of Europe's silkworms.

62   David, *Fashion Victims*, 197. See also Handley, *Nylon*, 20.

63   Handley, *Nylon*, 21.

64   Handley, *Nylon*, 22.

65   Handley, *Nylon*, 22.

66   Handley, *Nylon*, 24.

67   David, *Fashion Victims*, 198.

68   Handley, *Nylon*, 35.

69   Meikle, *American Plastic*, 132.

70   Handley, *Nylon*, 28, 33.

71   W. R. Ellis and G. H. Donovan, "Report for Week of July 9–15," NYWF Weekly Reports 1939, DPA, series II, part 2, box 35.

72   Jos. C. Matera quoted in W. Uffelman, "Report of the Week September 1–September 7, 1940," NYWF Weekly Reports 1940, DPA, series II, part 2, box 36. Meikle, *American Plastic*, 140, details DuPont's loss of control of the prerelease publicity for nylon; this concern is also evident in the company's papers at the Hagley Museum and Library. Some newspapers discussed the patent's use of cadaverine—a naturally occurring chemical found in animal and human corpses; however, what was meant in the patent was, in fact, the cadaverine found in sticky black tar when coal is melted. See news reports in DPA, series II, part 2, box 58, e.g., "Ladies' Hose and History," *Chicago Tribune*, September 26, 1938; and "Synthetic Silk May Hurt Japan," *Buffalo Evening News*, September 26, 1938.

73   Duane L. Greenfield, "Report for Week—May 19–25, 1940"; and Leonard Waller "Report for Week—October 6–October 13, 1940," both NYWF Weekly Reports 1940, DPA, series II, part 2, box 36.

74 W. Uffelman, "Report of the Week September 1–September 7, 1940," NYWF Weekly Reports 1940, DPA, series II, part 2, box 36.

75 See, for example, "New Hosiery Held Strong as Steel," NYT, October 28, 1938; see also news clippings in DPA, series II, part 2, boxes 58 and 59.

76 Paul Sampson, "Report for Week of June 18, 1939," NYWF Weekly Reports 1939, DPA, series II, part 2, box 35.

77 "Mannequins Dressed in Rayon at Early Rendition of Golden Gate International Exposition, San Francisco, 1939," in "Outline for Extension of Wonderworld of Chemistry Usefulness by Means of a Travelling Exhibit," DPA, 77.242.4–5.

78 Sastre-Juan, "'Science in Action,'" 161.

79 See B. S. Nicholson, "Supervisor's Report, June 24, 1939"; and Thomas Witherspoon, "Weekly Report, July 1–8, 1939," both NYWF Weekly Reports 1939, DPA, series II, part 2, box 35.

80 Mitton quoted in B. S. Nicholson, "Supervisor's Report, June 24, 1939," NYWF Weekly Reports 1939, DPA, series II, part 2, box 35.

81 See FAQs from Final NYWF Reports 1940; and Leonard Waller, "Report for Week of September 7, 1940," NYWF Weekly Reports 1940, both DPA, series II, part 2, box 36.

82 For more on corporate films during this period, see Bird, *Better Living*, 144–81.

83 Herbert Chason quoted in W. M. Hart, "Report for Week of July 9–15 1939," NYWF Weekly Reports 1939, DPA, series II, part 2, box 35.

84 John Boyko, "Lecturers' Reports, Shift A, July 22, 1939," NYWF Weekly Reports 1939, DPA, series II, part 2, box 35.

85 Regina Lee Blaszczyk, "Sold on Softness: DuPont Synthetics and Sensory Experience," paper presented at the Hagley Center for the History of Business, Technology, and Society, Wilmington, DE, November 6, 2021.

86 Herbert Chason quoted in W. M. Hart, "Report for Week of July 9–15, 1940," NYWF Weekly Reports 1939, DPA, series II, part 2, box 35.

87 C. E. Speakman quoted in "B. S. Nicholson Report, June 25—July 1, 1939," NYWF Weekly Reports 1939, DPA, series II, part 2, box 35.

88 See, for example, "New Hosiery Held Strong as Steel," NYT, October 28, 1938; see also news clippings in DPA, series II, part 2, boxes 58 and 59.

89 Herbert Chason quoted in W. M. Hart "Report for Week of July 9–15, 1940," NYWF Weekly Reports 1939, DPA, series II, part 2, box 35.

90 Classen, *Deepest Sense*, xi.

91 Herbert Chason quoted in W. M. Hart, "Report for Week of July 9–15, 1940," NYWF Weekly Reports 1939, DPA, series II, part 2, box 35.

92 DPAD, press release, April 24, 1940, NYWF folder, DPA, 1410, Public Affairs Department History Files, box 44.

93   DPAD, press release, April 24, 1940, NYWF folder, DPA, 1410, Public Affairs Department History Files, box 44.

94   DPAD, "Twenty-Fifth Anniversary of Nylon," February 1964, 1939 NYWF exhibit folder, DPA, 84.259 DP Textile Fibers Product Information Collection, box 27.

95   Leonard Waller, "Report September 7, 1940," NYWF Weekly Reports 1940, DPA, series II, part 2, box 36.

96   Leonard Waller, "Report September 7, 1940" and "October 12, 1940"; B. C. Hausdorf quoted in W. Uffelman, "Report of the Week September 1–September 7, 1940," all NYWF Weekly Reports 1940, DPA, series II, part 2, box 36.

97   Leonard Waller, "Report September 7, 1940" and "October 12, 1940," NYWF Weekly Reports 1940, DPA, series II, part 2, box 36.

98   Leonard Waller, "Report September 7, 1940," NYWF Weekly Reports 1940, DPA, series II, part 2, box 36.

99   Ndiaye, *Nylon and Bombs*, 156.

100   See also *Painted Lovelies*, a 1946 British Pathé film about nylon parachutes being recycled back into dresses after the war.

101   DPAD, press release, November 20, 1942, in "Papers on Nylon," DPA, acc. 1410, box 10.

102   Handley, *Nylon*, 48.

103   Elaine Cyrus, "Facts About Fashion," *Pittsburgh Courier*, July 7, 1945, 8.

104   See "Betty Grable in Nylons," DPA, acc. 1984.259, box 72.

105   Kakoudaki, "Pin-Up"; Buszek, *Pin-Up Grrrls*.

106   DuPont company product information photograph, *Man-Made Fibers, Nylon* (1942), DPA, acc. 1972.34, box 5.

107   For more on the politics of US World War II sacrifice, see Leff, "Politics of Sacrifice"; Henthorn, *From Submarines to Suburbs*.

108   Advertisement for American Lady Foundations, *Vogue*, November 1, 1943, 28.

109   Textron, "Things to Come, Streamlines Now," advertisement, *Vogue*, November 1, 1943, 8–9.

110   "Nylon Development" and "Angels Had Help with Nylon 'Miracle'" cited in DPAD, press release, October 1963, DPA, acc. 1410, box 57.

111   "Casino Burlesque," *Pittsburgh Courier*, September 15, 1945, 20.

112   Handley, *Nylon*, 48–50.

113   John L. Clark, "Wiley Avenue," *Pittsburgh Courier*, February 9, 1946, 28; Flip Mason, "Passing Parade," *Pittsburgh Courier*, February 2, 1946, 17.

114   "Nylon Development" and "Angels Had Help with Nylon 'Miracle'" cited in DPAD, press release, October 1963, DPA, acc. 1410, box 57.

115   "Nylon Development" and "Angels Had Help with Nylon 'Miracle'" cited in DPAD, press release, October 1963, DPA, acc. 1410, box 57.

Chapter 2. Nylon and the Test Tube Girl

Parts of a section of this chapter, "Redefining Nude: Nylon Futures Beyond Test Tube Girl Pink," appeared in "Redefining Nude: Unravelling Nylon's Unmarked Norms," in "Marginalized Identities," ed. Kelly Reddy-Best and Dyese Matthews, special issue, *Fashion, Style, and Popular Culture* 11, no. 1 (2024): 45–64.

1 Frederick Simpich, "Chemists Make a New World: Creating Hitherto Unknown Raw Materials, Science Disrupts Old Trade Routes and Revamps the World of Industry," *National Geographic*, November 1939, 601.

2 Hounshell and Smith, *Science and Corporate Strategy*; Meikle, *American Plastic*; Handley, *Nylon*; Ndiaye, *Nylon and Bombs*.

3 DuPont, "Nylon: Versatile Product of DuPont Chemistry," 1941, NMAH TLC.

4 DPAD, press release, April 22, 1940, NYWF folder, Public Affairs Department History Files, DPA, acc. 1410, box 44.

5 DPAD, "History of DuPont Exhibits 1935–1945," in notes on SFWF display, DPA, acc. 1410, boxes 58 and 59.

6 Gregory, *Black Corona*, 44; Rydell, *World of Fairs*, 192.

7 DPAD, "News for Publication," March 14, 1938, NYWF folder, Public Affairs Department History Files, DPA, acc. 1410, box 44.

8 DPAD, "News for Publication," March 14, 1938, NYWF folder, Public Affairs Department History Files, DPA, acc. 1410, box 44.

9 March, *Cassell Dictionary of Classical Mythology*, 57–60.

10 DPAD, "News for Publication," March 14, 1938, NYWF folder, Public Affairs Department History Files, DPA, acc. 1410, box 44.

11 See, for example, National Association of Manufacturers, "Fashions Out of Test Tubes," 1940, Hagley Museum and Library; DuPont, "Blazing the Trail to New Frontiers Through Chemistry," 1940, 14; and "Tall Tales and Fabulous Facts: Dow Corning Silicone News, New Frontier Edition," 1953, both SHI; and Dow Chemical Company, "Technology of Peace," 1946, NMAH TLC. See also Henthorn, *From Submarines to Suburbs*.

12 National Association of Manufacturers, "Fashions Out of Test Tubes," 1940, Hagley Museum and Library.

13 See Meikle, *American Plastic*; Handley, *Nylon*. The map is also on show at the Science History Institute in the museum's *Spinning the Elements* permanent display and featured in the Vitra Design Museum exhibition and catalog *Plastic: Remaking Our World*.

14 Meikle, *American Plastic*, 66.

15 Meikle, *American Plastic*, 66.

16 Meikle, *American Plastic*, 66. For more on colonialism and exploitation, see Tully, "Victorian Ecological Disaster"; Headrick, *Humans Versus Nature*.

17   "Our Motion Picture Library," *DuPont Magazine* 35, no. 7–8, (1941): 7, 24, Hagley Museum and Library. Screenings also include specialist groups, including engineers' groups, women's lunch clubs, farmers' clubs, and cleaners' associations. See "Mrs. Yarrington on Rotary Program," *Florence (AL) Herald*, April 19, 1946, 6; "Shriners See Picture," *Chattanooga (TN) Daily Times*, January 30, 1943, 2; "Cleaners Meeting Held Here Sunday," *Bristol (TN) Herald Courier*, September 1, 1941, 3; "Facts for Farmers" *McIntosh County Democrat* (Checotah, OK), June 29, 1944, 2; "DuPont Company's Varied Products Shown in Film," *Morning News* (Wilmington, DE), July 24, 1940, 14.

18   *A New World Through Chemistry*, 1939, DPA, DuPont Company films and commercials, acc. 1995.300, Hagley ID FILM_1995300_FC121.

19   DPAD, April 22, 1940, NYWF folder, DPA, Public Affairs Department History Files, acc. 1410, box 44.

20   Bird, *Better Living*.

21   The DPAD's aesthetic control over the 1939 "Wonder World of Chemistry" shows and publicity can be seen in a promotional photograph of a lab performance: The original male lecturer's face has been replaced with a superimposed image of another man. See "Nylon Hosiery Demonstration at 1939 New York World's Fair," DPA, AVD_2004268_P00001520, P-00001520, DuPont Company External Affairs Department photograph file, acc. 2004.268.

22   DuPont, "Show Business," *Better Living*, July–August 1952, 20–21, Hagley Museum and Library.

23   Nylon model introduced June 24, 1939, "Supervisor's Report—B. S. Nicholson," NYWF Weekly Reports 1939, DPA, series II, part 2, box 35.

24   Painter, *History of White People*, 367.

25   For Test Tube Girl, see DuPont, *Blazing the Trail to New Frontiers Through Chemistry*, 1940, SHI; for Princess Plastics, see DuPont Style News Service, "Test Tube Fashions," June 15, 1940, DPA, acc. 1410, box 44; for "Miss Chemistry," see DuPont, "The Chemical Girl of 1940," *DuPont Magazine* 34, no. 6 (June 1940): 16, Hagley Museum and Library.

26   "Outline for Extension of 'Wonder World of Chemistry' Usefulness by Means of Traveling Exhibit," 1939, DPA, 77.242.4–5, H/R 207601–609.

27   Marchand, "Designers Go to the Fair."

28   B. S. Nicholson, "Supervisor's Report," June 11, 1939, NYWF Weekly Reports 1939, DPA, series II, part 2, box 35.

29   Teague quoted in Marchand, "Designers Go to the Fair," 15.

30   Teague quoted in Marchand, "Designers Go to the Fair," 15.

31   DuPont, press release, September 19, 1940, NYWF folder, DPA, acc. 1410, box 44.

32   See DuPont, "The Chemical Girl of 1940," *DuPont Magazine* 34, no. 6 (June 1940): 16, Hagley Museum and Library; DuPont, *Blazing the Trail to New Frontiers Through Chemistry*, 1940, 14, SHI. Similar text and

imagery are also featured in the National Association of Manufacturers' promotional material, including a "Fashions Out of Test Tubes"–themed 1940 *Modern Pioneer* dinner program and stage show (see also Bird, *Better Living*, 134–35), and window display instructions, all Hagley Museum and Library. This continued into the postwar period, for example, with Spun-Lo advertising rayon in this way, as well as imagery in *Chemical Peddler*. See also the cover artwork for the pop music albums X-Ray Spex, *Germfree Adolescents* (1978) and Girls Aloud, *Chemistry* (2005).

33   DuPont Style News Service, "Test Tube Fashions," June 15, 1940, NYWF folder, DPA, acc. 1410, box 44.

34   DuPont, "Nylon: Versatile Product of DuPont Chemistry," 1941, 8, NMAH TLC.

35   See W. Uffelman, "Report of the Week September 1–September 7, 1940," NYWF Weekly Reports 1940, DPA, series II, part 2, box 36.

36   Simpich, "Chemists Make a New World," 601.

37   Scripts for Display, "Lectures for J. L. Hudson, 'Wonder World' Show," December 27, 1940, DPA, acc. 1410, box 58.

38   Frank Patton, "The Test Tube Girl," *Amazing Stories*, January 1942, 10.

39   Patton, "Test Tube Girl," 13.

40   Donald Deskey, "Radically New Dress System for Future Women Prophesies Donald Deskey," *Vogue*, February 1939, 137.

41   Deskey, "Radically New Dress System," 137.

42   Walter Dorwin Teague, "Nearly Nude Evening Dress Designed by Walter Dorwin Teague," *Vogue*, February 1939, 143.

43   George Sakier, "No Mechanistic Clothes for Future Women Predicts George Sakier," *Vogue*, February 1939, 144.

44   Raymond Loewy, "Raymond Loewy Designer of Locomotives and Lipsticks, Creates a Future Travel Dress," *Vogue*, February 1939, 141.

45   See Cogdell, *Eugenic Design*.

46   Cogdell, *Eugenic Design*, 4.

47   Loewy, *Never Leave Well Enough Alone*, 220.

48   See Cogdell, *Eugenic Design*, 50.

49   See Dyer, *White*; Painter, *History of White People*; Carter, *Heart of Whiteness*.

50   See Gilman, *Making the Body Beautiful*; Haiken, *Venus Envy*.

51   Painter, *History of White People*, 364.

52   Painter, *History of White People*, 364.

53   "Vogue Stars on the Screen," *Vogue*, February 1, 1939, 96–97.

54   *Eve, AD 2000!* (British Pathé, 1939), newsreel, "In the Year 2000: Fashion Predictions from 1939," 1 min., 18 sec., https://www.britishpathe.com/asset/68159/.

55   Hall, "Spectacle of the 'Other,'" 240.

56   Gregory, *Black Corona*, 44; Rydell, *World of Fairs*, 192.

57   Rydell, *World of Fairs*, 192.

58   "Good Form in America," *Vogue*, February 1939, 114.

59 "Fashion: La Belle Poitrine," *Vogue*, September 1944, 167.

60 Staiti, "Real Women, Normal Curves."

61 See Staiti, "Real Women, Normal Curves"; Carter, *Heart of Whiteness*; Dyer, *White*; Painter, *History of White People*; Cogdell, *Eugenic Design*; Urla and Swedland, "Anthropometry of Barbie"; Creadick, *Perfectly Average*; Stephens, "Normal Body on Display."

62 Staiti, "Real Women, Normal Curves."

63 See, for example, DuPont, "Experimental Stocking," 1937, 54G1/N20–40a; DuPont, "Experimental Stocking Section," 1937–38, 54G1/N20–44.6; DuPont, "Experimental Stocking," 1937, 54G1/N20–40b, all DuPont Museum Collection, Hagley Museum; various experimental nylon hose, 1938–39 in Labovsky box 11, Joseph X. Labovsky Nylon Collection, SH1.

64 Fields, *Intimate Affair*, 103.

65 Cosmetics, such as "invisible" face powders in shades of white and pink, emerged in the 1860s and share a similar history of color descriptions. See Peiss, *Hope in a Jar*, 58; Held, "Redefining Nude," Google Arts and Culture, June 2022, https://artsandculture.google.com/story/WwXxveh13cuaWQ.

66 Burris-Meyer, *Color and Design*, 90.

67 Textile Color Card Association of the United States, *Standard Color Card of America*.

68 Carter, *Heart of Whiteness*.

69 Blaszczyk, *Color Revolution*, 3.

70 Stott, "Dyeing of Nylon Fibers," 584; Stott, "Dyeing of Nylon Hosiery," 710.

71 Stott, "Dyeing of Nylon Fibers," 584.

72 Fields, *Intimate Affair*, 158, 164.

73 Fields, *Intimate Affair*, 113–73.

74 Fields, *Intimate Affair*, 114.

75 Burris-Meyer, *Color and Design*, 199.

76 Burris-Meyer, *Color and Design*, 199.

77 Sears catalog, Fall–Winter 1956, 258; Sears catalog, Spring–Summer 1954, 250.

78 Fifth Avenue Fashions, advertisement, *The Crisis*, August–September 1958, 463.

79 Basin Street Nylons, advertisement, *Tan Confessions*, November 1951, 55.

80 See, for example, "pecan" in Sears catalog, Spring–Summer 1958, 157, 174; Sears catalog, Spring–Summer 1959, 148, 151. See also "toast brown" and "spice brown" listed in Textile Color Card Association of the United States, *Standard Color Card of America*, 12.

81 For a more in-depth study of Sumner Hosiery and Basin Street Nylons, see Held, "Redefining Nude."

82 See, for example, Staiti, "Real Women, Normal Curves"; Creadick, *Perfectly Average*; Carter, *Heart of Whiteness*.

83 See the following Sears catalogs: Spring–Summer 1954, 250; Christmas Book 1954, 126; Spring–Summer 1956, 162; Christmas Book 1956, 48; Fall–Winter 1956, 259; Spring–Summer 1957, 194; Fall–Winter 1959, 148; Spring–Summer 1959, 148.

84 See, for example, Sears catalog, Christmas Book 1953, 136; Sears catalog, Christmas Book 1954, 127.

85 Sears catalog, Spring–Summer 1961, 187.

86 See, for example, SHI, "Big Mama Legsavers Pantyhose," ca. 1970s, https://digital.sciencehistory.org/works/c254dzj; Big Mama, advertisements in *Life*, November 19, 1971, 18; and *Life*, December 3, 1971, 22.

87 Betty-Bill, "Letter to Editor," *Letters from Female Impersonators* 15 (1963): 58, DTA.

88 Chris Ames, "Letter to Editor," *Letters from Female Impersonators* 9 (1962): 42–54, DTA.

89 Chris Ames, "Letter to Editor," *Letters from Female Impersonators* 6 (1962): 57, DTA.

90 Suli Montiel, "Letter to Editor," *Letters from Female Impersonators* 11 (1962): 62, DTA.

91 Bobbie Weaver, "Letter to Editor," *Letters from Female Impersonators* 6 (1962): 34, DTA.

92 Erica, "Letter to Editor," *New Trenns Magazine* 2, no. 6 (1971): 5, DTA.

93 See, for example, *Letters from Female Impersonators* 3 (1961): 30, 38; 9 (1962): 4, 15, 17; 15 (1963): 18, 36; 12 (1963): 59, all DTA.

94 Dorothy, "Letter to Editor," *Letters from Female Impersonators* 2 (1961): 9, DTA.

95 Betty-Bill, "Letter to Editor," *Letters from Female Impersonators* 15 (1963): 52, DTA; Diehl, "Zelda Wynn Valdes," 234.

96 Wilma, "Letter to Editor," *Letters from Female Impersonators* 4 (1961): 17, DTA.

97 Jakki J., "Letter to Editor," *Letters from Female Impersonators* 6 (1962): 9, DTA.

98 Betty Bill's sister Ruth, "Letter to Editor," *Letters from Female Impersonators* 11 (1962): 11, DTA.

99 Joyce, "Letter to Editor," *Letters from Female Impersonators*, Winter 1969, 66, DTA.

100 Lorraine Channing, "Letter to Editor," *Turnabout: A Magazine for Transvestitism* 2 (1963): 20, DTA.

101 Erica, "Letter to Editor," *New Trenns* 2, no. 6 (1971): 5, DTA.

Chapter 3. Soft Power

1 IG Farben, formed in 1925, was the result of a merger of six German chemical companies, which included Bayer, Agfa, and BASF. During World War II, IG Farben produced Zyklon B poison gas, which

killed over one million Jews. A factory was constructed close to the Auschwitz concentration camp, and it is estimated that over half of IG Farben's workforce of 330,000 individuals, working across its various subsidiaries in 1943, were enslaved or conscripts. The Allies seized IG Farben at the end of the war because of its involvement in the Holocaust, and in 1951, they split the company back into its original six entities; see Hayes, *Industry and Ideology*.

2   Bayer, "Polyurethanes."

3   See "Urethane Plastics," 1383. I have consciously decided to use "poly-urethane," as it is the term most commonly used in recent design and technology scholarship. See Pavitt, "Future Is Possibly Past"; Neushul and Westwick, "Blowing Foam."

4   Bayer, "Polyurethanes," 150.

5   Otto Bayer created these innovative materials by applying the princi-ple of polyaddition to liquid diisocyanates and existing polyester and polyether diols. This technique, referred to as the "basic diisocyanate polyaddition process," entails the reaction of two alcohol groups—isocyanates and polyols—to create urethane. See "Urethane Plastics"; Frisch and Saunders, *Plastic Foams*; Crawford, *Plastics and Rubber*.

6   Neushul and Westwick, "Blowing Foam," 58.

7   "Urethane Plastics."

8   "Urethane Plastics," 1383.

9   "Urethane Plastics."

10   "Urethane Plastics," 1384.

11   Dombow, *Polyurethanes*, 4; Frisch and Saunders, *Plastic Foams*; "Urethane Plastics," 1384; Crawford, *Plastics and Rubber*; Neushul and Westwick, "Blowing Foam," 58.

12   Neushul and Westwick, "Blowing Foam," 58; "Urethane Plastics"; Crawford, *Plastics and Rubber*.

13   Doriot, "Foreword," iii.

14   DeBell, Goggin, and Gloor, *German Plastics Practice*, 4.

15   Steen, *American Synthetic Organic Chemicals Industry*.

16   Leslie, *Synthetic Worlds*; Westermann, "Material Politics of Vinyl."

17   Colonel Romer and General Doriot, conversation transcript, August 14, 1945, War Department Files Correspondence: German Scientists and Experts, Georges Doriot Papers, box 8, LOC.

18   See correspondence in papers of Walter E. Gloor (SHI) and Georges Doriot (LOC).

19   See, for example, Gaston DuBois, letter to Walter E. Gloor, July 8, 1947, WEGP, box 1, folder 1, SHI.

20   John M. DeBell, William C. Goggin, and Walter E. Gloor, "HQ The-ater Service Forces, European Theater, Office of the Chief Quarter-master, Quartermaster Report: Technical Intelligence Reports; The German Plastics Industry," 1945, p. 1, WEGP, box 1, folder 9, SHI (here-after cited as DeBell, Goggin, and Gloor, "German Plastics Industry").

21   "Industrial Germany Today," 7, attachment in Gaston DuBois, letter to
      Walter E. Gloor, July 8, 1947, WEGP, box 1, folder 1, SHI.

22   John M. DeBell and Henry N. Richardson, "Notes on the Stimulation
      of Plastics Technology by German Disclosures," September 18, 1947,
      9, WEGP, box 1, folder 5, SHI. See also DeBell, Goggin, and Gloor,
      German Plastics Practice, 12.

23   Gaston DuBois, letter to Walter E. Gloor, July 8, 1947, WEGP, box 1,
      folder 1, SHI.

24   DeBell, Goggin, and Gloor, "German Plastics Industry," p. 1.

25   DeBell, Goggin, and Gloor, "German Plastics Industry," p. 1.

26   Kline, "Plastics in Germany" (part 1), 152F.

27   DeBell, Goggin, and Gloor, "German Plastics Industry," app. 8, p. 1.

28   DeBell, Goggin, and Gloor, "German Plastics Industry," app. 8, p. 2.

29   DeBell, Goggin, and Gloor, "German Plastics Industry," app. 8, p. 6,
      11.

30   DeBell, Goggin, and Gloor, "German Plastics Industry," app. 8, p. 1.

31   DeBell, Goggin, and Gloor, "German Plastics Industry," app. 8a,
      p. 22.

32   DeBell, Goggin, and Gloor, *German Plastics Practice*, 465.

33   DeBell, Goggin, and Gloor, *German Plastics Practice*, 465.

34   DeBell, Goggin, and Gloor, *German Plastics Practice*, 456–61.

35   John DeBell, letter to General Georges Doriot, October 23, 1945,
      WEGP, box 1, folder 1, SHI.

36   See W. C. Goggin of Dow Chemical Company, letter to Major Hob-
      son, Office of the Quartermaster General, January 29, 1946, WEGP,
      box 1, folder 1, SHI.

37   Kline, "Plastics in Germany" (part 1); Kline, "Plastics in Germany"
      (part 2).

38   Reports on German and Japanese industry, including FIAT reports,
      were also distributed in the United Kingdom by HM Stationery Office
      and could be viewed by appointment at the Technical Information
      and Document Unit. See *Classified List No. 19: Reports on German and
      Japanese Industry Published During the Period April 1 to August 31, 1948*
      (London: HMSO, 1948), WEGP, box 1, folder 1, SHI.

39   Kline, "Plastics in Germany" (part 1), 152A; Kline, "Plastics in Ger-
      many" (part 2), 212.

40   See O'Reagan, *Taking Nazi Technology*; Crim, *Our Germans*; Neufeld,
      "Nazi Aerospace Exodus."

41   Doriot, "Foreword," iii.

42   "Industrial Germany Today," attachment in Gaston DuBois, letter to
      Gloor, July 8, 1947, WEGP, box 1, folder 1, SHI.

43   Neushul and Westwick, "Blowing Foam"; "Urethane Plastics"; Frisch
      and Saunders, *Plastic Foams*; Crawford, *Plastics and Rubber*. For more
      on the military-industrial-academic complex, see Wisnioski, *Engineers
      for Change*, 24. For Lockheed, see Leslie, *Cold War and American*

Science. For DuPont, see Ndiaye, *Nylon and Bombs*, which points out that in comparison to DuPont, General Electric, McDonnell Douglas, and Lockheed all had far closer collaborative ties with the US government.

44   "Urethane Plastics," 1384.

45   Raider Winget, "Foam Rubber Business Has Hit Bonanza," *Sunday Times* (Cumberland, MD), March 19, 1950.

46   "Urethane Plastics," 1387.

47   Frisch and Saunders, *Plastic Foams*; "Urethane Plastics."

48   "Urethane Plastics."

49   Frisch and Saunders, *Plastic Foams*.

50   See, for example, Oreskes and Krige, *Science and Technology*; Crim, *Our Germans*; Neufeld, "Nazi Aerospace Exodus"; O'Reagan, *Taking Nazi Technology*; Leslie, *Cold War and American Science*; Epstein, *Torpedo*; Hitch and McKean, *Economics of Defense*.

51   See, for example, May, *Homeward Bound*; Henthorn, "Emblematic Kitchen"; Henthorn, "Commercial Fallout"; Henthorn, *From Submarines to Suburbs*; Hixson, *Parting the Curtain*; Castillo, *Cold War on the Home Front*; Colomina, *Domesticity at War*; Colomina et al., *Cold War Hothouses*; Oldenziel and Zachmann, *Cold War Kitchen*; Preciado, *Pornotopia*.

52   Forrestal, *Faith, Hope, and $5,000*, 160.

53   Forrestal, *Faith, Hope, and $5,000*, 160.

54   The photograph in figure 3.4 of Otto Bayer, labeled as being taken in Göttingen, is likely to have been taken at a foam demonstration to students in this famous university city.

55   For more on the carefully monitored use of images of the atomic bombings, see Maclear, *Beclouded Visions*; Boyer, *By the Bomb's Early Light*.

56   DuPont, "Uses of Organic Isocyanates" (ca. 1950–59), DPA, DuPont Company Product Information photographs, acc. 1972.341.

57   Barthes, "Plastic," 117.

58   "Foam Furniture" in "Tomorrow's Life Today—Man's New World: Part II," *Life*, November 11, 1957, 132.

59   See, for example, V. Kagan, *Upholstered Furniture*; *Designing with Foam* (Midland, MI: Dow Chemical Company, 1972), SHI.

60   Barthes, "Plastic," 117.

61   Colomina et al., *Cold War Hothouses*; Colomina, *Domesticity at War*; Castillo, *Cold War on the Home Front*; Preciado, *Pornotopia*.

62   Riesman, "Nylon War," 67.

63   See "First Woman on Trade Trip to Russia Leaves Tomorrow," *NYT*, July 4, 1963; Alice Hughes, "U.S. Clothiers Plan Tour of Russian Apparel Plants" *Poughkeepsie (NY) Journal*, June 27, 1963; Harry Golden, "Only in America: Uplift Mission," *New York Post*, November 18, 1963; Frank G. Siscoe, Department of State, letter to Ida Rosenthal, December 31, 1963; Hinda R. Kohn, "Olga and Sonja Have Style but Quality

	Is Still Lacking," *Women's Wear Daily*, July 29, 1963, all in Maidenform Collection 1922–1997, acc. 585, box 66, folder 7, NMAH Archives Center.
64	Ida Rosenthal quoted in Kohn, "Olga and Sonja Have Style but Quality Is Still Lacking."
65	See, for example, Henthorn, "Commercial Fallout"; Henthorn, "The Emblematic Kitchen"; Henthorn, *From Submarines to Suburbs*; Castillo, *Cold War on the Home Front*; Crowley and Pavitt, *Cold War Modern*; Colomina, "Cold War/Hothouses."
66	"50 Years of American Art" was organized in 1955 by the Museum of Modern Art, New York, and sent on tour to Paris, Zurich, Barcelona, Frankfurt, London, The Hague, Vienna, and Belgrade. See Pavitt, "Design and the Democratic Ideal," 83; Colomina, "Cold War/Hothouses"; Castillo, *Cold War on the Home Front*; de Grazia, *Irresistible Empire*.
67	See Oldenziel and Zachmann, *Cold War Kitchen*.
68	May, *Homeward Bound*.
69	May, *Homeward Bound*, 16.
70	Richard Nixon quoted in May, *Homeward Bound*, 17.
71	Wilson, *Mid-Century Modernism*, 3.
72	John H. Johnson, "Why Negroes Buy Cadillacs," *Ebony*, September 1949, 43, cited in Wilson, *Mid-Century Modernism*, 100; and Chambers, *Madison Avenue*, 44.
73	Bill Van Alstine, "Impact of the Negro on the Furniture Market," *Ebony*, April 1, 1963, 99–104.
74	Alstine, "Impact of the Negro," 99.
75	Castillo, *Cold War on the Home Front*.
76	Colomina, "Cold War/Hothouses," 14.
77	May, *Homeward Bound*, 106; Smith, "Bombshell"; Bailey, *From Front Porch to Back Seat*; Preciado, *Pornotopia*, 69–76.
78	Preciado, *Pornotopia*, 71.
79	Preciado, *Pornotopia*, 71. See also May, *Homeward Bound*, 92–113.
80	May, *Homeward Bound*, 19.
81	See, for example Henthorn, "Emblematic Kitchen"; Henthorn, "Commercial Fallout"; Oldenziel and Zachmann, *Cold War Kitchen*; May, *Homeward Bound*; Cowan, *More Work for Mother*; Lupton, *Mechanical Brides*; Hixson, *Parting the Curtain*; Castillo, *Cold War on the Home Front*; Colomina, *Domesticity at War*; Colomina et al., *Cold War Hothouses*.
82	Alstine, "Impact of the Negro," 99.
83	Reid, "'Our Kitchen Is Just as Good,'" 156.
84	Reid, "'Our Kitchen Is Just as Good,'" 154. See also Henthorn, *From Submarines to Suburbs*, 3.
85	Frisch and Saunders, *Plastic Foams*, 13.
86	Englander, Princess mattress advertisement, *Life*, March 12, 1965, 79.

87   Frisch and Saunders, *Plastic Foams*, 13.

88   See, for example, General Motors, "Big GMG Breakthrough" advertisement, *Life*, May 23, 1960, 100–101.

89   "Urethane Plastics," 1388.

90   See *Put the "Soft Sell" of Urethane Foam into Your Furniture Sales Story* (Pittsburgh: Mobay Chemical Company, 1959), Freda Diamond Collection 1945–1984, acc. 616, box 1, folder 9, NMAH Archives Center (hereafter cited as Mobay, *Soft Sell*).

91   Mobay, advertisement, *Interior Design*, March 1959, 33.

92   See, for example, the following Goodyear Airfoam advertisements in *Life*: July 31, 1950, 60; March 26, 1951, 3; October 31, 1955, 11; September 8, 1958, 86; February 29, 1960, 37.

93   See, for example, Goodyear, Airfoam advertisement, *Life*, March 26, 1951, 3.

94   Association of Motion Picture Producers Inc., "A Code Regulating Production of Motion Pictures," *Exhibitors Herald and Moving Picture World*, April 5, 1930, 13.

95   Mobay, *Soft Sell*.

96   For more on peacetime reconversion, gender, and consumption, see Henthorn, *From Submarines to Suburbs*, 86–100.

97   Mobay, *Soft Sell*.

98   It is important to note that the original Womb Chair was padded with latex foam, or what is described simply as a "layer of foam rubber," which is unlikely to have been polyurethane foam. Later editions and the contemporary version are padded with polyurethane foam. See McAtee, "Taking Comfort," 4; Pavitt, "Design and the Democratic Ideal," 83.

99   Florence Knoll quoted in Rita Reif, "Pioneer in Modern Furniture Is Charting Expansion Course," *NYT*, June 17, 1959.

100   McAtee, "Taking Comfort," 17.

101   McAtee, "Taking Comfort," 25.

102   McAtee, "Taking Comfort"; on the racialized gendered body, see Wilson, *Mid-Century Modernism*, 153–57.

103   Pavitt, "Design and the Democratic Ideal," 83.

104   V. Kagan, *Upholstered Furniture*.

105   See Johnson, "Why Negroes Buy Cadillacs"; Chambers, *Madison Avenue*; Wilson, *Mid-Century Modernism*.

106   "Hilltop Living," *Ebony*, February 1, 1960, 36–43; "New Ideas for Sectional Furniture," *Jet*, June 9, 1955, 38–40.

107   B. F. Goodrich, "Meet Mr. White," *Ebony*, August 1969, 30; Monsanto, "Engineers! Opportunity is Knocking . . . Will You Answer?," *Ebony*, August 1, 1973; "Monsanto's Designer 'Extraordinaire': Norma Curby Is Among Few Female Structural Design Engineers in U.S.", *Ebony*, October 1973, 135–40; Monsanto, "Success Is Not a Black or White Matter Here," *Ebony*, February 1, 1974, 7, and March 1, 1974, 90.

108	"Modern Living: Wrought Iron Furniture," *Jet*, February 5, 1953, 40.
109	Castro Convertibles, "Sleeping Beauties" advertisement, *Life*, September 15, 1961, 70.
110	Mobay, advertisement, *Interior Design*, March 1959, 33.
111	Mobay, *Soft Sell*.
112	V. Kagan, *Upholstered Furniture*.
113	Mobay, *Soft Sell*.
114	See Gaetano Pesce's 1969 UP seating series; and "The Plopupon Experiment in Casual Furniture" in *Designing with Foam* (Midland, MI: Dow Chemical Company, 1972), 12–13, SHI.
115	"B&B Italia's Revolutionary Sit-In," *Wallpaper*, October 25, 2019, archived October 28, 2019, at https://web.archive.org/web /20191028202852/https://www.wallpaper.com/w-bespoke/bb-italia -revolutionary-furniture.
116	Pavitt, "Future Is Possibly Past," 41.

Chapter 4. Outward, Upward, and Inward

Parts of a section of this chapter, "Foam Moved Upward: Queering Falsies," were published as "Shaping Foundations" in "Transgender Embodiment in Fashion and Beauty," special issue, *Critical Studies in Fashion and Beauty* 15, no. 1 (2024): 49–74.

1	Yalom, *History of the Breast*; Thesander, *Feminine Ideal*; Fields, *Intimate Affair*; Farrell-Beck and Gau, *Uplift*.
2	William Rosenthal, assignor to Maiden Form Brassiere Company, design for a brassiere, US Patent 112,238, published November 15, 1938. See also the selection of advertising materials and designs for the Maidenform Chansonette brassiere in a Maidenform foundation sales kit, 1937–38, Maidenform Collection 1922–1997, acc. 585, box 67, folder 1, NMAH Archives Center.
3	Farrell-Beck and Gau, *Uplift*.
4	"Dressing Through Two Decades," *Ebony*, November 1965, 213–18.
5	See Craig, *Ain't I a Beauty Queen?*; Matelski, *Reducing Bodies*, 108–28.
6	Jim Goodrich and A. S. "Doc" Young, "Why Hollywood Won't Glamorize Negro Girls," *Jet*, September 17, 1953, 56–61.
7	Craig, *Ain't I a Beauty Queen?*, 30.
8	Walker, *Style and Status*, 132.
9	See Matelski, *Reducing Bodies*, 108–28.
10	Brown, *Work!*, 208.
11	Brown, *Work!*, 208.
12	For more on *Tan Confessions*, see Ahad, "Confessions." See also Fields, *Intimate Affair*, 216.

13   For more on the 1930s emergence of "bra wardrobes," see Farrell-Beck and Gau, *Uplift*, 62; see also "Good Form in America" and "WACS Prove Good Market," *Corset and Underwear Review*, August 1943, 61.

14   "Maidenform Now Making Two New Types of 'Masquerade' Bust Pads," *Maidenform Mirror*, September 1946, 5, Maidenform Collection, NMAH. AC.0585, series 4, box 19, NMAH Archives Center (hereafter cited as "Maidenform Now Making Two New Types of 'Masquerade' Bust Pads").

15   See for example the following Sears catalogs: Fall–Winter 1940, 154; Spring–Summer 1941, 186; Fall–Winter 1941, 205; Spring–Summer 1942, 199.

16   Corset and Brassiere Association of America, *Foundations for Fashion*, 15.

17   "Maidenform Now Making Two New Types of 'Masquerade' Bust Pads," 5.

18   "Maidenform Now Making Two New Types of 'Masquerade' Bust Pads," 5.

19   Sears catalog, Fall–Winter 1951, 340–41; Bailey, *From Front Porch to Back Seat*, 74.

20   For more on the power of the pin-up and idealized visions of women during World War II, see also May, *Homeward Bound*; Kakoudaki, "Pin-Up"; Buszek, *Pin-up Grrrls*.

21   See Gottwald and Gottwald, *Frederick's of Hollywood*.

22   For more on this, see Held, "Redefining Nude"; Isabelle Held, "Redefining Nude," Google Arts and Culture, June 2022, https:// artsandculture.google.com/story/WwXxveh13cuaWQ.

23   See Farrell-Beck and Gau, *Uplift*.

24   Liftee, advertisement, *Ebony*, July 1947, 4.

25   Frederick's of Hollywood catalog 20, no. 56 (1960): 9, Hagley Museum and Library.

26   Frederick's of Hollywood catalog 20, no. 56 (1960): 17, Hagley Museum and Library.

27   Craig; *Ain't I a Beauty Queen?*; Walker, *Style and Status*; Matelski, *Reducing Bodies*.

28   "The World's Best Dressed Woman," *Jet*, November 29, 1951, 37.

29   Star, *"I Changed My Sex!,"* 99.

30   Maddock and Wheeler, *Sex Life of a Transvestite*, 76, DTA.

31   Maddock and Wheeler, *Sex Life of a Transvestite*, 75, DTA.

32   "Breast Works," *Female Impersonators* 1 (Winter 1969): 49; "Pudgy Roberts," *Female Mimics* 1, no. 8 (1965): 24, both DTA.

33   Pudgy Roberts interviewed in Gil Truman, "Girls Who Are and Can't," *Female Mimics* 1, no. 12 (1968): 21, DTA.

34   Roberts, *Female Impersonator's Handbook*, 48–52.

35   Roberts, *Female Impersonator's Handbook*, 24.

36   Ms. Bob and Carol Kleinmaier, "An Interview with Classic FI Pudgy Roberts," *TG Forum*, January 10, 2022, https://tgforum.com/an -interview-with-classic-fi-pudgy-roberts/.

37   Sally Douglas, "Breast Prosthetics for Ladies and Female Impersonators," *New Trenns* 4 (1970): 67, DTA.

38 Douglas, "Breast Prosthetics for Ladies and Female Impersonators," 67, DTA.

39 Douglas, "Breast Prosthetics for Ladies and Female Impersonators," 67, DTA.

40 "Pudgy Roberts Female Impersonator Guide," advertisement, *Female Impersonators* 2 (Summer 1969): 69, DTA.

41 Roberts, *Female Impersonator's Handbook*, 49.

42 Douglas, "Breast Prosthetics for Ladies and Female Impersonators," 73, DTA.

43 "The Truth About Female Impersonators," *Jet*, October 2, 1952, 27.

44 "2,500 Impersonators Frolic at Annual Ball in New York," *Jet*, December 10, 1953, 16–17.

45 "Female Impersonators Cavort at 'Fashionable' Chicago Ball," *Jet*, November 14, 1957, 23.

46 Miss Major quoted in Stryker, *Transgender History*, 75.

47 Toshio Meronek and Miss Major, *Miss Major Speaks*, 3.

48 Newton, "Selection from *Mother Camp*," 123.

49 See, for example, Farrell-Beck and Gau, *Uplift*; Fields, *Intimate Affair*.

50 *Female Mimics* 1 (1963): 35, 36, 55; *Female Impersonators on Parade* 3 (1961): 59, both DTA.

51 Ellison, "The Labor of Werqing It," 8.

52 Pudgy Roberts interviewed in Gil Truman, "Girls Who Are and Can't," *Female Mimics* 1, no. 12 (1968): 21, DTA.

53 Butler, "Imitation and Gender Insubordination," 313.

54 Roberts interviewed in Truman, "Girls Who Are and Can't," 21, DTA.

55 Brevard, *Woman I Was Not Born to Be*, 102.

56 Brevard, *Woman I Was Not Born to Be*, 79. See also Meyerowitz, *How Sex Changed*, 93. Susan Stryker, "Aleshia Brevard Crenshaw Interview Transcript," oral history, August 2, 1997, DTA.

57 *Female Impersonators on Parade* 3 (1961): 11, DTA.

58 Newton, "Selection from *Mother Camp*," 123.

59 Newton, "Selection from *Mother Camp*," 123.

60 "Hans Crystal," *Female Mimics* 1, no. 7 (1965): 34, DTA.

61 Letter from the editor to the cast of Finocchio Club, *Female Mimics* 1, no. 6 (1965): 67, DTA.

62 Farrell-Beck and Gau, *Uplift*, 75.

63 Gardner, "From Cotton to Silicone," 115.

64 Gardner, "From Cotton to Silicone."

65 Camp "Tru Life," promotional booklet, Walter Spohn Papers, Division of Medicine and Science, NMAH.

66 Camp "Tru Life," promotional booklet, Walter Spohn Papers, Division of Medicine and Science, NMAH.

67 "Society for the Second Self," brochure, 1980, DTA.

68 Stryker, *Transgender History*.

69 Hal Boyle, "Venus Was a Fat Pig: What's Behind the Bust Boom?," *Waterloo (IA) Daily Courier*, August 15, 1959, 8.

70 See, for example, Frederick's of Hollywood catalog 19, no. 87 (1965): 23, Hagley Museum and Library.

71 Frederick's of Hollywood catalog 19, no. 87 (1965): 16, Hagley Museum and Library.

72 "Maidenform Now Making Two New Types of 'Masquerade' Bust Pads," 5.

73 Joseph Coleman, assignor to Maiden Form Brassiere Company Inc., design for a hip pad, US Patent 2,481,291, published September 6, 1949.

74 "Maidenform Now Making Two New Types of 'Masquerade' Bust Pads," 5.

75 For more on neoclassical sculpture and its role in the creation of whiteness, see Painter, *History of White People*; Dyer, *White*, 207–23.

76 Frederick's of Hollywood catalog 19, no. 87 (1965): 20–21, Hagley Museum and Library.

77 Franklyn, *On Developing Bosom Beauty*, 73.

78 The pathologization of cis women's bodies, particularly breasts, and the way this was used by US surgeons and health insurance companies to legitimize the cost of cosmetic surgery, is covered in the established scholarship. See, for example, Haiken, *Venus Envy*; Gilman, *Making the Body Beautiful*; Jacobson, *Cleavage*; Matelski, *Reducing Bodies*.

79 Serlin, *Replaceable You*; Gilman, *Making the Body Beautiful*; Haiken, *Venus Envy*; Ott, "Sum of Its Parts."

80 "Plastic Surgery Perils 'Falsie' Trade," *The Austin Daily Herald*, November 15, 1949, 12.

81 "Building Up Bosoms," *Time*, November 18, 1957, RAF.

82 La Roe, *Breast Beautiful*; Franklyn, *On Developing Bosom Beauty*.

83 Franklyn, *On Developing Bosom Beauty*, 22–27; La Roe, *Breast Beautiful*, 71–85.

84 Franklyn, *On Developing Bosom Beauty*, 22–27; La Roe, *Breast Beautiful*, 71–85. See also Kind and Moreck, *Gefilde der Lust*; Niemoeller, *Complete Guide to Bust Culture*, 9–18.

85 For more on this, see also Gilman, *Making the Body Beautiful*.

86 "Edith Lances" note, n.d., HCP, box 7, "Prostheses 1946–53" file.

87 "Nubrest" clipping, n.d., HCP, box 7, "Prostheses 1946–53" file.

88 Ocularist Walter Spohn's Papers at the Smithsonian National Museum of American History also hold information on mastectomy prostheses.

89 "Campbell Company" card, HCP, box 7, "Prostheses 1946–53" file.

90 Franklyn, *On Developing Bosom Beauty*, 76–79; Franklyn, *Augmentation Mammaplasty*.

91 Franklyn, *On Developing Bosom Beauty*, 76.

92 Franklyn, *On Developing Bosom Beauty*, 83–84.

93 See Walter Spohn Papers, NMAH; HCP; Milton Edgerton Papers, Alan Mason Chesney Medical Archives, Johns Hopkins University.

94   Herbert Conway correspondence, Science Service Collection, folder 68, broadcast August 29, 1953, RU 7091, box 400, Smithsonian Institution Archives.

95   Herbert Conway interviewed in Gobind Behari Lal, "Surgery Pledges Beauty," *Los Angeles Examiner*, October 7, 1960.

96   Franklyn, *On Developing Bosom Beauty*, 140.

97   Franklyn, *Augmentation Mammaplasty*, 72–73.

98   Hollander, *Seeing Through Clothes*, 447.

99   Entwistle, *Fashioned Body*.

100   Minor, "Operation Hollywood," 14, RAF.

101   See, for example, Edgerton and McClary, "Augmentation Mammaplasty"; Edgerton et al., "Augmentation Mammaplasty II"; Franklyn, *On Developing Bosom Beauty*, *Beauty Surgeon* (1960, 1961), and *Augmentation Mammaplasty*; JPW, box 208, box 228, folder 3; La Roe, *Breast Beautiful*.

102   Star, *"I Changed My Sex!,"* 99.

103   JPW, box 208, folder 3.

104   For letters from prospective patients, see JPW and AMA Archives.

105   Franklyn, *On Developing Bosom Beauty*, 8.

106   Minor, "Operation Hollywood," RAF; Victor Warren Quale, "Beauty and the Bust," *Esquire*, June 1954, 85–109.

107   See Haiken, *Venus Envy*, 212–13; Matelski, *Reducing Bodies*, 109–10; Serlin, "Engineering Masculinity."

108   "Plastic Surgery," *Ebony*, May 1949, 19–23.

109   Michele Burgen, "Plastic Surgery: A Lift for the Face, Figure and Spirit," *Ebony*, January 1977, 40–50.

110   Burgen, "Plastic Surgery," 40.

111   "The Case of the Counterfeit Bride," in Franklyn, *Beauty Surgeon* (1960), 63–71.

112   "The Case of the Negro Nurse," in Franklyn, *On Developing Bosom Beauty*, 131–33.

113   "The Case of the Counterfeit Bride," in Franklyn, *Beauty Surgeon* (1960), 63–71, 71.

114   "The Case of the Counterfeit Bride," in Franklyn, *Beauty Surgeon* (1960), 63–71, 71.

115   Darlene Gray at 27 min., 18 sec., in *Mondo Topless*, directed by Russ Meyer (Eve Productions, 1966).

116   Jean Howard, "'Charlotte' in Second Surgery," *New York Journal–American*, August 14, 1955, 9, JPW, box 199, folder 4.

117   See also Meyerowitz, "'Fierce and Demanding' Drive," 372.

118   Watson Crews Jr., "Sex Change Breaks Up Old Gang of His (Hers)," *Sunday News* (New York), March 22, 1964, 24.

119   Crews, "Sex Change Breaks Up Old Gang of His (Hers)," 24.

120   See, for example, Roberts, *Female Impersonators' Handbook*, 40; *Female Impersonators* 1 (Winter 1969): 49.

121   Skidmore, "Constructing the 'Good Transsexual'"; Snorton, *Black on Both Sides*.

122   Brown told *Jet* she had arranged for gender-affirming surgery with a doctor in Bonn, Germany. See "Male Shake Dancer Plans to Change Sex, Wed GI in Europe," *Jet*, June 18, 1953, 24–25; "Male Dancer Becomes Danish," *Jet*, June 25, 1953, 26; "Jail Male Shake Dancer for Posing as Woman in Boston," *Jet*, July 9, 1953, 20; "Tax Snag Halts Male Dancer's Trip for Sex Change," *Jet*, October 15, 1953, 19.

123   See also Snorton, *Black on Both Sides*.

124   Delisa Newton, "From Man to Woman: Part II," *Sepia*, May 1966, 66.

125   Franklyn, *On Developing Bosom Beauty*, 74.

126   Robert Alan Franklyn in Henry Lee, "Breastplasty: The Operation That Remolds Flat-Chested Women," *Pageant*, August 25, 1953, 73.

127   Grindlay and Waugh, "Plastic Sponge"; Pangman and Wallace, "Use of Plastic Prosthesis"; Conway, "Mammaplasty."

128   Grindlay and Waugh, "Plastic Sponge."

129   Pangman and Wallace, "Use of Plastic Prosthesis"; Conway and Smith, "Breast Plastic Surgery."

130   Oppenheimer et al., "Sarcomas Induced in Rodents"; Kiskadden, "Operations on Bosoms Dangerous."

Chapter 5. Bombshells, Bombers, and Bumpers

1   McLuhan, *Mechanical Bride*, 13.

2   McLuhan, *Mechanical Bride*, 93–97.

3   Partridge, *Dictionary of Slang*, 39.

4   "Blonde Missile," *Ogden (UT) Standard Examiner*, March 2, 1958, 11A.

5   Yalom, *History of the Breast*, 177.

6   See, for example, "The Two-Way Stretch," *Chicago Tribune*, August 3, 1963.

7   Haiken, *Venus Envy*; Gilman, *Making the Body Beautiful*; Serlin, *Replaceable You*; Matelski, *Reducing Bodies*.

8   Haiken, *Venus Envy*, 12.

9   Haiken, *Venus Envy*, 7; Serlin, *Replaceable You*, 4, 9; Plemons, *Look of a Woman*, 13.

10   See, for example, Haiken, *Venus Envy*, 12; Serlin, *Replaceable You*, 4.

11   Serlin, *Replaceable You*, 4.

12   See, for example, Weldon Wallace, "Spare Parts for the Human Machine," *Baltimore Sun*, September 2, 1953, 6.

13   American architect and design theorist Richard Buckminster Fuller coined the term "livingry" to mean the opposite of weaponry; see Fuller, *Critical Path*, xxv.

14   See Franklyn, *On Developing Bosom Beauty*, 73–74; Franklyn, *Beauty Surgeon* (1960, 1961).

15   See, for example, Matelski, *Reducing Bodies*, 44. Matelski recounts Franklyn's plastic foam origin story in a brief section that summarizes his career but does not provide context or explore the subject further.

16   Franklyn's active pursuit of prospective patients is one reason he was blacklisted by the AMA. AMA-associated doctors are not allowed to advertise their services. The AMA's files on him contain cuttings from Yellow Pages directories across the United States, listing his services. See also Franklyn, "Augmentation Mammaplasty," and the accompanying Surgifoam advertorial, RAF.

17   See, for example, Franklyn, *Beauty Surgeon* (1960, 1961); *On Developing Bosom Beauty*. For self-authorship and self-actualization, see Menon, *Refashioning Race*.

18   Franklyn, *Beauty Surgeon* (1961). Other titles in the series included cookbooks and hairstyling guides, further highlighting Franklyn's role as a contributor to publishers of self-help books for women.

19   See Franklyn, "Augmentation Mammaplasty," and the accompanying Surgifoam advertorial, RAF. This number is repeated throughout his publications but difficult to verify.

20   Ralph Lee Smith, "All the Twiggies Want to Be Sophia," *True*, November 1967, 81–82.

21   William Johnson, "Dr. Beauty Buys a Beast," *Sports Illustrated*, February 19, 1968, 51.

22   Franklyn, *Clinical Atlas of Cosmetic Plastic Surgery*, 3.

23   Franklyn, *On Developing Bosom Beauty*, 73. Correspondence held in the AMA Archives (RAF) shows that Franklyn told prospective patients he was a chief plastic surgeon for the US Armed Forces. According to his draft registration form, Franklyn changed his name from Frank Mark Eisenberg in 1941. I have found no records verifying that he was a surgeon in the US Armed Forces or that he served at all.

24   For more information about this photograph, see Ed Das, "Warprizes," Canadian Aerospace, December 14, 2012, https://canadianaerospace .weebly.com/german-wwii-jets-in-canada.html.

25   Franklyn, *Beauty Surgeon* (1961), 14. When it came to synthetic rubber replacements during World War II, Germany was in the lead: Buna, the first mass-produced synthetic rubber, made in 1929, became the major replacement for natural rubber. See Westermann, *Plastik und Politische Kultur*; Rubin, *Synthetic Socialism*, 20–21.

26   Franklyn, *Beauty* Surgeon (1961), 14.

27   See Henthorn, *From Submarines to Suburbs*, 86–100.

28   Franklyn, *Beauty Surgeon* (1961), 14.

29   Don Brown, "Rare Souvenirs of War Consigned to Scrap Heap," *Ottawa Citizen*, September 18, 1957, 3.

30   Franklyn, *Beauty Surgeon* (1961), 15.

31   Franklyn, *On Developing Bosom Beauty*, 73.

32   Franklyn, *On Developing Bosom Beauty*, 73.

33   Franklyn, *On Developing Bosom Beauty*, 73–74.

34   Franklyn, *On Developing Bosom Beauty*, 74.

35   DeBell, Goggin, and Gloor, *German Plastics Practice*.

36   Franklyn, *On Developing Bosom Beauty*, 73.

37   Franklyn, *Beauty Surgeon* (1960), 18.

38   Franklyn, *On Developing Bosom Beauty*, 73.

39   Franklyn, *Beauty Surgeon* (1960), 21.

40   Franklyn, *On Developing Bosom Beauty*, 74.

41   Franklyn, *Beauty Surgeon* (1961), 17.

42   Surgifoam, advertisement, *General Practice: The Medical Journal of the West* 20, no. 11 (1957): 27; Franklyn, "Augmentation Mammaplasty," both RAF.

43   Polyester foam: Franklyn, *On Developing Bosom Beauty*, 74. Polyurethane foam: Franklyn, "Kosmetische Korrektur," 1752; Franklyn, *Beauty Surgeon* (1960), 21; Franklyn, *Beauty Surgeon* (1961), 17. Plastic foam: Franklyn, *Augmentation Mammaplasty*, 46.

44   See Franklyn, "Augmentation Mammaplasty," and the accompanying Surgifoam advertorial; Franklyn, "Breastplasty," both RAF. Webster and Conway's archives both hold clippings relating to Franklyn. See JPW, 1888–1974, series 9: Newspaper Clippings, box 199, folder 1; HCP, box 33, series V: Miscellaneous, 1948–1968.

45   See RAF, particularly files for 1954 and 1959 (0288-01, 0289-01).

46   Henry Lee, "Breastplasty: The Operation That Remolds Flat-Chested Women," *Pageant*, August 25, 1953, 68–75. This includes the same images used in Franklyn, *On Developing Bosom Beauty*.

47   Lee, "Breastplasty," 68.

48   Franklyn, "Breastplasty," 72.

49   Lee, "Breastplasty"; Franklyn, *On Developing Bosom Beauty*, 79.

50   Franklyn quoted in Lee, "Breastplasty," 75.

51   Franklyn quoted in Lee, "Breastplasty," 71.

52   Lee, "Breastplasty," 74.

53   Franklyn, *On Developing Bosom Beauty*, 74.

54   Franklyn, "Der neue Kunststoff 'Surgifoam.'"

55   Lee, "Breastplasty." Other plastic foam materials that were implanted include Ivalon foam, which was made from polyvinyl alcohol formaldehyde polymer sponge or a polyethylene sac with Ivalon.

56   Ivalon was found to be toxic when implanted. For a more detailed description of the Ivalon sterilization process, see Armand Hartley, "Surgery's New Miracle—Permanent Curves," *Confidential*, New York, January 1956, 18–19.

57   Franklyn quoted in Lee, "Breastplasty," 74–75.

58   "The Business of Bolstering Bosoms," *JAMA* Bureau of Investigation, November 28, 1953, 1200.

59   Franklyn quoted in AMA news release, November 27, 1953, RAF, Correspondence and Articles.

60   See AMA news release, November 27, 1953, RAF, Correspondence and Articles; "Business of Bolstering Bosoms."

61   "Business of Bolstering Bosoms," 1200.

62   Robert M. Yoder, "What Are American Women Made Of?," *Pageant*, May 1953, 66–70.

63   Minor, "Operation Hollywood," 15, RAF.

64   Minor, "Operation Hollywood," 14, RAF.

65   See Tanquero and Rawlinson, *Remember Me, Vicki Starr*, 268.

66   For more on the star factory, the golden age of Hollywood, and consumption of celebrity culture, see Dyer, *Stars*; Dyer, *Heavenly Bodies*; Stacey, "Feminine Fascinations"; Gundle, *Glamour*, 172–99.

67   Rodriguez, *Heroes, Lovers, and Others*, 76–78. See also McLean, "'I'm a Cansino.'"

68   Gina Lance and Bijoux Deluxe, "Bombshell: Interview with Mamie Van Doren," *Girl Talk* 4, no. 1 (2002): 16–20.

69   Franklyn, *On Developing Bosom Beauty*, 84.

70   Minor, "Operation Hollywood," 14, RAF.

71   "Is Bra Secret to a Bigger Bustline?," *Westport (CT) News*, July 18, 1975, 56; Gottwald and Gottwald, *Frederick's of Hollywood*.

72   Franklyn, *On Developing Bosom Beauty*, 128.

73   Franklyn, *On Developing Bosom Beauty*, 84.

74   See Hawes, "Making the Modern Consumer"; Gartman, "Harley Earl and the Art and Color Section."

75   Oldsmobile, advertisement, *Life*, January 28, 1952, 62–63.

76   Antonelli, *Safe*, 29.

77   Hudson Hornet V8, advertisement, *Life*, November 12, 1956, 75.

78   Ingrassia, *Engines of Change*.

79   Henry Ford claimed he made this comment during a meeting in 1909; see Ford and Crowther, *My Life and Work*, 72.

80   For more on the Art and Color Section, see Hawes, "Making the Modern Consumer"; Gartman, "Harley Earl and the Art and Color Section."

81   *Maidenform Mirror*, April 1952, 2, Maidenform Collection, NMAH, AC.0585, series 4, box 19, NMAH Archives Center.

82   See "Raquel Welch Pillow" and "Life-Like Lady's Legs" in Demarais, *Mail-Order Mysteries*, 87, 135.

83   Franklyn, *On Developing Bosom Beauty*, 11.

84   Minor, "Operation Hollywood," 14, RAF.

85   For a visual history of designed objects resembling women's bodies, see Strang, *Working Women*.

86   Lippincott, *Design for Business*, 54–55.

87   Wallace, "Spare Parts," 6.

88   "Spare Parts for Humans," *Ebony*, April 1953, 16, quoted in Serlin, *Replaceable You*, 4.

89   Wallace, "Spare Parts," 6.

90   Edgerton et al., "Augmentation Mammaplasty II," 284.

91   Noel Robbins, Robbins Instruments Company, letter to Herbert Conway, January 29, 1959, HCP, box 7.

92   "Etheron Foam Physical Properties," attached to Noel Robbins, letter to Herbert Conway, January 29, 1959, HCP, box 7.

93   Noel Robbins, letter to Herbert Conway, October 8, 1959, HCP, box 7.

94   See correspondence between Noel Robbins and Herbert Conway, 1959, HCP, box 7.

95   D. H. Bryan, Mobay district sales manager, letter to Herbert Conway, November 18, 1960, HCP, box 7.

96   D. H. Bryan, Mobay district sales manager, letter to Herbert Conway, November 18, 1960, HCP, box 7.

97   Herbert Conway, letter to M. J. Sanger, head of cellular material development, General Tire and Rubber Company, August 20, 1960, HCP, box 7.

98   Conway and Dietz, "Augmentation Mammaplasty," 574.

99   "Building Up Bosoms," *Time*, November 18, 1957, RAF.

100   Conway and Dietz, "Augmentation Mammaplasty," 576.

101   Conway, "Mammaplasty."

102   Conway and Dietz, "Augmentation Mammaplasty," 576.

103   For more on how medical actors pathologized women's bodies, particularly in relation to the taxonomy of the youthful breast, see, for example, Kind and Moreck, *Gefilde der Lust*. See also Gilman, *Making the Body Beautiful*, 206–257; Haiken, *Venus Envy*; Jacobson, *Cleavage*; Berney, "Streamlining Breasts."

104   Conway and Dietz, "Augmentation Mammaplasty," 579.

105   Franklyn commented, "[This] amused me as it gave a picture of a lot of discontented Russian women seeking improvement where Nature had failed." Quoted in Whitney Bolton's column, *Philadelphia Inquirer*, February 17, 1958, 7.

106   Francisco Duran Acosta, letter to Herbert Conway, October 12, 1962, HCP, box 7.

107   Herbert Conway, letter to Francisco Duran Acosta, October 19, 1962, HCP, box 7.

108   Herbert Conway, letter to Francisco Duran Acosta, October 19, 1962, HCP, box 7.

109   Alexa Klossowski, letter to Herbert Conway, September 3, 1962, HCP, box 7.

110   Herbert Conway, letter to Alexa Klossowski, September 7, 1962, HCP, box 7.

Chapter 6. Silicones on the Surface

1   Kenneth A. Thompson, "Question: What's More Perfect Than Any Lady?," *Detroit Free Press*, September 9, 1952, 18.

2   *What's a Silicone?* (Midland, MI: Dow Corning, 1952), NMAH TLC and SHI.

3	"The Silicones: Cornerstone of a New Industry," *Fortune*, May 1947, 104.
4	"The Silicones," 104–5.
5	"The Silicones," 104.
6	McWhan, *Sand and Silicon*.
7	McWhan, *Sand and Silicon*.
8	McWhan, *Sand and Silicon*.
9	See Warrick, *Forty Years*; Dow Corning trade literature at NMAH TLC and SHI.
10	Meals and Lewis, *Silicones*.
11	See, for example, Fordham, *Silicones*; Morris, *Polymer Pioneers*; Oakes, *A to Z of STS Scientists*, 163–64; Simona Morini, "Beauty and Health," *Vogue*, March 15, 1971, 84–87, 114–15; Ashley et al., "Present Status of Silicone Fluid," 419.
12	See, for example, Fordham, *Silicones*; Morini, "Beauty and Health," 115; Ashley et al., "Present Status of Silicone Fluid," 419.
13	J. Franklin Hyde, oral history interview, 1986, SHI, 22, citing Andrianov, "Synthesis of Alkyl-Substituted Ortho-Esters of Silic Acid."
14	Wöhler, Liebig, and Kopp, *Annalen der Chemie*, 263, cited in J. Franklin Hyde, oral history interview, 1986, SHI, 22.
15	Noll, *Chemistry and Technology of Silicones*, presents a more global and independent perspective on silicone R & D than his American peers, even addressing work in Japan. Noll, like his peers, does not mention silicone's use in cosmetic surgery.
16	Noll, *Chemistry and Technology of Silicones*, 19.
17	"The Silicones," 224.
18	Fordham, *Silicones*, 1.
19	Fordham, *Silicones*.
20	J. Franklin Hyde, oral history interview, 1986, SHI, 21.
21	Braley, "Use of Silicones."
22	J. Franklin Hyde, oral history interview, 1986, SHI, 19.
23	J. Franklin Hyde, oral history interview, 1986, SHI, 24.
24	"The Silicones," 106–7.
25	Noll, *Chemistry and Technology of Silicones*, 19.
26	Oral histories with chemists from both companies record that in late 1937, a pivotal meeting occurred between representatives at CGW and GE, after which the latter also began silicone research. Here accounts diverge: Hyde and Warrick presented this visit and what happened after as akin to industrial espionage (see J. Franklin Hyde, oral history interview, 1986, 24; Earl Warrick, oral history interview, 1986, 11, both SHI). It should be noted here that CGW and GE had a history of collaboration; for instance, GE chemist Eugene Rochow described CGW as "virtually a branch—a department—of General Electric" (Rochow, oral history interview, 1995, 24, SHI). Rochow and Charles Reed of GE offer a different take on the story, as does German chemist Walter

Noll. See Charles Reed, oral history interview, 1986, SHI; and Noll, *Chemistry and Technology of Silicones*.

27   "The Silicones," 107.

28   In his oral history, GE chemistry engineer and MIT graduate Charles Reed describes the US synthetic rubber program's collaborative efforts between MIT, Dow, and Bradley Dewey, which advanced synthetic rubber R & D to support the wartime push for greater material independence. Charles Reed, oral history interview, 1986, 14, SHI.

29   Eugene G. Rochow, assignor to General Electric Company, designs related to polymeric methyl silicone, US Patents 2,258,218–2,258,222, issued October 7, 1941, cited in J. Franklin Hyde, oral history interview, 1986, 26, SHI.

30   J. Franklin Hyde, oral history interview, 1986, 26–27, SHI.

31   J. Franklin Hyde, oral history interview, 1986, SHI. Warrick reflected that in the end, the two companies decided that, since they had so many overlapping patents, "or that were in fields of each other's, that neither one would be able to operate without license from the other." Earl Warrick, oral history interview, 1986, 23, SHI.

32   This argument is often made in STS. See, for example, Pinch and Bijker, "Social Construction of Facts and Artefacts"; Wisnioski, *Engineers for Change*; Leslie, *Cold War and American Science*.

33   "The Silicones," 109.

34   "The Silicones," 110.

35   J. Franklin Hyde, oral history interview, 1986, 26, SHI. Harold Boeschenstein, who headed the Owens-Corning merger, was responsible for interesting the navy in electric motor insulation, according to Earl Warrick, oral history interview, 1986, 10, SHI.

36   Warrick, *Forty Years*, 21; Coe, *Unlikely Victory*, 28.

37   Braley, "Use of Silicones," 281.

38   Warrick, *Forty Years*, 10.

39   Allen and Polmar, *Rickover*; Warrick, *Forty Years*, 21.

40   Coe, *Unlikely Victory*, 28. It is worth noting that the British government had collaborated with the US Navy to address the issue of high-altitude ignition loss in jet aircraft, demonstrating the urgency of the issue and the joint transatlantic Allied effort to find a solution. However, the problem was ultimately solved in the United States with the development of the DC-4 compound. After the war, Dow Corning's Shailer Bass was invited by the British government to share developments and applications in silicone, hoping these might resolve issues the British faced with their own equipment, including radio and radar systems developed during the war. See Warrick, *Forty Years*, 103.

41   "The Silicones," 110.

42   *What's a Silicone?*, 32.

43   Coe, *Unlikely Victory*, 28.

44   William R. Collings, "From Me to You," letter to Dow Corning Staff, February 12, 1953, Dow Chemical Company Buildings and Grounds, Dow Collection, box 2, file 00175, SHI.

45   Coe, *Unlikely Victory*, 27; Charles Reed, oral history interview, 1986, 30, SHI.

46   Charles Reed, oral history interview, 1986, 30, SHI.

47   For more details on the unusual circumstances of this government-sanctioned joint venture, see Warrick, *Forty Years*, 43.

48   "The Silicones," 104.

49   *Fortune* advertisement, *Life* magazine, May 12, 1947, 151.

50   Fordham, *Silicones*, 2. For advertising, see, for example, the 1956 Dow Corning promotional film *The Invisible Protectors*, 2 min., 59 sec., posted April 8, 2012, Internet Archive, https://archive.org/details /0570_Invisible_Protectors_The.

51   "The Silicones," 226.

52   Fordham, *Silicones*; Warrick, *Forty Years*.

53   "Bouncing Putty," *Life*, January 29, 1945, 63.

54   Coe, *Unlikely Victory*, 28.

55   Chasan, "History of Injectable Silicone Fluids," 2035.

56   Dow Corning, *What's a Silicone?* See also Dow Corning trade literature at NMAH TLC and SHI.

57   "The Silicones," 104.

58   "What Are Silicones?," *Philadelphia Inquirer*, April 21, 1952, 27.

59   *Electronics*, February 1953, 37; "Tall Tales in DC Silicone News," *Electronics*, October 1953, and *Scientific American*, March 1956, 49; "Tall Tales," *Scientific American*, October 1954, 47.

60   Dow Corning, *What's a Silicone?*, 1.

61   Bix, *Girls Coming to Tech!*, 115.

62   See Pearson, *Beyond Small Numbers*, 16, 174; Brown, *African American Women Chemists*.

63   Pearson, *Black Scientists, White Society*.

64   Dow Corning, *What's a Silicone?*, 33.

65   Dow Corning, *Tall Tales and Fabulous Facts: Dow Corning Silicone News—New Frontier Edition* (Midland, MI: Dow Corning, 1953), SHI.

66   For an American cultural history of space through the lens of frontier, see Weitekamp, *Space Craze*.

67   For more on midcentury Westernwear and atomic culture, see Abrego, *Westernwear*, 71–112.

68   *What's a Silicone?*, 33.

69   *What's a Silicone?*, 1.

70   "The Silicones," 109.

71   Brown, *Work!*, 63.

72   *What's a Silicone?*, 2.

73   Davis, *Plastic Matter*, 47.

74	For more on the concept of bodily betrayals in relation to ageing, see Howson, *Body in Society*, 193–200.
75	For scholarship on aging and gender in Western discourse, see Susan Sontag, "The Double Standard of Ageing," *Saturday Review*, September 23, 1972, 29–38; Gibson, "No-One Expects Me Anywhere"; Twigg, *Fashion and Age*; King, *Discourses of Ageing*.
76	*What's a Silicone?*, 2.
77	Morini, "Beauty and Health," 86.
78	Braley, "Use of Silicones," 281.
79	For more on the politics and commercial power of Cold War spin-offs, see Leslie, *Cold War and American Science*; Henthorn, "Commercial Fallout"; Henthorn, *From Submarines to Suburbs*, 86–88; Cowan, *Social History of American Technology*; for more on military-industrial connections and the body, see Serlin, *Replaceable You*; de Monchaux, *Spacesuit*; Hersch, "High Fashion."
80	Henthorn, *From Submarines and Suburbs*, 86.
81	Henthorn, *From Submarines and Suburbs*, 86.
82	"Changes for 1954," *Vogue*, January 1, 1954, 127.
83	See, for example, Lentheric, "On Hand Protective Lotion," advertisement, *Life*, February 1, 1954, 48; Cara Nome silicone lotion, *Life*, February 6, 1956, 55; Wolco Glas spray cleaner, *Life*, October 19, 1953, 114; O-Cedar Dri-Glo polish, *Life*, March 31, 1952, 61.
84	See the Dow Corning promotional film *The Invisible Protectors*.
85	For more on the war on insects, see Russell, *War and Nature*; on germs, see Lupton and Miller, *Bathroom, the Kitchen*.
86	Russell, *War and Nature*, 168.
87	Henthorn, *From Submarines to Suburbs*.
88	Henthorn, *From Submarines to Suburbs*; Harris, *Little White Houses*; Wilson, *Mid-Century Modernism*; Lupton, *Bathroom, the Kitchen*.
89	Pond's Angel Face, advertisement, in the following issues of *Life*: November 30, 1959, 45; February 17, 1961, 54; August 15, 1960, 62; September 19, 1960, 70–71.
90	Pond's Angel Face, advertisement, in the following issues of *Life*: November 30, 1959, 45; February 17, 1961, 54.
91	Pond's Angel Face, advertisement, *Life*, January 20, 1961, 62.
92	Peiss, *Hope in a Jar*, 151.
93	"East-West Twain Find a Meeting in MacLaine," *Life*, February 17, 1961, 91.
94	Pond's Angel Face, advertisement, *Life*, November 30, 1959, 45.
95	Peiss, *Hope in a Jar*, 258.
96	Posner's Special Silicone Formula, advertisement, *Ebony*, June 1960, 84.
97	Apex, advertisement, *Ebony*, September 1961, 117.
98	American Health and Beauty Aids Institute, "Black Clout," *Ebony*, March 1988, 160–61; Nulox Sheen, advertisement, in the following issues of *Ebony*: May 1962, 86; August 1962, 51.

99	See also Long Aid, advertisement, *Ebony*, July 1976, 14; Persulan, advertisement, *Ebony*, February, 1965, 102; Drake Laboratories, Persulan Synthesized Bergamot Hair Conditioner, ca. 1970, Acc. 1985.0475, Division of Medicine and Science, NMAH, https://americanhistory.si.edu/collections/object/nmah_209597.
100	Craig, *Ain't I a Beauty Queen?*, 93.
101	Posner's Bergamot, advertisement, *Ebony*, February 1965, 129.
102	Henthorn, *From Submarines to Suburbs*, 285.
103	See Painter, *History of White People*, 366–73; Henthorn, *From Submarines to Suburbs*, 205–12; Harris, *Little White Houses*; Wilson, *Mid-Century Modernism*.
104	Warrick, *Forty Years*, 166.
105	*What's a Silicone?*, 33; *Tall Tales and Fabulous Facts*, 24.
106	Rob McGregor quoted in Braley, "Use of Silicones," 281.
107	Braley, "Use of Silicones," 283. See also Warrick, *Forty Years*, 185.
108	Braley, "Use of Silicones," 283.
109	See, for example, Ott, "Sum of Its Parts," 20; Ott, "Hard Wear and Soft Tissue"; Serlin, "Engineering Masculinity."
110	Warrick, *Forty Years*, 185.
111	Warrick, *Forty Years*, 187.
112	Warrick, *Forty Years*, 186.
113	Braley, "Use of Silicones," 281.

Chapter 7. Silicones Beneath the Surface

1	For a selection, see Haiken, *Venus Envy*, 246; Williams, *Breasts*, 65; Matelski, *Reducing Bodies*, 33; Byrne, *Informed Consent*, 41; Angell, *Science on Trial*, 35; Parker, *Women, Doctors and Cosmetic Surgery*, 196; Zimmermann, *Silicone Survivors*, 23; Jacobson, *Cleavage*, 80; Davis, *Reshaping the Female Body*, 20; Toni Kosover, "Fill Her Up," *W Magazine*, November 1972, 20; Deborah Larned, "A Shot—or Two or Three—in the Breast," *Ms.*, September 1977, 55; Simona Morini, "Beauty and Health," *Vogue*, March 15, 1971, 84–87, 114–15; Schalk, "The History of Augmentation Mammaplasty," 88; Mimi Swartz, "Silicone City," *Texas Monthly*, August 1995, 69–100; John Byrne, "Beauty and the Breast: How Industry Sold Implants to Women," *Ms.*, June 1996, 45–46; Phillip J. Hilts, "Strange History of Silicone Held Many Warning Signs," *NYT*, January 18, 1992, 1, 8; Riordan, *Inventing Beauty*, 111. Miller critiques Byrne and Haiken in "Mammary Mania in Japan," 281–82; and *Beauty Up*, 82–83.
2	Byrne, *Informed Consent*, 41.
3	Within histories of cosmetic surgery, Sander L. Gilman is the exception, as he mentions a Japanese medical study that covers some of

Japan's silicone breast augmentation research and development. See *Making the Body Beautiful*, 103.

4   Braley, "Illegal, Immoral and Dangerous," 173.

5   See, for example, "Bosoms Can Be Increased 2 Inches," *Confidential*, February 1959, 30–31; Wallace Turner, "Silicone Inquiry Shows Wide Use: Illegal Breast Injections Are Reported on Coast," *NYT*, April 28, 1968; "Liquid Silicone Used in Reconstructive Surgery," *NYU Medical Center News*, February 1968, JMP, box 1, folder 33; Ashley et al., "Present Status of Silicone Fluid"; Braley, "Silicone Fluids with Added Adulterants."

6   Zimmermann, *Silicone Survivors*, 23.

7   See Zimmermann, *Silicone Survivors*; Byrne, *Informed Consent*; Haiken, *Venus Envy*.

8   Miller, "Mammary Mania in Japan," 279.

9   Miller, "Mammary Mania in Japan," 281–82; Miller, *Beauty Up*, 82–83.

10   See note 1.

11   Bryne, *Informed Consent*, 43; Haiken, *Venus Envy*, 247; H. Kagan, "Sakurai Injectable Silicone Formula." In 1958 German MD R. Kaden, at the Haut Klinik im Rudolf Virchow-Krankenhaus, Berlin, also began experimenting with silicone injections, using DC 200/100. See Kaden, "Verwendung des Silikonöls."

12   See, for example, Narins and Beer, "Liquid Injectable Silicone"; "Liquid Silicone Used in Reconstructive Surgery," *NYU Medical Center News*, February 1968, JMP, box 1, folder 33.

13   Franklyn, *Art of Staying Young*, 118–19.

14   In 1939, Lenz's citizenship was revoked, and he fled to Cairo, where he continued his practice and research. After the war, he practiced in both Egypt and Baden-Baden, Germany.

15   Franklyn, *Art of Staying Young*; H. Kagan, "Sakurai Injectable Silicone Formula."

16   Symmers, "Silicone Mastitis in 'Topless' Waitresses," 20.

17   Gilman, *Making the Body Beautiful*, 251–57; Haiken, *Venus Envy*, 235–36.

18   Boo-Chai, "Complications of Augmentation Mammaplasty by Silicone Injection"; Haiken, *Venus Envy*; Jacobson, *Cleavage*.

19   Haiken, *Venus Envy*, 236.

20   Symmers, "Silicone Mastitis in 'Topless' Waitresses," 19.

21   Webb, "Cleopatra's Needle."

22   Webb, "Cleopatra's Needle."

23   "Plan for Execution of Industrialization Experiments: Yosuke Suzuki, Seisakusho Ltd," RG 331 (Allied Operational and Occupation Headquarters, World War II), Supreme Commander for the Allied Powers, Economic and Scientific Division, Industrial Production and Construction Branch, topic file 1945–1950, box 7173, National Archives, Washington, DC, declassified authority 775018.

24   "Plan for Execution of Industrialization Experiments: Yosuke Suzuki, Seisakusho Ltd," RG 331 (Allied Operational and Occupation Headquarters, World War II), Supreme Commander for the Allied Powers, Economic and Scientific Division, Industrial Production and Construction Branch, topic file 1945–1950, box 7173, National Archives, Washington, DC, declassified authority 775018.

25   "Plan for Execution of Industrialization Experiments: Yosuke Suzuki, Seisakusho Ltd," RG 331 (Allied Operational and Occupation Headquarters, World War II), Supreme Commander for the Allied Powers, Economic and Scientific Division, Industrial Production and Construction Branch, topic file 1945–1950, box 7173, National Archives, Washington, DC, declassified authority 775018.

26   "History," Shin-Etsu Silicone, accessed February 27, 2024, https://www.shinetsusilicone-global.com/info/development.shtml.

27   Warrick, *Forty Years*, 108, 239.

28   "Tokyo MD Claims 'Youth' Injections," *Honolulu Star-Bulletin*, January 24, 1965, 17.

29   "Tokyo MD Claims 'Youth' Injections"; H. Kagan, "Sakurai Injectable Silicone Formula," 55.

30   H. Kagan, "Sakurai Injectable Silicone Formula," 55.

31   "Tokyo MD Claims 'Youth' Injections."

32   "History," Shin-Etsu Silicone, accessed February 27, 2024, https://www.shinetsusilicone-global.com/info/development.shtml.

33   Warrick, *Forty Years*, 108.

34   Warrick, *Forty Years*, 239.

35   Japan was a key country in Dow Corning's efforts to establish international joint venture operations. In 1965, Paul Sawada of Shin-Etsu met with Bill May at Dow Corning's Tokyo office to begin negotiations for what was to become their Japanese government-sanctioned joint venture, Shin-Etsu Hadotai, specializing in semiconductor silicons. In 1966, Dow Corning also set up a joint venture and licensing agreement with Toray Silicone Co. Ltd, becoming one of the first American companies to receive an equal partnership in postwar Japan. See Warrick, *Forty Years*, 240. Eugene Rochow also visited Shin-Etsu in 1964 and spent some days there working on silicone. See Rochow, oral history interview, 1995, 105, SHI. These were two of the major silicone companies in Japan; Tokyo Shibaura Electrical Co., which worked with General Electric, and Fuji Kobunshi also made silicones. See Noll, *Chemistry and Technology of Silicones*, 20.

36   Correspondence between Herbert Conway and Takeya Shirakabe, 1960–61, HCP, box 7, series I.

37   Herbert Conway, letter to Silas Braley, December 19, 1960, HCP, box 7, series I.

38   F. D. Smith, manager, Monsanto Chemical Company, letter to Conway, January 4, 1961, HCP, box 7, series I.

39   Yakuo Fujimoto, letter to Herbert Conway, November 9, 1960, HCP, box 7, series I.

40   See materials listed in Tsuguo Naruke, "Breast Cancer in a Patient After Injection of Synthetic Material into the Breast for Cosmetic Purposes," *Nippon Iji Shinpo*, January 17, 1970, translated from the Japanese for Dow Corning and sent out in correspondence from Silas Braley to the Silicone Committee, June 14, 1971, JMP, box 1, folder 33.

41   Herbert Conway, letter to Rob McGregor, Dow Corning, October 24, 1960, HCP, box 7, series I.

42   Herbert Conway, letter to Rob McGregor, Dow Corning, October 24, 1960, HCP, box 7, series I.

43   Herbert Conway, letter to Rob McGregor, Dow Corning, October 24, 1960, HCP, box 7, series I.

44   Herbert Conway, letter to Rob McGregor, Dow Corning, October 24, 1960, HCP, box 7, series I.

45   Rob McGregor, letter to Herbert Conway, October 29, 1959, HCP, box 7, series I.

46   Morini, "Beauty and Health," 86. See also "The Silicone Injection Story Updated," *Harper's Bazaar*, May 1967, 148; "Yes, You Can Have a Bigger Bosom!," *Cosmopolitan*, January 1970, 66.

47   Morini, "Beauty and Health"; Morini, *Body Sculpture*.

48   Braley, "Use of Silicones," 288.

49   Rob McGregor, letter to Herbert Conway, October 29, 1959; Silas Braley, letter to Herbert Conway, November 10, 1960, all HCP, box 7, series I.

50   Silas Braley, letter to Herbert Conway, November 10, 1960, HCP, box 7, series I.

51   Braley, "Use of Silicones," 283.

52   See correspondence in JMP and HCP, as well as Badura et al., "Investigations of Intravenous Application of Silicone Oils"; Kaden, "Verwendung des Silikonöls," 217; Ben-Hur et al., "Local and Systemic Effects."

53   Ferreira et al., "Changes in the Lung."

54   Warrick, *Forty Years*, 103.

55   Ashley et al., "Present Status of Silicone Fluid."

56   See Leslie, *Cold War and American Science*; Wisnioski, *Engineers for Change*.

57   Ashley et al., "Present Status of Silicone Fluid," 411.

58   William R. Collings, "From Me to You," letter to Dow Corning staff, February 12, 1953, Dow Chemical Company Buildings and Grounds, Dow Collection, box 2, file 00175, SHI.

59   Clarke, *Tupperware*, 27.

60   Wayne Koning, supervisor, Hospital/Surgical Products Division at Dow Corning, letter to Joseph Murray, January 13, 1969, JMP, box 1, folder 33.

61   A broad culture of sending out plastic samples played an important part in the rapid distribution and types of applications in postwar America. For further examples, see Walter Spohn Papers, Division of Medicine and Science, NMAH, for medical plastic samples; and NMAH TLC for trade catalogs, many of which include actual plastic samples. In a March 2018 conversation with the California sculptor David Best, he stated that in the 1960s, he and his fellow artist friends were regularly sent plastic pellets to experiment with by DuPont. Eva Hesse is another artist who worked with plastics samples. See Barger, "Thoughts on Replication."

62   As previously noted, the "Sakurai" formula, attributed to Japanese medical doctor Rin Sakurai, originated in 1946 in Japan. See H. Kagan, "Sakurai Injectable Silicone Formula," 53; Jacobson, *Cleavage*, 80. Organogen, largely made of petroleum, was advertised as a "fleshy injection" in 1950s Japan.

63   See also Miller, "Mammary Mania in Japan"; Miller, *Beauty Up*.

64   "Bosoms Can Be Increased 2 Inches," *Confidential*, February 1959, 30–31.

65   Silas Braley, letter to Herbert Conway, January 4, 1961, HCP, box 7, series I.

66   Silas Braley, letter to Herbert Conway, January 4, 1961, HCP, box 7, series I.

67   Herbert Conway, letter to Silas Braley, January 13, 1961, HCP, box 7, series I.

68   Herbert Conway, letter to Silas Braley, January 13, 1961, HCP, box 7, series I.

69   Charles Vinnik, a vocal opponent of silicone shots, noted that Dow Corning distanced itself from breast augmentation shots by stressing its other applications. See Vinnik, "The Hazards of Silicone Injections"; and Vinnik, cited April 21, 1975, Public Health Service, FDA, Agenda, Special Drugs Advisory Committee Minutes, JMP, box 1, folder 33. In *Informed Consent*, 43, Byrne similarly claims that Dow Corning would later publicly deny having any knowledge of breast injections prior to 1963, when Harvey D. Kagan, an osteopath by training, shared his research findings at the weeklong meeting of the American Otorhinological Society for Plastic Surgery. It should also be noted that Byrne is frequently cited in the established scholarship on breast augmentation; see, for example, Zimmermann, *Silicone Survivors*; Webb, "Cleopatra's Needle"; Jacobson, *Cleavage*; Miller, *Beauty Up*.

70   Symmers, "Silicone Mastitis in 'Topless' Waitresses," 21.

71   Silas Braley, letter to William St. Clair Symmers, August 1, 1968, JMP, box 1, folder 33.

72   Conway and Goulian, "Experience with an Injectable Silastic RTV," 294.

73   Silas Braley, letter to William St. Clair Symmers, August 1, 1968, JMP, box 1, folder 33.

74	Kopf et al., "Complications of Silicone Injections."
75	As openly documented in Conway and Goulian, "Experience with an Injectable Silastic RTV."
76	Silas Braley, letter to William St. Clair Symmers, August 1, 1968, JMP, box 1, folder 33.
77	See, for example, "First Annual Report Dow Corning Center for the Aid to Medical Research," August 15, 1960, HCP, and Dow Corning reports held in Edgerton, Murray, and Spohn papers. Of the bulletins I have accessed there is one exception to this—Fourth Annual Report of the DCCAMR, August 15, 1963, which lists developments in "prostheses and materials for mammary augmentation" (HCP) but does not explicitly mention silicone injections for breast augmentation.
78	For more on these claims, see Byrne, *Informed Consent*, 43. For evidence of prior knowledge, see the Conway Dow Corning correspondence, cited throughout, particularly the following letters: Rob McGregor to Herbert Conway, October 29, 1959; Silas Braley to Herbert Conway, November 10, 1960; Silas Braley to Herbert Conway, January 4 1961, HCP, box 7, series I.
79	Rob McGregor, letter to Herbert Conway, October 29, 1959, HCP, box 7, series I.
80	Dow Corning has previously been accused of altering records and paperwork: "A review of the Dow Corning Corporation's silicone gel breast implant business found that some documents about the manufacturing of the implants in the 1980s were altered, the company said today." See "Dow Corning Says Records on Implants Were Altered," *NYT*, November 3, 1992, 3. Haiken, *Venus Envy*, 343, further notes, "Dow Corning has conceded that it should have made more of an effort to ensure that its product was not being used in humans." See *San Francisco Chronicle*, January 20, 1971, 4.
81	Conway and Goulian, "Experience with an Injectable Silastic RTV." It is worth noting this article is not referenced by cosmetic surgery scholars, including Haiken. It does, however, get a brief mention in Zimmermann, *Silicone Survivors*, 24–25.
82	Conway and Goulian, "Experience with an Injectable Silastic RTV," 294.
83	Symmers, "Silicone Mastitis in 'Topless' Waitresses."
84	Conway and Goulian, "Experience with an Injectable Silastic RTV," 295. See also "Silicones Now Reaching Out to Medical Market," January 22, 1962, in Dow Chemical Company, "Oil, Drug, and Oil Reporter Subject Clips," box 1, SHI.
85	Conway and Goulian, "Experience with an Injectable Silastic RTV," 295.
86	Conway and Goulian, "Experience with an Injectable Silastic RTV," 295.
87	Conway and Goulian, "Experience with an Injectable Silastic RTV," 295.
88	Hollings, *Welcome to Mars*.

89	Miller, "Mammary Mania in Japan," 279. For more on sex workers in Japan during this period, see Sanders, "Panpan"; Kovner, "Base Cultures."
90	Snorton, *Black on Both Sides*, 46.
91	See *Women of the World*, directed by Paolo Cavara, Gualtiero Jacopetti, and Franco Prosperi (Cineriz, 1963), esp. 47 min., 32 sec., through 47 min., 42 sec.
92	For scholarship on eyelid surgery in East Asia, see Miller, *Beauty Up*, 118–22; Menon, *Refashioning Race*, 67–68, 70, 72–74; Shirakabe et al., "Double-Eyelid Operation in Japan"; Nguyen et al., "Asian Blepharoplasty."
93	Doda quoted in Sam Hudson, "The Silicone Bosom Revisited," *Sir!*, March 1968, 4, 61.
94	Mutou, "Study of 317 Akiyama's DMPS Cases."
95	Mutou, "Study of 317 Akiyama's DMPS Cases."
96	AMA Archives, AMA Department of Investigation records, box 94, Breast, Bust Developers, Correspondence, 1956–68, 0094–05. See the original inquiry, attached to the AMA response letter from Robert A Youngerman, September 7, 1967.
97	AMA Archives, AMA Department of Investigation records, box 94, Breast, Bust Developers, Correspondence, 1956–68, 0094–05, AMA response letter from Robert A Youngerman, September 7, 1967.
98	Miller, "Mammary Mania in Japan."
99	See Klein, *Cold War Orientalism*; Serlin, *Replaceable You*, 57–110; Bardsley, "Girl Royalty"; Barker, *Hiroshima Maidens*.
100	Klein, *Cold War Orientalism*, 299.
101	Dower, *Japan in War and Peace*; Klein, *Cold War Orientalism*.
102	Klein, *Cold War Orientalism*, 149.
103	Serlin, *Replaceable You*.
104	Bardsley, "Girl Royalty," 375.
105	Bardsley, "Girl Royalty," 375.
106	Klein *Cold War Orientalism*; Serlin, *Replaceable You*; Bardsley, "Girl Royalty"; Shibusawa, *America's Geisha Ally*.
107	It is worth noting here that Japanese beauty pageants have a longer history. For a critical study of 1930s Miss Nippon, see Robertson, "Japan's First Cyborg?"
108	See Bardsley, "Girl Royalty," 386.
109	See also Kovner, *Occupying Power*; Kovner, "Base Cultures."
110	Herbert Conway, letter to K. L. Pickrell, January 22, 1962, HCP, box 7, series I.
111	"Dinner in Honor of Milton T. Edgerton" menu, June 18, 1970, at Elkridge Country Club, in Edgerton Milton T. Jr Bio Files, Alan Mason Chesney Medical Archives, Johns Hopkins University.
112	For more on the influence of classical white sculpture on the creation of whiteness, see Painter, *History of White People*; Dyer, *White*.

Chapter 8. Queering Silicones

1   Shteir, *Striptease.*
2   Alan Levy, "A Morality Play in Three Acts," *Life*, March 11, 1966, 80.
3   "Miss Carol Doda: Originator of the Topless," *San Francisco Chronicle*, October 31, 1964, 34.
4   Levy, "Morality Play in Three Acts," 80; "Nude Discotheque with Carol Doda," *Playboy*, April 1965; "Topless," *Playboy*, September 1966, 160–67.
5   Carol Doda as Sally Silicone at 34 min., 30 sec., in *Head*, directed by Bob Rafelson (Columbia Pictures, 1968); Wolfe, "Put-Together Girl."
6   "Topless Insured for $1 Million," *Jet*, March 16, 1972, 30; Sam Roberts, "Carol Doda, Pioneer of Topless Entertainment, Dies at 78," *NYT*, November 11, 2015.
7   David Perlman, "Analysis of the Bust Injections," *San Francisco Chronicle*, November 15, 1965, 15.
8   "Topless Insured for $1 Million," 30.
9   Roberts, "Carol Doda."
10 Herb Caen quoted in Levy, "Morality Play in Three Acts," 80.
11 Carol Doda quoted in "Escalation," *Newsweek*, October 25, 1965, 110.
12 Carol Doda quoted in "Escalation," 110. See also Hal Schaefer, "The Owl Steps Out," *San Francisco Chronicle*, June 20, 1964, 32; "Nude Discotheque with Carol Doda"; Levy, "Morality Play in Three Acts"; "The Broadway Boys All Sing This Song—Doda! Doda!" *San Francisco Chronicle*, August 1, 1965, 116; Perlman, "Analysis of the Bust Injections."
13 Plastic surgeon Norman Anderson, in his December 18, 1990, testimony, before the US House of Representatives, described some fifty thousand women in the United States who had been injected with silicone during this period. Cited in Webb, "Cleopatra's Needle"; Jacobson, *Cleavage.* Narins and Beer, "Liquid Injectable Silicone," estimates that between twenty thousand and forty thousand women were injected in the United States by 1965.
14 Simona Morini, "Beauty and Health: A New Aid to Plastic Surgery: Silicone," *Vogue*, January 15, 1971, 84.
15 Morini, "Beauty and Health," 86.
16 See Morini, "Beauty and Health"; Morini, *Body Sculpture*; Haiken, *Venus Envy*, 254; "The Silicone Injection Story Updated," *Harper's Bazaar*, May 1967, 148; Al Reinert, "Dr. Jack Makes His Rounds," *Esquire*, May 1975, 114–16.
17 Morini, "Beauty and Health."
18 Vinnik, "Hazards of Silicone Injections," 959.
19 Morini, "Beauty and Health," 86.
20 Warrick, *Forty Years*, 232.
21 Braley, "Use of Silicones," 284.

22 Braley, "Use of Silicones," 284.

23 See, for example, Morini, "Beauty and Health,"; Haiken, *Venus Envy*, 247; Byrne, *Informed Consent*; Jacobson, *Cleavage*.

24 See HCP, for example, Silas Braley, letter to Dicran Goulian Jr., April 24, 1963, which details how ten pounds had "unexplainably" gone missing and needed to be resent.

25 Charles Vinnik cited April 21, 1975, Public Health Service, FDA, Agenda, Special Drugs Advisory Committee Minutes, 14, JMP, box 1, folder 33.

26 "Dow Corning 3 Officers Indicted on Silicone Fluid," *Wall Street Journal*, August 17, 1967, HCP. Regulating interstate commerce was one of the forms of jurisdiction that the FDA held prior to the enactment of 1976 Medical Device Regulation Act.

27 Haiken, *Venus Envy*, 250.

28 Al Reinert, "Dr. Jack Makes His Rounds"; Morini, "Beauty and Health," 84; "Silicone Injection Story Updated."

29 For more on Nancy Reagan as a patient of Orentreich, see "FDA Slow to Protect from Beauty Products," *Santa Cruz Sentinel*, November 23, 1992, 7.

30 Morini, "Beauty and Health," 115.

31 Morini, "Beauty and Health," 115.

32 Franklyn, *Art of Staying Young*, 116.

33 Morini, "Beauty and Health," 115; Anne Louise Bardach, "The Dark Side of Plastic Surgery," NYT, April 17, 1988.

34 See, for example, Milton Edgerton, letter to Kelman Cohen, June 24, 1975, JMP, box 1, folder 33.

35 FDA inspection records reveal that Orentreich continued to order industrial-grade silicone and process it in his clinic for subcutaneous injection until 1991. In November of that year, Orentreich signed a consent decree of permanent injunction to agree to stop manufacturing, distributing, promoting, and administering fluid injectable silicone in surgical procedures in humans to this order. See FDA, HHS news release, February 28, 1992, and FDA quarterly reports, January to March 1992, 32–33.

36 John Furlong at 4 min., 3 sec.–4 min., 47 sec., in *Mondo Topless*, directed by Russ Meyer (Eve Productions, 1966).

37 "Dressing Through Two Decades," *Ebony*, November 1965, 213.

38 Levy, "Morality Play," 80.

39 Levy, "Morality Play," 87.

40 Shteir, *Striptease*, 321.

41 Levy, "Morality Play."

42 See Chester Higgins, "Bare-Bosomed Waitresses," *Jet*, June 10, 1965, 36, 60; Levy, "Morality Play," 87.

43 See Mary Ann Schildknecht at 54 min., 20 sec., in *Carol Doda Topless at the Condor*, directed by Marlo McKenzie and Jonathan Parker (Picturehouse, 2023).

44   Higgins, "Bare-Bosomed Waitresses"; Levy, "Morality Play," 87.

45   *Female Mimics* 1, no. 12 (1968) 59, DTA; *Bay Area Reporter*, December 1, 1971, 16.

46   *Carol Doda Topless at the Condor*, dir. McKenzie and Parker.

47   Otálvaro-Hormillosa, *Erotic Resistance*, 76.

48   Otálvaro-Hormillosa, *Erotic Resistance*, 76.

49   Mattioli, *Three Nights*, 108.

50   Earl Wilson, "The Almost Topless Gown," *San Francisco Examiner*, June 29, 1964, 16.

51   Doda quoted in Wolfe, "Put-Together Girl," 91.

52   Larry Still, "'Choice' Movie May Not Be," *Jet*, November 5, 1964, 20–25.

53   Stryker, *Transgender History*, 85.

54   Levy, "Morality Play," 82.

55   Shteir, *Striptease*, 321.

56   Shteir, *Striptease*, 321.

57   *Mondo Topless*, dir. Meyer, at 19 min., 43 sec.

58   Mattioli, *Three Nights*, 143.

59   Mattioli, *Three* Nights, 159; "Topless," *Playboy*, September 1966, 163.

60   Mattioli, *Three Nights*, 162.

61   Haiken, *Venus Envy*, 249.

62   "Nude Discotheque," *Playboy*, April 1965, 75; Mattioli, *Three Nights*, 10; "Rascal Go-Go Girls," *Rascal*, March 1966.

63   See also "Dancer Finds Boa's Stereo Hideaway," *Redwood City Tribune*, July 26, 1967, 2.

64   "Spoonful of Topless Hauls in the Sugar," *Salt Lake Tribune*, January 19, 1967, 16. Topless dancing became the dominant practice, replacing the traditional burlesque tease, which had required performers to keep their breasts covered under censorship laws. For more on this, see Shteir, *Striptease*; Scott, *Costumes of Burlesque*.

65   Sam Hudson, "The Silicone Bosom Revisited," *Sir!*, March 1968, 4; "Escalation," *Newsweek*, October 25, 1965, 110.

66   Hudson, "Silicone Bosom Revisited," 4, 61; "Escalation," 110.

67   Franklyn, *Art of Staying Young*.

68   Symmers, "Silicone Mastitis in 'Topless' Waitresses."

69   Symmers, "Silicone Mastitis in 'Topless' Waitresses," 19.

70   Symmers, "Silicone Mastitis in 'Topless' Waitresses," 19.

71   Harry Nelson, "Silicone Injections Health Hazard, Study Indicates," *Los Angeles Times*, November 2, 1975.

72   Ruth Ponce interviewed in Linda Witt, "What Is a Woman Without Breasts?," *Today's Health*, April 1974, 35.

73   Kopf et al., "Complications of Silicone Injections"; Witt, "What Is a Woman Without Breasts?"

74   Judy Mamou at 30 min., 49 sec., in *Carol Doda Topless at the Condor*, dir. McKenzie and Parker.

75   "Escalation."
76   "Deaths Tied to Silicone Breast Shots!," *Drag* 6, no. 21 (1976): 5, DTA.
77   Witt, "What Is a Woman Without Breasts?," 33.
78   Witt, "What Is a Woman Without Breasts?," 33.
79   See Silicone Committee correspondence, JMP. For example, Dow
      Corning medical meeting notes, New York, September 1975, 11–13.
80   Dow Corning medical meeting notes, New York, September 1975, 15,
      JMP.
81   Bob Emmons quoted in Byrne, *Informed Consent*, 44.
82   Morini, "Beauty and Health."
83   "Escalation." Teenagers are also mentioned in the AMA correspon-
      dence files; Symmers, "Silicone Mastitis in 'Topless' Waitresses";
      Webb, "Cleopatra's Needle."
84   "Escalation."
85   "Deaths Tied to Silicone Breast Shots!," 5, DTA.
86   "Escalation."
87   Michele Burgen, "Plastic Surgery: A Lift for the Face, Figure and
      Spirit," *Ebony*, January 1977, 40–50.
88   Burgen, "Plastic Surgery," 40.
89   Burgen, "Plastic Surgery," 42.
90   See also "Topless Insured for $1 Million," 30; "Larger and Smaller
      Busts Offered at Naval Hospital," *Jet*, December 26, 1974, 7.
91   Byrne, *Informed Consent*, 45.
92   Haiken, "Modern Miracles," 183.
93   See Ortiz-Monasterio and Trigos, "Management of Patients with
      Complications from Injections."
94   Ruth Ponce quoted in Witt, "What Is a Woman Without Breasts?," 35.
95   *SILASTIC Demonstration Mammary Sizer Implant Seamless Design
      350 CC* (Midland, MI: Dow Corning, 1970), SHI.
96   *SILASTIC Demonstration Mammary Sizer Implant Seamless Design
      350 CC*, SHI.
97   *SILASTIC Demonstration Mammary Sizer Implant Seamless Design
      350 CC*, SHI.
98   Braley, "Use of Silicones," 286.
99   Dow Corning, "Dow Corning Seeks Approval for Silicone Fluid Injec-
      tions," press release, September 3, 1975, JMP, box 1, folder 34.
100   Dow Corning, "Dow Corning Seeks Approval for Silicone Fluid Injec-
      tions," press release, September 3, 1975, JMP, box 1, folder 34.
101   A. H. Rathjen, "Testimony Presented to the Assembly Committee
      on Criminal Justice in Sacramento," January 7, 1976, 2, circulated
      to Silicone Committee members, February 16, 1976, JMP, box 1,
      folder 34.
102   See Serlin, "Christine Jorgensen and the Cold War Closet"; Skidmore,
      "Constructing the 'Good Transsexual'"; Stryker, *Transgender History*;
      Snorton, *Black on Both Sides*.

103   Exceptions I have come across: Ellenbogen and Rubin, "Injectable
       Fluid Silicone Therapy"; Mutou, "Study of 317 Akiyama's DMPS."

104   See, for example, the New York City Trans Oral History Project,
       https://nyctransoralhistory.org/. Most of these accounts on silicone
       focus on medical gatekeeping and community organizing around
       harm reduction in the 1990s and ensuing decades.

105   Dorian Corey interviewed in Michael Cunningham, "The Slap of Love,"
       Open City 6 (1995), https://opencity.org/archive/issue-6/the-slap-of-love.

106   Gill-Peterson, "DIY."

107   See, for example, Edgerton and McClary, "Augmentation Mamma-
       plasty"; Edgerton et al., "Augmentation Mammaplasty II."

108   Gill-Peterson, *Histories of the Transgender Child*, 132; Stryker, *Trans-
       gender History*, 97, 117.

109   Meyerowitz, *How Sex Changed*, 222; Stryker, *Transgender History*, 117.

110   "Sex Reassignment in Transsexualism: Research Justification and
       Protocol—Follow Up Questionnaire," John Money Collection, Gen-
       der Identity Clinic 1966–1967, box 503600, Alan Mason Chesney
       Medical Archives, Johns Hopkins University.

111   Notes from Gender Identity Clinic meeting, May 21, 1971, John
       Money Collection, 1971–1973, box 503600, Alan Mason Chesney
       Medical Archives, Johns Hopkins University.

112   Stryker, *Transgender History*, 104.

113   Stryker, *Transgender History*, 104.

114   See, for example, Skidmore, "Constructing the 'Good Transsexual'";
       Stryker, *Transgender History*; Snorton, *Black on Both Sides*. For DIY
       forms of transition, see Gill-Peterson, "DIY."

115   Gill-Peterson, "DIY."

116   "Statement of STAR's Political Platform," in Lewis, "Trans History,"
       76–77.

117   Ronald Kotulak and George Bliss, "County Office Probes Illegal Sili-
       cone Injections," *Chicago Tribune*, Sunday December 7, 1975, JMP, box
       1, folder 34.

118   Cook County case 76CR3340, July 26, 1975, Cook County
       Court Archives, Chicago (hereafter cited as Cook County case
       76CR3340).

119   Kotulak and Bliss, "County Office Probes Illegal Silicone Injections."

120   Kotulak and Bliss, "County Office Probes Illegal Silicone Injections."

121   Cook County case 76CR3340, 4.

122   Kotulak and Bliss, "County Office Probes Illegal Silicone Injections."

123   Cook County case 76CR3340, 3.

124   Cook County case 76CR3340, 3.

125   Cook County case 76CR3340, 3.

126   Cook County case 76CR3340, 5.

127   "Silicone Injections Are Dangerous," *Erickson Educational Foundation
       Newsletter* 6 no. 2 (1973): 2, DTA.

128   "Deaths Tied to Silicone Breast Shots!"; "Silicone Shots Busted Wide Open," *Drag* 6, no. 22 (1976): 9, both DTA.

129   "The Meeting Place," *Female Impersonator* 5, no. 8 (1975): 36, DTA.

130   "News," *Drag* 3, no. 10 (1973): 4–5, DTA.

131   "Deaths Tied to Silicone Breast Shots!"; Linda Lee, "Changing Your Life," *Female Mimics International* 11, no. 1 (1980): 16, both DTA.

132   "Lee G. Brewster's Mardi Gras '73," advertisement, *Drag* 3, no. 9 (1973): 32, DTA.

133   Stryker, "Lee Greer Brewster."

134   "Lee Brewster: The First Lady of Drag," *Female Impersonator* 5, no. 8 (1975): 60–63, DTA.

135   "Lee's Mardi Gras Enterprises Inc.," advertisement, *Drag* 6, no. 21 (1976): back cover, DTA.

136   "The Passing Parade," *Drag* 1, no. 3 (1971): 33, DTA.

137   Stryker, "Lee Greer Brewster."

138   Lee, "Changing Your Life"; Sally Douglas, "Hormones and Me," *New Trenns Magazine* 2, no. 6 (1971): 10–21, both DTA.

139   "New Discovery!," *Drag* 4, no. 13 (1973): 43, 45, DTA.

140   "Treasure Chest," advertisement, *Female Impersonator* 5, no. 8 (1975): 51, DTA.

141   Surgitek Inc., "Sales Brochure," 1973, Walter Spohn Papers, NMAH, Division of Science and Medicine.

142   Lee, "Changing Your Life," 14; see also "Society for the Second Self," brochure, 1980, both DTA.

143   Lee, "Changing Your Life," 14; see also Douglas, "Hormones and Me," both DTA.

144   Pedersen, *Bra*, 77.

145   Ashley Altadonna, "A Brief History of Breast Forms," Trans Tool Shed, December 23, 2019, https://transtoolshed.com/blogs/news/a-brief-history-of-breast-forms.

146   Gardner, "Hiding the Scars," 321.

147   "Mirage on the Horizon," *Female Mimics International* 13, no. 6 (1984): 5, DTA.

148   Linda Lee, "Changing Your Life," 14; Douglas, "Hormones and Me," both DTA.

149   "The Perfect Illusion," *Female Mimics International* 11, no. 4 (1980): 15, DTA.

150   See, for example, "Hormones DO Make a Difference," *Drag* 1, no. 3 (1971): 14–15; "Carolyn Summers Discovers Miss Stephanie," *Drag* 4, no. 14 (1974): 24; Stefanie in "Out of the Closet," *Female Impersonator* 5, no. 8 (1975): 12; Douglas, "Hormones and Me," all DTA.

151   Douglas, "Hormones and Me," DTA.

152   Douglas, "Hormones and Me," DTA.

153   Georgina, "Letter to the Editor," *Female Mimics International* 12, no. 3 (1982): 11, DTA.

154 Davis, *Finer Specimen of Womanhood*, 26.

155 Linda Lee, "Linda Lee's Pages," *Female Mimics International* 13, no. 6 (1984): 9, DTA.

156 Stone, "Empire Strikes Back," 229.

157 Stone, "Empire Strikes Back," 232.

158 A. H. Rathjen, "Testimony Presented to the Assembly Committee on Criminal Justice in Sacramento," January 7, 1976, 5, circulated to Silicone Committee members, February 16, 1976, JMP, box 1, folder 34.

Epilogue

1 Firestone, *Dialectic of Sex*; Murphy, *Seizing the Means of Reproduction*; Hamraie, *Building Access*; Kafer, *Feminist, Queer, Crip*; Stryker, *Transgender History*; Haraway, "Cyborg Manifesto"; Stone, "Empire Strikes Back."

2 cárdenas, "Android Goddess," 30.

3 "Chromat × Tourmaline ss22," Chromat, September 15, 2021, https://chromat.co/blogs/news/ss22.

4 Alex Jenny, "Inside Chromat and Tourmaline's 'Lifesaving' Swimwear Show at Riis," *Them*, September 16, 2021, https://www.them.us/story/chromat-x-tourmaline-swimwear-show-riis-beach?.

5 "Chromat × Tourmaline ss22."

6 Bex McCharen, "Tourmaline on the Power of Being Your Full, Sexy Self in Public," *Them*, February 13, 2023, https://www.them.us/story/tourmaline-chromat-becca-mccharen-tran-beach.

7 Jenny, "Inside Chromat and Tourmaline's 'Lifesaving' Swimwear Show at Riis."

8 "Cruz Suit-Red Ribbed," Chromat, accessed August 7, 2024, https://chromat.co/collections/all/products/cruz-suit-red.

9 Reddy-Best et al., "Visibly Queer- and Trans-Fashion Brands."

10 Grant et al., *Injustice at Every Turn*, 22.

11 Reddy-Best et al., "Visibly Queer-and Trans-Fashion Brands," 47.

12 For a frontline account of the impact of plastics pollution on the world's oceans and environment, as well as its role in environmental racism, see Cirino, *Thicker Than Water*.

13 "About Us," Chromat, accessed August 7, 2024, https://chromat.co/pages/about-us.

14 Kinoshita et al., "Utilization of a Cyclic Dimer."

15 Bombshell Club, Instagram profile, accessed August 7, 2024, https://www.instagram.com/bombshell_club/. For more on Essa Noche and Bombshell Brunch, see Noche's Instagram profile, accessed August 7, 2024, https://www.instagram.com/essanoche/.

16 Miss Rosen, "Inside Ethan James Green's New Book of New York Bombshells," *Dazed Digital*, September 13, 2022, https://www

.dazeddigital.com/art-photography/article/56940/1/ethan-james
-green-bombshell-photography-interview.

17   Green, *Bombshell*.

18   Green interviewed in Rosen, "Inside Ethan James Green's New Book
of New York Bombshells."

19   Rosen, "Inside Ethan James Green's New Book of New York Bombshells."

20   Deirdre Simonds, "Kim Kardashian Is a Blonde Bombshell in a New
Series of Photos with a Cryptic Message: 'Time Will Always Tell,'"
Daily Mail, August 23, 2021, https://www.dailymail.co.uk/tvshowbiz
/article-11139039/Kim-Kardashian-blonde-bombshell-new-series-new
-photos-cryptic-message.html; Rebecca Calderwood, "Kim Kardashian Transforms into 80s Bombshell with Huge Ice Blonde Wig and
Silver Swimwear," *Mirror*, July 3, 2022, https://www.mirror.co.uk/3am
/celebrity-news/kim-kardashian-transforms-80s-bombshell-27386280.

21   Due to the impact of COVID-19, it is difficult to compare the American Society of Plastic Surgeons' annual statistics for 2020 to previous years. See *Plastic Surgery Statistics Report* (Arlington Heights,
IL: American Society of Plastic Surgeons, 2020), https://www
.plasticsurgery.org/documents/News/Statistics/2020/plastic-surgery
-statistics-full-report-2020.pdf.

22   Strings, *Fearing the Black Body*, 62–64; Gilman, *Making the Body
Beautiful*, 210–18.

23   For more on the Kardashians and hyperfemininity, see Monteverde,
"Kardashian Komplicity."

24   Ida Simon, "The Bum," archived March 20, 2023, at https://web
.archive.org/web/20230320182428/https://ida-simon.com/The-Bum.

25   Amanda Fortini, "Break the Internet: Kim Kardashian," *Paper*,
November 12, 2014, https://www.papermag.com/break-the-internet
-kim-kardashian-cover-1427450475.html.

26   Hatti Rex, "You Can Try Kim K's Bum on for Size—Thanks to This
Creative Trio," *Dazed*, February 19, 2020, https://www.dazeddigital
.com/beauty/article/48019/1/kim-kardashian-wearable-bum-ida
-jonsson-simon-saarinen-beate-karlsson-nyfw.

27   Beate Karlsson, "3 x Kim K Wearable Bum Replica," Instagram, February 20, 2022, https://www.instagram.com/p/CaNS1QfLR0K/?utm
_source=ig_web_copy_link&igsh=MzRlODBiNWFlZA==.

28   Bucy, "How to Get a Perfect Drag Body? Hip Padding Tutorial,"
posted May 17, 2022, YouTube, https://youtu.be/w66lV6Gybbg, at
11 min., 37 sec.

29   Connor Lennox, "ASMR 1 HOUR Foam Snipping (Making Drag Queen
Hip Pads)," archived June 4, 2022, https://web.archive.org/web
/20220604203947/https://youtu.be/2_U-B_-p3n8, at 0 min., 37 sec.

30   Patrick Starrr, "How to Make Drag Queen Body Padding," posted
October 8, 2018, YouTube, https://youtu.be/bSsajoeYauc, at 2 min.,
9 sec.

31   Sharon Waniz, "How Hip Pads Are Made. Tailors Techniques Revealed. Curvy Instagram Models," posted September 9, 2019, YouTube, https://youtu.be/zuK4FVVMS-I, at 7 min., 20 sec.

32   Waniz, "How Hip Pads Are Made," at 9 min., 50 sec.

33   Liboiron, *Pollution Is Colonialism*, 5.

34   US Environmental Protection Agency, "Flexible Polyurethane Foam Fabrication Operations: National Emission Standards for Hazardous Air Pollutants," last modified March 26, 2025, https://www.epa.gov/stationary-sources-air-pollution/flexible-polyurethane-foam-fabrication-operations-national.

35   Shaw et al., "Halogenated Flame Retardants." For more on endocrine-disrupting compounds, see Liboiron, "Plasticizers"; Liboiron, *Pollution Is Colonialism*, 81–112.

36   Cambria Bold, "Is a Sofa Made with Polyurethane Foam Okay or Unhealthy?," *Apartment Therapy*, March 14, 2011, archived January 26, 2017, https://web.archive.org/web/20170126051315/https://www.apartmenttherapy.com/is-a-sofa-made-with-polyuretha-141503; "Non-Toxic Sofas: Do They Exist and How to Find Them," *State Home*, August 15, 2020, https://blog.thestatedhome.com/non-toxic-sofas/.

37   "About Certified Foam," CertiPUR-US, accessed August 7, 2024, https://certipur.us/about-the-seal/.

38   Irina Webb, "Polyurethane Foam: What Is Inside Our Furniture?," *I Read Labels For You*, April 30, 2021, https://ireadlabelsforyou.com/polyurethane-foam-what; Joanne Chen, "How to Choose a Mattress," *Wirecutter*, updated April 2, 2025, https://www.nytimes.com/wirecutter/guides/buying-a-mattress/.

39   "Cotton," World Wildlife Fund, accessed August 7, 2024, https://www.worldwildlife.org/industries/cotton. For research on sustainable fashion practices, see the Centre for Sustainable Fashion at London College of Fashion, https://www.sustainable-fashion.com.

40   See Liboiron, *Pollution Is Colonialism*; Davis, *Plastic Matter*; Anna Tsing, "Feral Atlas: The More-Than-Human Anthropocene," Stanford University Press, 2021, https://feralatlas.supdigital.org/?cd=true&bdtext=introduction-to-feral-atlas; Cirino, *Thicker Than Water*.

41   *Plastic Surgery Statistics Report*.

42   Danielle Nett, "For Trans Women, Silicone 'Pumping' Can Be a Blessing and a Curse," NPR, September 1, 2019, https://www.npr.org/sections/codeswitch/2019/09/01/755629721/for-trans-women-silicone-pumping-can-be-a-blessing-and-a-curse.

43   Safer et al., "Barriers to Healthcare for Transgender Individuals."

44   Wilson et al., "Use and Correlates of Illicit Silicone or 'Fillers'"; Sonnenblick et al., "MRI Features of Free Liquid Silicone."

45   Sonnenblick et al., "MRI Features of Free Liquid Silicone."

46   Wilson et al., "Use and Correlates of Illicit Silicone or 'Fillers'"

47   Wilson et al., "Use and Correlates of Illicit Silicone or 'Fillers'";
      Sonnenblick et al., "MRI Features of Free Liquid Silicone."

48   Wilson et al., "Use and Correlates of Illicit Silicone or 'Fillers'"; Madeline B. Deutsch and Barry Zevin, "Free Silicone and Other Filler Use," UCSF, June 17, 2016, https://transcare.ucsf.edu/guidelines/silicone-filler.

49   Andrea James, "Injected Silicone," accessed December 12, 2022, https://www.transgendermap.com/medical/injected-silicone/; *SJI Occupational Health and Safety Handbook*, 3rd ed. (San Francisco: St. James Infirmary, 2010); Chyten-Brennan, "Surgical Transition," 269–70.

50   Nick Gorton, "Silicone Pumping and Harm Reduction in the Transgender Community," TransLine, June 15, 2016, https://transline.zendesk.com/hc/en-us/article_attachments/213758928.

51   Callen-Lorde Community Health Center, "Safer Silicone," June 2016, https://web.archive.org/web/20221102024142/http://callen-lorde.org/graphics/2016/06/HOTT-Safer-Silicone-Brochure.pdf.

52   *SJI Occupational Health and Safety Handbook*, 41.

53   "What We Believe," St. James Infirmary, accessed December 12, 2022, https://www.stjamesinfirmary.org/about.

54   Ruby Corado cited in Nett, "For Trans Women, Silicone 'Pumping' Can Be a Blessing and a Curse."

BIBLIOGRAPHY

Abrego, Sonya. *Westernwear: Postwar American Fashion and Culture.* London: Bloomsbury Visual Arts, 2022.

Ahad, Badia. "Confessions." In *Rethinking Therapeutic Culture*, edited by Timothy Richard Aubry and Trysh Travis, 85–95. Chicago: University of Chicago Press, 2015.

Allen, Thomas B., and Norman Polmar. *Rickover: Father of the Nuclear Navy.* Military Profiles. Washington, DC: Potomac Books, 2007.

Anderson, Karen. *Wartime Women: Sex Roles, Family Relations, and the Status of Women During World War II.* Contributions in Women's Studies. Westport, CT: Greenwood Press, 1981.

Andrianov, K. A. "Synthesis of Alkyl-Substituted Ortho-Esters of Silic Acid." *Journal of General Chemistry* 8 (1938): 552–56.

Angell, Katelyn, and K. R. Roberto. "Cataloging." *TSQ: Transgender Studies Quarterly* 1, no. 1–2 (2014): 53–56.

Angell, Marcia. *Science on Trial: The Clash of Medical Evidence and the Law in the Breast Implant Case.* New York: W. W. Norton, 1996.

Angus, Siobhan. *Camera Geologica.* Durham, NC: Duke University Press, 2024.

Antonelli, Paola. *Safe: Design Takes on Risk.* New York: Museum of Modern Art, 2005.

Anzaldúa, Gloria. *Borderlands/La Frontera: The New Mestiza.* San Francisco: Aunt Lute Books, 1987.

Ashley, F. L., S. Braley, T. D. Rees, D. Goulian, and D. L. Ballantyne. "The Present Status of Silicone Fluid in Soft Tissue Augmentation." *Plastic and Reconstructive Surgery* 39, no. 4 (1967): 411–20. https://doi.org/10.1097/00006534-196704000-00012.

Badura, R., A. Buczek, J. Kotz, J. Utzis, and J. Wasilewski. "Investigations of Intravenous Application of Silicone Oils." *Medycyna Weterynaryjna* 3 (1968): 151–54.

Bailey, Beth L. *From Front Porch to Back Seat: Courtship in Twentieth-Century America.* 1988. Reprint, Baltimore: Johns Hopkins University Press, 1989.

Balsamo, Anne. *Technologies of the Gendered Body: Reading Cyborg Women.* Durham, NC: Duke University Press, 1995.

Bardsley, Jan. "Girl Royalty: The 1959 Coronation of Japan's First Miss Universe." *Asian Studies Review* 32, no. 3 (2008): 375–91. https://doi.org/10.1080/10357820802302512.

Barger, Michelle. "Thoughts on Replication and the Work of Eva Hesse." *Tate Papers*, no. 8 (Autumn 2007). https://www.tate.org.uk/research/tate-papers /08/thoughts-on-replication-and-the-work-of-eva-hesse.

Barker, Rodney. *The Hiroshima Maidens: A Story of Courage, Compassion, and Survival*. New York: Viking, 1985.

Barthes, Roland. "Plastic." In *Mythologies*, translated by Annette Lavers, 117–20. 1972. Reprint, London: Vintage Classics, 2009.

Bayer, Otto. "Polyurethanes." Translated by I. G. Callomon and G. M. Kline. *Modern Plastics* 24:10 (1947): 149–52.

Ben-Hur, N., D. L. Ballantyne, T. D. Rees, and I. Seidman. "Local and Systemic Effects of Dimethylpolysiloxane Fluid in Mice." *Plastic and Reconstructive Surgery* 39, no. 4 (1967): 423–26. https://doi.org/10.1097 /00006534-196704000-00014.

Berney, Adrienne. "Streamlining Breasts: The Exaltation of Form and Disguise of Function in 1930s' Ideals." *Journal of Design History* 14, no. 4 (2001): 327–42. https://doi.org/10.1093/jdh/14.4.327.

Bird, William L. *Better Living: Advertising, Media and the New Vocabulary of Business Leadership, 1935–1955*. Evanston, IL: Northwestern University Press, 1999.

Bix, Amy Sue. *Girls Coming to Tech! A History of American Engineering Education for Women*. Engineering Studies Series. Cambridge, MA: MIT Press, 2013.

Blaszczyk, Regina Lee. *The Color Revolution*. Cambridge, MA: MIT Press, 2012.

Boo-Chai, K. "The Complications of Augmentation Mammaplasty by Silicone Injection." *British Journal of Plastic Surgery* 22, no. 3 (1969): 281–85. https:// doi.org/10.1016/s0007-1226(69)80120-7.

Boyer, Paul. *By the Bomb's Early Light: American Thought and Culture at the Dawn of the Atomic Age*. Chapel Hill: University of North Carolina Press, 1994.

Boyer, Paul. "The United States, 1941–1963: A Historical Overview." In *Vital Forms: American Art and Design in the Atomic Age, 1940–1960*, edited by Brooke Kamin Rapaport and Kevin Stayton, 38–77. New York: Brooklyn Museum of Art in association with H. N. Abrams, 2001.

Braley, Silas. "Illegal, Immoral and Dangerous." *Science News* 93, no. 7 (1968): 173. https://doi.org/10.2307/3952909.

Braley, Silas. "Silicone Fluids with Added Adulterants." *Plastic and Reconstructive Surgery* 45, no. 3 (1970): 288. https://doi.org/10.1097 /00006534-197003000-00014.

Braley, Silas. "The Use of Silicones in Plastic Surgery: A Retrospective View." *Plastic and Reconstructive Surgery* 51, no. 3 (1973): 280–88. https://doi.org/10.1097 /00006534-197303000-00006.

Brevard, Aleshia. *The Woman I Was Not Born to Be: A Transsexual Journey*. Philadelphia: Temple University Press, 2001.

Bright, Christopher J. *Continental Defense in the Eisenhower Era: Nuclear Antiaircraft Arms and the Cold War*. Palgrave Studies in the History of Science and Technology. New York: Palgrave Macmillan, 2010.

Brown, Elspeth H. *Work! A Queer History of Modeling*. Durham, NC: Duke University Press, 2019.

Brown, Jeannette E. *African American Women Chemists*. New York: Oxford University Press, 2012.

Burris-Meyer, Elizabeth. *Color and Design in the Decorative Arts*. Retailing Series. New York: Prentice Hall, 1937.

Buszek, Maria Elena. *Pin-Up Grrrls: Feminism, Sexuality, Popular Culture*. Durham, NC: Duke University Press, 2006.

Butler, Judith. "Imitation and Gender Insubordination." In *The Lesbian and Gay Studies Reader*, edited by Henry Abelove, Michèle Aina Barale, and David M. Halperin, 307–20. New York: Routledge, 1993.

Butler, Judith. *Who's Afraid of Gender?* New York: Farrar, Straus and Giroux, 2024.

Byrne, John A. *Informed Consent*. New York: McGraw-Hill, 1996.

cárdenas, micha. "The Android Goddess Declaration: After Man(ifestos)." In *Bodies of Information*, edited by Elizabeth Losh and Jacqueline Wernimont, 25–38. Intersectional Feminism and the Digital Humanities. Minneapolis: University of Minnesota Press, 2018.

Carter, Julian. *The Heart of Whiteness: Normal Sexuality and Race in America, 1880–1940*. Durham, NC: Duke University Press, 2007.

Castillo, Greg. *Cold War on the Home Front: The Soft Power of Midcentury Design*. Minneapolis: University of Minnesota Press, 2010.

Chambers, Jason. *Madison Avenue and the Color Line: African Americans in the Advertising Industry*. Philadelphia: University of Pennsylvania Press, 2008.

Chasan, Paul E. "The History of Injectable Silicone Fluids for Soft-Tissue Augmentation." *Plastic and Reconstructive Surgery* 120, no. 7 (2007): 2034–40. https://doi.org/10.1097/01.prs.0000267580.92163.33.

Chyten-Brennan, Jules. "Surgical Transition." *Trans Bodies, Trans Selves: A Resource for the Transgender Community*, edited by Laura Erickson-Schroth, 265–90. New York: Oxford University Press, 2014.

Cirino, Erica. *Thicker Than Water: The Quest for Solutions to the Plastic Crisis*. Washington, DC: Island Press, 2021.

Clarke, Alison J. *Tupperware: The Promise of Plastic in 1950s America*. Washington, DC: Smithsonian Institution, 2014.

Classen, Constance. *The Deepest Sense: A Cultural History of Touch*. Studies in Sensory History. Urbana: University of Illinois Press, 2012.

Coe, Jerome T. *Unlikely Victory: How General Electric Succeeded in the Chemical Industry*. New York: American Institute of Chemical Engineers, 2000.

Cogdell, Christina. *Eugenic Design: Streamlining America in the 1930s*. Philadelphia: University of Pennsylvania Press, 2004.

Cole, Thomas G., II. "(The) Bikini: EmBodying the Bomb." *Genders, 1998–2013*, March 1, 2011. https://www.colorado.edu/gendersarchive1998-2013/2011/03/01/bikini-embodying-bomb.

Colomina, Beatriz. "Cold War/Hothouses." In Colomina, Brennan, and Kim, *Cold War Hothouses*, 10–21.

Colomina, Beatriz. *Domesticity at War*. Illustrated ed. Cambridge, MA: MIT Press, 2007.

Colomina, Beatriz, Annmarie Brennan, and Jeannie Kim, eds. *Cold War Hothouses: Inventing Postwar Culture, from Cockpit to Playboy*. New York: Princeton Architectural Press, 2004.

Conway, Herbert. "Mammaplasty: Analysis of 110 Consecutive Cases with End-Results." *Plastic and Reconstructive Surgery* 10, no. 5 (1952): 303–15.

Conway, Herbert, and George Dietz. "Augmentation Mammaplasty." *Surgery, Gynecology and Obstetrics* 114 (May 1962): 573.

Conway, Herbert, and Dicran Goulian Jr. "Experience with an Injectable Silastic RTV as a Subcutaneous Prosthetic Material: A Preliminary Report." *Plastic and Reconstructive Surgery* 32 (1963): 294–302. https://doi.org/10.1097/00006534-196309000-00002.

Conway, Herbert, and James Smith. "Breast Plastic Surgery: Reduction Mammaplasty, Mastopoxy, Augmentation Mammaplasty, and Mammary Construction; Analysis of Two Hundred and Forty-Five Cases." *Plastic and Reconstructive Surgery and the Transplantation Bulletin* 21, no. 1 (1958): 8–19. https://doi.org/10.1097/00006534-195801000-00002.

Corset and Brassiere Association of America. *Foundations for Fashion: Facts, Figures, Feature Material*. New York: Corset and Brassiere Association of America, 1948.

Cowan, Ruth Schwartz. *More Work for Mother: The Ironies of Household Technology from the Open Hearth to the Microwave*. New York: Basic Books, 1983.

Cowan, Ruth Schwartz. *A Social History of American Technology*. Oxford: Oxford University Press, 1997.

Cowan, T. L., and Jasmine Rault. "Introduction: Metaphors as Meaning and Method in Technoculture." *Catalyst: Feminism, Theory, Technoscience* 8, no. 2 (2022): 1–23. https://doi.org/10.28968/cftt.v8i2.39036.

Craig, Maxine Leeds. *Ain't I a Beauty Queen? Black Women, Beauty, and the Politics of Race*. Oxford: Oxford University Press, 2002.

Crawford, R. *Plastics and Rubber: Engineering Design and Applications*. London: MEP, 1985.

Creadick, Anna G. *Perfectly Average: The Pursuit of Normality in Postwar America*. Culture, Politics, and the Cold War. Amherst: University of Massachusetts Press, 2010.

Crenshaw, Kimberlé. "Beyond Racism and Misogyny: Black Feminism and 2 Live Crew." In *Words That Wound: Critical Race Theory, Assaultive Speech, and the First Amendment*, edited by Mari J. Matsuda, Charles R. Lawrence III, Richard Delgado, and Kimberlé Crenshaw, 111–32. Boulder, CO: Westview, 1993.

Crenshaw, Kimberlé. "Demarginalizing the Intersection of Race and Sex: A Black Feminist Critique of Antidiscrimination Doctrine, Feminist Theory and Antiracist Politics." *University of Chicago Legal Forum* 1989, no. 1 (1989): 139–66.

Crim, Brian E. *Our Germans: Project Paperclip and the National Security State*. Baltimore: Johns Hopkins University Press, 2018.

Crowley, David, and Jane Pavitt, eds. *Cold War Modern: Design, 1945–1970*. London: V & A Publishing, 2008.

David, Alison Matthews. *Fashion Victims: The Dangers of Dress Past and Present.* London: Bloomsbury Visual Arts, 2015.

Davidann, Jon Thares. "'A Certain Presentiment of Fatal Danger': The Sino-Japanese War and U.S.-Japanese Relations, 1937–1939." In *Cultural Diplomacy in U.S.-Japanese Relations, 1919–1941*, edited by Jon Thares Davidann, 179–203. New York: Palgrave Macmillan US, 2007. https://doi.org/10.1057/9780230609730_11.

Davis, Heather M. *Plastic Matter.* Elements. Durham, NC: Duke University Press, 2022.

Davis, Kathy. *Reshaping the Female Body: The Dilemma of Cosmetic Surgery.* New York: Routledge, 1995.

Davis, Sharon. *A Finer Specimen of Womanhood: A Transsexual Speaks Out.* New York: Vantage Press, 1985.

DeBell, John M., William C. Goggin, and Walter E. Gloor. *German Plastics Practice: A Record, Rewritten and Amplified, from the Quartermaster Reports.* Cambridge, MA: Murray Printing Press, 1945.

de Grazia, Victoria. *Irresistible Empire: America's Advance Through Twentieth-Century Europe.* Cambridge, MA: Harvard University Press, 2009.

Demarais, Kirk. *Mail-Order Mysteries: Real Stuff from Old Comic Book Ads!* San Rafael, CA: Insight Editions, 2011.

de Monchaux, Nicholas. *Spacesuit: Fashioning Apollo.* Cambridge, MA: MIT Press, 2011.

Denton, Sally, and Roger Morris. *The Money and the Power: The Making of Las Vegas and Its Hold on America, 1947–2000.* New York: Alfred A. Knopf, 2001.

Diehl, Nancy. "Zelda Wynn Valdes: Uptown Modiste." In *The Hidden History of American Fashion: Rediscovering 20th-Century Women Designers*, edited by Nancy Diehl, 223–36. London: Bloomsbury Academic, 2018.

Dombow, Bernard. *Polyurethanes.* New York: Reinhold Publishing, 1957.

Doriot, Georges. "Foreword." In DeBell, Goggin, and Gloor, *German Plastics Practice*, iii.

Dower, John W. *Japan in War and Peace: Selected Essays.* New York: New Press, 1995.

Dyer, Richard. *Heavenly Bodies: Film Stars and Society.* 2nd ed. London: Routledge, 2004.

Dyer, Richard. *Stars.* New ed. London: BFI, 1998.

Dyer, Richard. *White.* London: Routledge, 1997.

Edgerton, Milton, and Allan McClary. "Augmentation Mammaplasty; Psychiatric Implications and Surgical Indications; (with Special Reference to Use of the Polyvinyl Alcohol Sponge Ivalon)." *Plastic and Reconstructive Surgery* 21, no. 4 (1958): 279–305.

Edgerton, Milton, E. Meyer, and W. E. Jacobson. "Augmentation Mammaplasty II: Further Surgical and Psychiatric Evaluation." *PRS* 27 (March 1961): 279–302. https://doi.org/10.1097/00006534-196103000-00005.

E. I. Du Pont de Nemours and Company. *Du Pont: The Autobiography of an American Enterprise.* Wilmington, DE: E. I. du Pont de Nemours and Company, 1952. Distributed by Scribner's Sons.

Ellenbogen, R., and L. Rubin. "Injectable Fluid Silicone Therapy: Human Morbidity and Mortality." *JAMA* 234, no. 3 (1975): 308–9.

Ellison, Treva. "The Labor of Werqing It: The Performance and Protest Strategies of Sir Lady Java." In Gossett, *Trap Door*, 1–22.

Entwistle, Joanne. *The Fashioned Body: Fashion, Dress, and Modern Social Theory.* Cambridge: Polity Press, 2000.

Epstein, Katherine C. *Torpedo: Inventing the Military-Industrial Complex in the United States and Great Britain.* Cambridge, MA: Harvard University Press, 2014.

Farrell-Beck, Jane, and Colleen Gau. *Uplift: The Bra in America.* Philadelphia: University of Pennsylvania Press, 2002.

Ferreira, M. C., V. Spina, and K. Iriya. "Changes in the Lung Following Injections of Silicone Gel." *British Journal of Plastic Surgery* 28, no. 3 (1975): 173–76. https://doi.org/10.1016/0007-1226(75)90124-1.

Fields, Jill. *An Intimate Affair: Women, Lingerie, and Sexuality.* Berkeley: University of California Press, 2007.

Firestone, Shulamith. *The Dialectic of Sex: The Case for Feminist Revolution.* 1970. Reprint, London: Verso, 2015.

Ford, Henry, and Samuel Crowther, *My Life and Work.* London: William Heinemann, 1922.

Fordham, Stanley. *Silicones.* London: George Newnes, 1960.

Forrestal, Dan. *Faith, Hope, and $5,000.* New York: Simon and Schuster, 1977.

Franklyn, Robert Alan. *The Art of Staying Young.* New York: Frederick Fell, 1964.

Franklyn, Robert Alan. "Augmentation Mammaplasty." *General Practice: The Medical Journal of the West* 20, no. 11 (1957): 11–12.

Franklyn, Robert Alan. *Augmentation Mammaplasty.* 3rd ed. Rome: International Academy of Cosmetic Surgery, 1976.

Franklyn, Robert Alan. *Beauty Surgeon.* Long Beach, CA: White Horn Publishing, 1960.

Franklyn, Robert Alan. *Beauty Surgeon.* New York: Pyramid Royal, 1961.

Franklyn, Robert Alan. "Breastplasty." *Southern General Practitioner of Medicine and Surgery* 4 (April 1953): 71–76.

Franklyn, Robert Alan. *The Clinical Atlas of Cosmetic Plastic Surgery: A Teaching Manual.* Rome: International Academy of Cosmetic Surgery, 1976.

Franklyn, Robert Alan. "Der neue Kunststoff 'Surgifoam' und seine Chirurgische Nutzbarkeit." *Zentralblatt für Chirurgie* 29 (1956): 1192–93.

Franklyn, Robert Alan. "Kosmetische Korrektur Unästhetischer Postoperativer Endergebnisse der Mastoplastik." *Zentralblatt für Chirurgie* 41 (1957): 1752–53.

Franklyn, Robert Alan. *On Developing Bosom Beauty.* New York: Frederick Fell, 1959.

Franklyn, Robert Alan, with Marcia Borie. *Instant Beauty.* New York, Frederick Fell, 1967.

Frisch, Kurt, C., and James H. Saunders. *Plastic Foams, Part 1.* New York: Marcel Decker, 1972.

Fuller, R. Buckminster. *Critical Path*. New York: St. Martin's Press, 1981.

Gabrys, Jennifer, Gay Hawkins, and Mike Michael, eds. *Accumulation: The Material Politics of Plastic*. London: Routledge, 2013.

Gardner, Kirsten, E. "From Cotton to Silicone: Breast Prosthesis before 1950." In Ott, Serlin, and Mihm, *Artificial Parts, Practical Lives*, 102–18.

Gardner, Kirsten, E. "Hiding the Scars: History of Breast Prostheses After Mastectomy Since 1945." In *Beauty and Business: Commerce, Gender, and Culture in Modern America*, edited by Philip Scranton, 309–26. Hagley Perspectives on Business and Culture. New York: Routledge, 2001.

Gartman, David. "Harley Earl and the Art and Color Section: The Birth of Styling at General Motors." *Design Issues* 10, no. 2 (1994): 3–26. https://doi.org/10.2307/1511626.

Gibson, Pamela Church. "No-One Expects Me Anywhere: Invisible Women, Ageing and the Fashion Industry." In *Fashion Cultures: Theories, Explorations, and Analysis*, edited by Stella Bruzzi and Pamela Church Gibson, 79–90. London: Routledge, 2000.

Gill-Peterson, Jules. *Histories of the Transgender Child*. Minneapolis: University of Minnesota Press, 2018.

Gill-Peterson, Jules. "DIY." In *The SAGE Encyclopedia of Trans Studies*, edited by Abbie E. Goldberg and Genny Beemyn. Los Angeles: SAGE Reference, 2021. https://doi.org/10.4135/9781544393858.n70.

Gill-Peterson, Jules. *A Short History of Trans Misogyny*. London: Verso, 2024.

Gilman, Sander L. *Making the Body Beautiful: A Cultural History of Aesthetic Surgery*. Princeton, NJ: Princeton University Press, 2001.

Glickman, Lawrence B. "'Make Lisle the Style': The Politics of Fashion in the Japanese Silk Boycott, 1937–1940." *Journal of Social History* 38, no. 3 (2005): 573–608. https://doi.org/10.1353/jsh.2005.0032.

Gossett, Reina, Eric A. Stanley, and Johanna Burton, eds. *Trap Door: Trans Cultural Production and the Politics of Visibility*. Critical Anthologies in Art and Culture. Cambridge, MA: MIT Press, 2017.

Gottwald, Laura, and Janusz Gottwald. *Frederick's of Hollywood, 1947–1973: 26 Years of Mail Order Seduction*. Secaucus, NJ: Castle Books, 1973.

Grant, Jaime M., Lisa Mottet, Justin Edward Tanis, Jack Harrison, Jody Herman, and Mara Keisling. *Injustice at Every Turn: A Report of the National Transgender Discrimination Survey*. Washington, DC: National Center for Transgender Equality/National Gay and Lesbian Task Force, 2011. https://transequality.org/sites/default/files/docs/resources/NTDS_Report.pdf.

Green, Ethan James. *Bombshell*. London: Baron Books, 2024.

Gregory, Steven. *Black Corona: Race and the Politics of Place in an Urban Community*. Princeton Studies in Culture/Power/History. Princeton, NJ: Princeton University Press, 1998.

Grindlay, John, and John Waugh. "Plastic Sponge Which Acts as Framework for Living Tissue." *American Medical Association Archives of Plastic Surgery* 63 (1951): 288–97.

Gundle, Stephen. *Glamour: A History*. Oxford: University Press, 2009.

Haiken, Elizabeth. "Modern Miracles: The Development of Cosmetic Prosthetics." In Ott, Serlin, and Mihm, *Artificial Parts, Practical Lives*, 171–98.

Haiken, Elizabeth. *Venus Envy: A History of Cosmetic Surgery*. Baltimore: Johns Hopkins University Press, 1997.

Halberstam, Jack. *Trans*: A Quick and Quirky Account of Gender Variability*. Berkeley: University of California Press, 2018.

Hall, Stuart. "The Spectacle of the 'Other.'" In *Representation*, edited by Stuart Hall, Jessica Evans, and Sean Nixon, 2nd ed., 215–87. London: Open University, 2013.

Hamraie, Aimi. *Building Access: Universal Design and the Politics of Disability*. Minneapolis: University of Minnesota Press, 2017.

Handley, Susannah. *Nylon: The Story of a Fashion Revolution*. London: Bloomsbury, 1999.

Haraway, Donna. "A Cyborg Manifesto: Science, Technology, and Socialist-Feminism in the Late Twentieth Century." In *Simians, Cyborgs, and Women: The Reinvention of Nature*, 149–82. London: Free Association Books, 1991.

Harris, Dianne Suzette. *Little White Houses: How the Postwar Home Constructed Race in America*. Architecture, Landscape, and American Culture Series. Minneapolis: University of Minnesota Press, 2013.

Hawes, Esme. "Making the Modern Consumer." In *Cars: Accelerating the Modern World*, edited by Brendan Cormier, Elizabeth Bisley, and Esme Hawes, 94–115. London: V&A Publishing, 2019.

Hayes, Peter. *Industry and Ideology: IG Farben in the Nazi Era*. Cambridge: Cambridge University Press, 1987.

Headrick, Daniel, R. *Humans Versus Nature: A Global Environmental History*. Oxford: Oxford University Press, 2020.

Held, Isabelle. "Redefining Nude: Unravelling Nylon's Unmarked Norms," *Fashion, Style, and Popular Culture* 11, no. 1 (2024): 45–64. https://doi.org/10.1386/fspc_00225_1.

Held, Isabelle. "Shaping Foundations: Trans Feminine Self-fashioning, DIY and Community Building in the Postwar United States," *Critical Studies in Fashion and Beauty* 15, no. 1 (2024): 49–74. https://doi.org/10.1386/csfb_00069_1.

Henthorn, Cynthia Lee. "Commercial Fallout: The Image of Progress and the Feminine Consumer from World War II to the Atomic Age, 1942–1962." In *The Writing on the Cloud: American Culture Confronts the Atomic Bomb*, edited by Alison M. Scott and Christopher D. Geist, 24–44. Lanham, MD: University Press America, 1997.

Henthorn, Cynthia Lee. "The Emblematic Kitchen: Labor-Saving Technology as National Propaganda, the United States, 1939–1959." *Knowledge and Society* 12 (2000): 153–87.

Henthorn, Cynthia Lee. *From Submarines to Suburbs: Selling a Better America, 1939–1959*. Athens: Ohio University Press, 2006.

Hersch, Matthew H. "High Fashion: The Women's Undergarment Industry and the Foundations of American Spaceflight." *Fashion Theory* 13, no. 3 (2009): 345–70. https://doi.org/10.2752/175174109X438118.

Hill Collins, Patricia, and Sirma Bilge. *Intersectionality*. Cambridge: Polity Press, 2016.

Hitch, Charles Johnston, and Roland N. McKean. *The Economics of Defense in the Nuclear Age*. RAND Corporation Research Study. Cambridge, MA: Harvard University Press, 1960.

Hixson, Walter L. *Parting the Curtain: Propaganda, Culture, and the Cold War, 1945–1961*. New York: St. Martin's Griffin, 1998.

Hollander, Anne. *Seeing Through Clothes*. Berkeley: University of California Press, 1993. First published 1978.

Hollings, Ken. *Welcome to Mars: Politics, Pop Culture, and Weird Science in 1950s America*. Berkeley, CA: North Atlantic Books, 2014.

Hounshell, David A., and John K. Smith. *Science and Corporate Strategy: Du Pont R & D, 1902–1980*. Studies in Economic History and Policy. New York: Cambridge University Press, 1988.

Howson, Alexandra. *The Body in Society: An Introduction*. 2nd ed. Cambridge: Polity Press, 2013.

Ingrassia, Paul. *Engines of Change: A History of the American Dream in Fifteen Cars*. New York: Simon and Schuster, 2013.

Jacobson, Nora. *Cleavage: Technology, Controversy, and the Ironies of the Man-Made Breast*. New Brunswick, NJ: Rutgers University Press, 2000.

Kaden, R. "Verwendung des Silikonöls bei Faltenbildung." *Ästhetische Medizin* 17, no. 10 (1968): 217–20.

Kafer, Alison. *Feminist, Queer, Crip*. Bloomington: Indiana University Press, 2013.

Kagan, H. "Sakurai Injectable Silicone Formula." *Archives of Otolaryngology* 78 (November 1963): 53–58. https://doi.org/10.1001/archotol.1963.00750020677009.

Kagan, Vladimir. *Upholstered Furniture: Designer's Portfolio of Urethane Foam Construction Techniques*. New York: Vladimir Kagan Design Associates and Mobay Chemical Company, 1964.

Kakoudaki, Despina. "Pin-Up: The American Secret Weapon in World War II." In *Porn Studies*, edited by Linda Williams, 335–69. Durham, NC: Duke University Press, 2004.

Kind, Alfred, and Curt Moreck. *Gefilde der Lust: Morphologie, Physiologie und Sexual-Psychologische Bedeutung der Sekundären Geschlechtsmerkmale des Weibes*. Wien: Verlag für Kulturforschung, 1930.

King, Jeannette. *Discourses of Ageing in Fiction and Feminism: The Invisible Woman*. New York: Palgrave Macmillan, 2013.

Kinoshita, Shinichi, Sadao Kageyama, Kazuhiko Iba, Yasuhiro Yamada, and Hirosuke Okada. "Utilization of a Cyclic Dimer and Linear Oligomers of E-Aminocaproic Acid by *Achrornobacter Guttatus* Ki 72." *Agricultural and Biological Chemistry* 39, no. 6 (1975): 1219–23. https://doi.org/10.1080/00021369.1975.10861757.

Kiskadden, W. S. "Operations on Bosoms Dangerous." *Plastic and Reconstructive Surgery* 15, no. 1 (1955): 79–80. https://doi.org/10.1097/00006534-195501000-00012.

Klein, Christina. *Cold War Orientalism: Asia in the Middlebrow Imagination, 1945–1961.* Berkeley: University of California Press, 2003.

Kline, Gordon. "Plastics in Germany, 1939–1945." *Modern Plastics* 23 (1945): 152A–152P.

Kline, Gordon. "Plastics in Germany, 1939–1945, Part 2." *Modern Plastics* (1945): 212–27.

Knepp, Donn. *Las Vegas, the Entertainment Capital.* Sunset Pictorial. Menlo Park, CA: Lane Publishing, 1987.

Koistinen, Paul A. C. *The Military-Industrial Complex: A Historical Perspective.* Praeger Special Studies. New York: Praeger, 1980.

Kopf, Edward. "Injectable Silicones," *Rocky Mountain Medical Journal* 63, no. 3 (1966): 34–36.

Kopf, Edward, Charles Vinnik, Joseph Bongiovi, and Donald Dombrowski. "Complications of Silicone Injections." *Rocky Mountain Medical Journal* 73, no. 2 (1976): 77–80.

Kovner, Sarah. "Base Cultures: Sex Workers and Servicemen in Occupied Japan." *Journal of Asian Studies* 68, no. 3 (2009): 777–804.

Kovner, Sarah. *Occupying Power: Sex Workers and Servicemen in Postwar Japan.* Stanford, CA: Stanford University Press, 2012.

Lambert, Susan. "Introduction." In *Provocative Plastics: Their Value in Design and Material Culture,* edited by Susan Lambert, 1–27. Cham, Switzerland: Palgrave Macmillan, 2020.

La Roe, Else K. *The Breast Beautiful.* New York: Froben Press, 1947.

Leff, Mark. "The Politics of Sacrifice on the American Home Front in World War II," *Journal of American History* 77, no. 4 (1991): 1296–318.

Leslie, Esther. *Synthetic Worlds: Nature, Art and the Chemical Industry.* London: Reaktion, 2005.

Leslie, Stuart W. *The Cold War and American Science: The Military-Industrial-Academic Complex at MIT and Stanford.* New York: Columbia University Press, 1993.

Lewis, Abram J. "Trans History in a Moment of Danger: Organizing Within and Beyond 'Visibility' in the 1970s." In Gossett, Stanley, and Burton, *Trap Door,* 57–90.

Liboiron, Max. "Plasticizers: A Twenty-First Century Miasma." In *Accumulation: The Material Politics of Plastic,* edited by Jennifer Gabrys, Gay Hawkins, and Mike Michael, 134–49. London: Routledge, 2013.

Liboiron, Max. *Pollution Is Colonialism.* Durham, NC: Duke University Press, 2021.

Lippincott, Joshua Gordon. *Design for Business.* Chicago: P. Theobald, 1947.

Lockwood, William W. "Japanese Silk and the American Market." *Far Eastern Survey* 5, no. 4 (1936): 31–36. https://doi.org/10.2307/3021472.

Loewy, Raymond. *Never Leave Well Enough Alone.* 1951. Reprint, Baltimore: Johns Hopkins University Press, 2002.

Loftin, Craig M. "Unacceptable Mannerisms: Gender Anxieties, Homosexual Activism, and Swish in the United States, 1945–1965." *Journal of Social History* 40, no. 3 (2007): 577–96.

Lorde, Audre. "Age, Race, Class and Sex: Women Redefining Difference." In *Sister Outsider: Essays and Speeches*, 114–23. Freedom, CA: Crossing Press, 1984.

Lupton, Ellen. *Mechanical Brides: Women and Machines from Home to Office*. New York: Princeton Architectural Press, 1993.

Lupton, Ellen, and J. Abbott Miller. *The Bathroom, the Kitchen, and the Aesthetics of Waste: A Process of Elimination*. New York: Princeton Architectural Press, 1996.

Maclear, Kyo. *Beclouded Visions: Hiroshima-Nagasaki and the Art of Witness*. Albany: State University of New York Press, 1998.

Maddock, Larry, and Leonard Wheeler. *Sex Life of a Transvestite*. Hollywood, CA: KDS Publishing, 1964.

March, Jenny. *Cassell Dictionary of Classical Mythology*. London: Cassell, 2000.

Marchand, Roland. "The Designers Go to the Fair: Walter Dorwin Teague and the Professionalization of Corporate Industrial Exhibits, 1933–1940." *Design Issues* 8, no. 1 (1991): 4–17. https://doi.org/10.2307/1511449.

Matelski, Elizabeth. *Reducing Bodies: Mass Culture and the Female Figure in Postwar America*. New York: Routledge, 2017.

Mattioli, Benita. *Three Nights at the Condor: A Coal Miner's Son, Carol Doda, and the Topless Revolution*. California: Keisho Publications, 2018.

May, Elaine Tyler. *Homeward Bound: American Families in the Cold War Era*. New York: Basic Books, 1988.

McAtee, Cammie. "'Taking Comfort in the Age of Anxiety: Eero Saarinen's Womb Chair.'" In Schuldenfrei, *Atomic Dwelling*, 3–25.

McLean, Adrienne L. "'I'm a Cansino': Transformation, Ethnicity, and Authenticity in the Construction of Rita Hayworth, American Love Goddess." *Journal of Film and Video* 44, no. 3–4 (1992): 8–26.

McLuhan, Marshall. *The Mechanical Bride*. London: Routledge and Kegan Paul, 1967.

McWhan, Denis. *Sand and Silicon: Science That Changed the World*. Oxford: Oxford University Press, 2012.

Meals, Robert N., and Frederick Minton Lewis. *Silicones*. Reinhold Plastics Applications Series. New York: Reinhold Publishing, 1959.

Meikle, Jeffrey L. *American Plastic: A Cultural History*. New Brunswick, NJ: Rutgers University Press, 1995.

Menon, Alka V. *Refashioning Race: How Global Cosmetic Surgery Crafts New Beauty Standards*. Berkeley: University of California Press, 2023.

Meronek, Toshio, and Miss Major. *Miss Major Speaks: Conversations with a Black Trans Revolutionary*. London: Verso, 2023.

Meyerowitz, Joanne J. "Beyond the Feminine Mystique: A Reassessment of Postwar Mass Culture, 1946–1958." In *Not June Cleaver: Women and Gender in Postwar America 1945–1960*, edited by Joanne J. Meyerowitz, 229–62. Philadelphia: Temple University Press, 1994.

Meyerowitz, Joanne J. "A 'Fierce and Demanding' Drive." In Whittle and Stryker, *Transgender Studies Reader*, 362–86.

Meyerowitz, Joanne J. *How Sex Changed: A History of Transsexuality in the United States*. Cambridge, MA: Harvard University Press, 2002.

Miller, Laura. *Beauty Up: Exploring Contemporary Japanese Body Aesthetics.*
Berkeley: University of California Press, 2006.

Miller, Laura. "Mammary Mania in Japan." *Positions: East Asia Cultures Critique* 11,
no. 2 (2003): 271–300.

Monteverde, Giuliana. "Kardashian Komplicity: Performing Post-Feminist Beauty."
Critical Studies in Fashion and Beauty 7, no. 2 (2016): 153–72. https://doi
.org/10.1386/csfb.7.2.153_1.

Moraga, Cherríe, and Gloria Anzaldúa, eds. *This Bridge Called My Back: Writings by
Radical Women of Color.* Watertown, MA: Persephone Press, 1981.

Morini, Simona. *Body Sculpture: Plastic Surgery from Head to Toe.* New York: Dela-
corte Press, 1972.

Morris, Peter J. T. *Polymer Pioneers: A Popular History of the Science and Technology
of Large Molecules.* Philadelphia: Center for History of Chemistry, 1986.

Murphy, Michelle. *Seizing the Means of Reproduction: Entanglements of Feminism,
Health, and Technoscience.* Experimental Futures. Durham, NC: Duke University Press, 2012.

Murphy, Michelle. *Sick Building Syndrome and the Problem of Uncertainty: Environmental Politics, Technoscience, and Women Workers.* Durham, NC: Duke
University Press, 2006.

Mutou, Yasuo. "A Study of 317 Akiyama's DMPS (Cross-Linked Dimethylpolysiloxane, Branded as Elicon) Cases Between 1960 to April 1964." *Journal of
Japanese Practical Surgical Society* 26, no. 1 (1965): 25–35.

Narins, Rhoda S., and Kenneth Beer. "Liquid Injectable Silicone: A Review of
Its History, Immunology, Technical Considerations, Complications, and
Potential." *Plastic and Reconstructive Surgery* 118, no. S3 (2006): S77–S84.
https://doi.org/10.1097/01.prs.0000234919.25096.67.

Ndiaye, Pap A. *Nylon and Bombs: DuPont and the March of Modern America.* Studies
in Industry and Society. Baltimore: Johns Hopkins University Press, 2007.

Neufeld, Michael J. "The Nazi Aerospace Exodus: Towards a Global, Transnational
History." *History and Technology* 28, no. 1 (2012): 49–67. https://doi.org/10
.1080/07341512.2012.662338.

Neushul, Peter, and Peter Westwick. "Blowing Foam and Blowing Minds." In
Groovy Science: Knowledge, Innovation and American Counterculture, edited
by David Kaiser and Patrick McCray, 51–69. Chicago: University of Chicago Press, 2016.

Newton, Esther. "Selection from *Mother Camp*." In Whittle and Stryker, *Transgender Studies Reader*, 121–30.

Nguyen, Marilyn Q., Patrick W. Hsu, and Tue A. Dinh. "Asian Blepharoplasty."
Seminars in Plastic Surgery 23, no. 3 (2009): 185–97. https://doi.org/10.1055
/s-0029-1224798.

Niemoeller, A. F. *The Complete Guide to Bust Culture.* New York: Harvest House,
1939.

Noll, Walter. *Chemistry and Technology of Silicones.* New York: Academic Press, 1968.

Oakes, Elizabeth H. *A to Z of STS Scientists.* Notable Scientists. New York: Facts on
File, 2002.

Oldenziel, Ruth, and Karin Zachmann, eds. *Cold War Kitchen: Americanization, Technology, and European Users.* Inside Technology. Cambridge, MA: MIT Press, 2009.

Oppenheimer, B. S., E. T. Oppenheimer, and A. P. Stout. "Sarcomas Induced in Rodents by Imbedding Various Plastic Films." *Proceedings of the Society for Experimental Biology and Medicine* 79, no. 3 (1952): 366–69. https://doi.org/10.3181/00379727-79-19380.

O'Reagan, Douglas M. *Taking Nazi Technology: Allied Exploitation of German Science After the Second World War.* Baltimore: Johns Hopkins University Press, 2019.

Oreskes, Naomi, and John Krige, eds. *Science and Technology in the Global Cold War.* Cambridge, MA: MIT Press, 2014.

Ortiz-Monasterio, F., and I. Trigos. "Management of Patients with Complications from Injections of Foreign Materials into the Breasts." *Plastic and Reconstructive Surgery* 50, no. 1 (1972): 42–47. https://doi.org/10.1097/00006534-197207000-00007.

Otálvaro-Hormillosa, Gigi. *Erotic Resistance: The Struggle for the Soul of San Francisco.* Oakland: University of California Press, 2024.

Ott, Katherine. "Hard Wear and Soft Tissue: Craft and Commerce in Artificial Eyes." In Ott, Serlin, and Mihm, *Artificial Parts, Practical Lives,* 147–70.

Ott, Katherine. "The Sum of Its Parts: An Introduction to Modern Histories of Prosthetics." In Ott, Serlin, and Mihm, *Artificial Parts, Practical Lives,* 1–43.

Ott, Katherine, David Serlin, and Stephen Mihm, eds. *Artificial Parts, Practical Lives: Modern Histories of Prosthetics.* New York: New York University Press, 2002.

Painter, Nell Irvin. *The History of White People.* New York: W. W. Norton, 2011.

Pak, Susie. "Complex International Alliances: Japan." In *Gentlemen Bankers: The World of J. P. Morgan,* 160–91. Harvard Studies in Business History. 2013. Reprint, Cambridge, MA: Harvard University Press, 2014.

Pangman, John, and Robert Wallace. "The Use of Plastic Prosthesis in Breast Plastic and Other Soft Tissue Surgery." *Western Journal of Surgery, Obstetrics and Gynecology* 8, no. 63 (1955): 503–12.

Parker, Rhian. *Women, Doctors and Cosmetic Surgery: Negotiating the "Normal" Body.* Basingstoke: Palgrave Macmillan, 2010.

Partridge, Eric. *A Dictionary of Slang and Unconventional English.* Edited by Paul Beale. 8th ed. London: Routledge, 1981.

Pavitt, Jane. "Design and the Democratic Ideal." In Crowley and Pavitt, *Cold War Modern,* 73–93.

Pavitt, Jane. "The Future Is Possibly Past: The Anxious Spaces of Gaetano Pesce." In Schuldenfrei, *Atomic Dwelling,* 26–44.

Pearson, Willie, Jr. *Beyond Small Numbers: Voices of African American PhD Chemists.* Leeds, UK: Emerald Publishing, 2005.

Pearson, Willie, Jr. *Black Scientists, White Society, and Colorless Science: A Study of Universalism in American Science.* Millwood, NY: Associated Faculty Press, 1985.

Pedersen, Stephanie. *Bra: A Thousand Years of Style, Support and Seduction.* Newton Abbot, UK: David and Charles, 2004.

Peiss, Kathy Lee. *Hope in a Jar: The Making of America's Beauty Culture*. New York: Metropolitan Books, 1998.

Pinch, Trevor J., and Wiebe E. Bijker. "The Social Construction of Facts and Artefacts: Or How the Sociology of Science and the Sociology of Technology Might Benefit Each Other." *Social Studies of Science* 14, no. 3 (1984): 399–441.

Plemons, Eric. *The Look of a Woman: Facial Feminization Surgery and the Aims of Trans- Medicine*. Durham, NC: Duke University Press, 2017.

Preciado, Paul B. *Pornotopia: An Essay on Playboy's Architecture and Biopolitics*. New York: Zone Books, 2014.

Puar, Jasbir K. "'I Would Rather Be a Cyborg Than a Goddess': Becoming-Intersectional in Assemblage Theory." *philosOPHIA* 2, no. 1 (2012): 49–66. https://doi.org/10.1353/phi.2012.a486621.

Pursell, Carroll W. *The Military-Industrial Complex*. New York: Harper and Row, 1972.

Rawson, K. J. "Introduction: 'An Inevitably Political Craft.'" In "Archives and Archiving," special issue, *TSQ: Transgender Studies Quarterly* 2, no. 4 (2015): 544–52. https://doi.org/10.1215/23289252-3151475.

Reddy-Best, Kelly L., Kyra Streck, and Jennifer Farley Gordon. "Visibly Queer- and Trans-Fashion Brands and Retailers in the Twenty-First Century." *Dress* 48, no. 1 (2022): 33–53. https://doi.org/10.1080/03612112.2021.1967606.

Reid, Susan E. "'Our Kitchen Is Just as Good': Soviet Responses to the American National Exhibitions in Moscow, 1959." In Crowley and Pavitt, *Cold War Modern*, 154–62.

Riesman, David. "The Nylon War." In *"Abundance for What?" and Other Essays*, 67–75. Piscataway, NJ: Transaction Publishers, 1964. First published 1951.

Riordan, Teresa. *Inventing Beauty: A History of the Innovations That Have Made Us Beautiful*. New York: Broadway Books, 2004.

Roberts, Pudgy. *Female Impersonator's Handbook*. Newark, NJ: Capri Publishers, 1967.

Robertson, Jennifer. "Japan's First Cyborg? Miss Nippon, Eugenics and Wartime Technologies of Beauty, Body and Blood." *Body and Society* 7, no. 1 (2001): 1–34. https://doi.org/10.1177/1357034X01007001001.

Rodriguez, Clara E. *Heroes, Lovers, and Others: The Story of Latinos in Hollywood*. Washington, DC: Smithsonian Books, 2004.

Rubin, Eli. *Synthetic Socialism: Plastics and Dictatorship in the German Democratic Republic*. Chapel Hill: University of North Carolina Press, 2014.

Russell, Edmund. *War and Nature: Fighting Humans and Insects with Chemicals from World War I to Silent Spring*. Studies in Environment and History. Cambridge: Cambridge University Press, 2001.

Russell, Legacy. *Glitch Feminism: A Manifesto*. London: Verso, 2020.

Rydell, Robert W. *World of Fairs: The Century-of-Progress Expositions*. Chicago: University of Chicago Press, 1993.

Safer, Joshua D., Eli Coleman, Jamie Feldman, Robert Garofalo, Wylie Hembree, Asa Radix, and Jae Sevelius. "Barriers to Health Care for Transgender

Individuals." *Current Opinion in Endocrinology, Diabetes, and Obesity* 23, no. 2 (2016): 168–71. https://doi.org/10.1097/MED.0000000000000227.

Sanders, Holly. "Panpan: Streetwalking in Occupied Japan." *Pacific Historical Review* 81, no. 3 (2012): 404–31. https://doi.org/10.1525/phr.2012.81.3.404.

Sastre-Juan, Jaume. "'Science in Action': The Politics of Hands-On Display at the New York Museum of Science and Industry." *History of Science* 59, no. 2 (2021): 155–78. https://doi.org/10.1177/0073275317725239.

Schalk, Deborah N. "The History of Augmentation Mammaplasty." *Plastic and Aesthetic Nursing* 8, no. 3 (1988): 88–90.

Schuldenfrei, Robin, ed. *Atomic Dwelling: Anxiety, Domesticity, and Postwar Architecture*. London: Routledge, 2012.

Schuller, Kyla, and Jules Gill-Peterson. "Introduction: Race, the State, and the Malleable Body." In "The Biopolitics of Plasticity," special issue, *Social Text* 38, no. 2 (143) (2020): 1–17. https://doi.org/10.1215/01642472-8164716.

Scott, Coleen. *The Costumes of Burlesque, 1866–2018*. New York: Routledge, 2019.

Serano, Julia. *Whipping Girl: A Transsexual Woman on Sexism and the Scapegoating of Femininity*. 2007. Reprint, Berkeley: Seal Press, 2016.

Serlin, David. "Christine Jorgensen and the Cold War Closet." *Radical History Review* 1995, no. 62 (1995): 137–65. https://doi.org/10.1215/01636545-1995-62-137.

Serlin, David. "Engineering Masculinity: Veterans and Prosthetics After World War II." In Ott, Serlin, and Mihm, *Artificial Parts, Practical Lives*, 45–74.

Serlin, David. *Replaceable You: Engineering the Body in Postwar America*. Chicago: University of Chicago Press, 2004.

Seu, Mindy. *Cyberfeminism Index*. Los Angeles: Inventory Press, 2022.

Shaw, Susan D., Arlene Blum, Roland Weber, Kurunthachalam Kannan, David Rich, Donald Lucas, Catherine P. Koshland, Dina Dobraca, Sarah Hanson, and Linda S. Birnbaum. "Halogenated Flame Retardants: Do the Fire Safety Benefits Justify the Risks?" *Reviews on Environmental Health* 25, no. 4 (2010): 261–305. https://doi.org/10.1515/reveh.2010.25.4.261.

Shibusawa, Naoko. *America's Geisha Ally: Reimagining the Japanese Enemy*. 2006. Reprint, Cambridge, MA: Harvard University Press, 2010.

Shirakabe, Y., T. Kinugasa, M. Kawata, T. Kishimoto, and T. Shirakabe. "The Double-Eyelid Operation in Japan: Its Evolution as Related to Cultural Changes." *Annals of Plastic Surgery* 15, no. 3 (1985): 224–41. https://doi.org/10.1097/00000637-198509000-00006.

Shteir, Rachel. *Striptease: The Untold History of the Girlie Show*. New York: Oxford University Press, 2004.

Skidmore, Emily. "Constructing the 'Good Transsexual': Christine Jorgensen, Whiteness, and Heteronormativity in the Mid-Twentieth-Century Press." *Feminist Studies* 37, no. 2 (2011): 270–300.

Smith, Stephanie A. "Bombshell." In *Household Words: Bloomers, Sucker, Bombshell, Scab, Nigger, Cyber*, 69–96. Minneapolis: University of Minnesota Press, 2006.

Snorton, C. Riley. *Black on Both Sides: A Racial History of Trans Identity*. Minneapolis: University of Minnesota Press, 2017.

Sonnenblick, Emily B., Shivani Chaudhry, Karen A. Lee, Shabnam Jaffer, Frank Fang, Jess Ting, and Laurie R. Margolies. "MRI Features of Free Liquid Silicone in the Transgender Female Breast." *Journal of Breast Imaging* 3, no. 3 (2021): 322–31. https://doi.org/10.1093/jbi/wbab016.

Spillers, Hortense J. "Mama's Baby, Papa's Maybe: An American Grammar Book." *Diacritics* 17, no. 2 (1987): 65–81. https://doi.org/10.2307/464747.

Stacey, Jackie. "Feminine Fascinations: Forms of Identification in Star-Audience Relations." In *Stardom: Industry of Desire*, edited by Christine Gledhill, 141–63. London: Routledge, 1991.

Staiti, Alana. "Real Women, Normal Curves, and the Making of the American Fashion Mannequin, 1932–1946." *Configurations* 28, no. 4 (2020): 403–31. https://dx.doi.org/10.1353/con.2020.0029.

Star, Hedy Jo. *"I Changed My Sex!"* Chicago: Novel Books, 1963.

Steen, Kathryn. *The American Synthetic Organic Chemicals Industry: War and Politics, 1910–1930*. Chapel Hill: University of North Carolina Press, 2014.

Stephens, Elizabeth. "The Normal Body on Display: Public Exhibitions of the Norma and Normman Statues." In *The Routledge Companion to Media, Sex and Sexuality*, edited by Clarissa Smith, Feona Attwood, and Brian McNair, 7–18. London: Routledge, 2017.

Stone, Sandy. "The Empire Strikes Back: A Posttranssexual Manifesto." In Whittle and Stryker, *Transgender Studies Reader*, 221–35.

Stott, Philip. "The Dyeing of Nylon Fibers: A Preliminary Survey." *American Dyestuff Reporter*, October 2, 1939, 582–89.

Stott, Philip. "The Dyeing of Nylon Hosiery." *American Dyestuff Reporter*, December 22, 1941, 710–14.

Strang, Jessica. *Working Women: An Appealing Look at the Appalling Uses and Abuses of the Feminine Form*. New York: H. N. Abrams, 1984.

Strings, Sabrina. *Fearing the Black Body: The Racial Origins of Fat Phobia*. New York: New York University Press, 2019.

Stryker, Susan. "(De)Subjugated Knowledges: An Introduction to Transgender Studies." In Whittle and Stryker, *Transgender Studies Reader*, 1–18.

Stryker, Susan. "Lee Greer Brewster." In *Encyclopedia of Lesbian, Gay, Bisexual, and Transgender History in America*, edited by Marc Stein, 166–67. New York: Charles Scribner's Sons, 2004.

Stryker, Susan. *Transgender History: The Roots of Today's Revolution*. New York: Seal Press, 2017.

Sullivan, Steve. *Bombshells: Glamour Girls of a Lifetime*. New York: St. Martin's Griffin, 1998.

Symmers, William St. Clair. "Silicone Mastitis in 'Topless' Waitresses and Some Other Varieties of Foreign-Body Mastitis." *BMJ* 3, no. 5609 (1968): 19–22.

Taha, Hebatalla. "Atomic Aesthetics: Gender, Visualization and Popular Culture in Egypt." *International Affairs* 98, no. 4 (2022): 1169–87. https://doi.org/10.1093/ia/iiac115.

Tanquero, Albert, and Lewis Rawlinson. *Remember Me, Vicki Starr: The Visual History of a Trans Renegade*. N.p.: Albert Tanquero, 2021.

Textile Color Card Association of the United States. *Standard Color Card of America*. New York: Textile Color Card Association of the United States, 1941.

Thesander, Marianne. *The Feminine Ideal*. London: Reaktion Books, 1997.

Titus, A. Constandina. *Bombs in the Backyard: Atomic Testing and American Politics*. Nevada Studies in History and Political Science. Reno: University of Nevada Press, 1986.

Tully, John. "A Victorian Ecological Disaster: Imperialism, the Telegraph, and Gutta-Percha," *Journal of World History* 20, no. 4 (2009): 559–79.

Twigg, Julia. *Fashion and Age: Dress, the Body and Later Life*. London: Bloomsbury Academic, 2013.

"Urethane Plastics—Polymers of Tomorrow." *Industrial and Engineering Chemistry* 48 (1956): 1383–91.

Urla, Jacqueline, and Alan Swedland. "The Anthropometry of Barbie: Unsettling Ideals of the Feminine Body in Popular Culture." In *Deviant Bodies: Critical Perspectives on Difference in Science and Popular Culture*, edited by Jennifer Terry and Jacqueline Urla, 277–313. Race, Gender, and Science. Bloomington: Indiana University Press, 1995.

Vinnik, Charles. "The Hazards of Silicone Injections." *JAMA* 236, no. 8 (1976): 959.

Voyles, Traci Brynne. "Anatomic Bombs: The Sexual Life of Nuclearism, 1945–57." *American Quarterly* 72, no. 3 (2020): 651–73. https://doi.org/10.1353/aq .2020.0039.

Walker, Susannah. *Style and Status: Selling Beauty to African American Women, 1920–1975*. Lexington: University Press of Kentucky, 2007.

Wark, McKenzie. "Blog-Post for Cyborgs." *Public Seminar*, September 24, 2015. https://publicseminar.org/2015/09/blog-post-for-cyborgs/.

Wark, McKenzie. "Introduction." In *When Monsters Speak: A Susan Stryker Reader*, by Susan Stryker, edited by McKenzie Wark, 1–19. Durham, NC: Duke University Press, 2024.

Warrick, Earl L. *Forty Years of Firsts: The Recollections of a Dow Corning Pioneer*. New York: McGraw-Hill, 1990.

Webb, Sharon, M. "Cleopatra's Needle: The History and Legacy of Silicone Injections." Unpublished paper, Harvard University, 1997.

Weitekamp, Margaret A. *Space Craze: America's Enduring Fascination with Real and Imagined Spaceflight*. Washington, DC: Smithsonian Books, 2022.

Weitekamp, Margaret A. "Technology: The Spacesuit Unpicked." *Nature* 475, no. 7356 (2011). https://doi.org/10.1038/475294a.

Westermann, Andrea. "The Material Politics of Vinyl: How the State, Industry and Citizens Created and Transformed West Germany's Consumer Democracy." In *Accumulation: The Material Politics of Plastic*, edited by Jennifer Gabrys, Gay Hawkins, and Mike Michael, 68–86. London: Routledge, 2013.

Westermann, Andrea. *Plastik und politische Kultur in Westdeutschland*. Zurich: Chronos Verlag, 2007.

Whittle, Stephen, and Susan Stryker. *The Transgender Studies Reader*. New York: Routledge, 2006.

Williams, Cristan. "Transgender." *TSQ: Transgender Studies Quarterly* 1, no. 1–2 (2014): 232–34.

Williams, Florence. *Breasts: A Natural and Unnatural History*. New York: W. W. Norton, 2012.

Wilson, Erin, Jenna Rapues, Harry Jin, and H. Fisher Raymond. "The Use and Correlates of Illicit Silicone or 'Fillers' in a Population-Based Sample of Transwomen, San Francisco, 2013." *Journal of Sexual Medicine* 11, no. 7 (2014): 1717–24. https://doi.org/10.1111/jsm.12558.

Wilson, Kristina. *Mid-Century Modernism and the American Body: Race, Gender, and the Politics of Power in Design*. Princeton, NJ: Princeton University Press, 2021.

Wisnioski, Matthew. *Engineers for Change: Competing Visions of Technology in 1960s America*. Cambridge, MA: MIT Press, 2012.

Wöhler, Friedrich, Justus Liebig, and Hermann Kopp. *Annalen der Chemie und Pharmacie*, Leipzig: C. F. Winter'sche Verlagshandlung, 1863.

Wolfe, Tom. "The Put-Together Girl." In *The Pump House Gang*, 81–96. New York: Farrar, Straus and Giroux, 1968.

Xiang, Sunny. "Bikinis and Other Atomic Incidents: The Synthetic Life of the Nuclear Pacific." *Radical History Review* 2022, no. 142 (2022): 37–56. https://doi.org/10.1215/01636545-9397030.

Yalom, Marilyn. *A History of the Breast*. London: Pandora, 1998.

Zimmermann, Susan M. *Silicone Survivors: Women's Experiences with Breast Implants*. Philadelphia: Temple University Press, 1998.

INDEX

hormone therapy, 71, 138, 151, 153–54, 259–61, 263, 267–68, 279, 280

Horne, Lena, 14, 125

housewives, 7, 110–13, 115, 179, 204–5, 256, 259

Hudson Motor Car Company, 176

Hunter, Mel, 211

Hyde, Franklin, 192–95, 317n26

hyperfemininity, 13, 112, 275. *See also* femininity

IG Farben, 91–93, 95–97, 168, 301–2n1

implants. *See* breast implants

Industrial and Engineering Chemistry, 103

injectables. *See* paraffin injections; silicone injections

insecticides, 205

Instagram, 276–78

Institut für Sexualwissenschaft, 217

Interior Design (magazine), 113

intersectionality, 3, 18, 21, 122, 274, 279–80

Interview (magazine), 274–75

isocyanates, 96–97, 103, 105, 302n5

Ivalon, 144, 155, 157, 172–73, 314n55–56

Jacopetti, Gualtiero, 231. See also *Women of the World* (dir. Cavara, Jacopetti, and Prosperi; 1963)

Japan: Americanization of, 109; Kyoto, 219, 229; Osaka, 220; and Pearl Harbor attacks, 194, 196; silicone injections from, 25, 214–19, 231–34, 246–47, 258; silicone technology transfer from, 219–31; silk from, 23, 30, 36–40, 43; Tokyo, 196, 220, 235, 237, 246, 323n35; women's beauty standards in, 234–37; Yokohama Harbor, 222, 214

Japanese Americans, 207, 240, 252–53

Japanese Society of Plastic and Reconstructive Surgery, 222

Jaru Inc., 264–65

Jenny, Alex, 271–72

Jet (magazine): advertisements in, 125–26, 128; on the conical bustline, *8*, 123–24; on foam padding, 129; on furniture, 117; on the topless craze, 241, 251; on trans feminine people, 135, 154, 259

Jewel Box Revue, 131, 135, 265

Johns Hopkins Hospital, 244, 259–60; Gender Identity Clinic, 260

Johnson, John Harold, 110

Johnson, Marsha P., 261, 264, 272

Jones, Shirley May (Miss Leonard), 130

Jonsson, Ida: "The Bum" project, 275–76

Jorgensen, Christine, 14, 16, 154

Journal of the American Medical Association (*JAMA*), 165

Kagan, Harry, 216, 245, 325n69

Kardashian, Kim, 275–76

Karefa-Johnson, Gabriella, 274

Karlsson, Beate, 276

Keystone Cosmetics: Hi-Brown Liquid Face Powder and Leg Make-Up, 51, *52*

Khrushchev, Nikita S., 109–10

Kipping, Frederic Stanley, 191, 193

Kitchen Debate (1959), 109–12

Kitt, Eartha, 14

Klein, Christina, 234–35

Kline, Gordon, 100, 102

Klossowski, Alexa, 184–85

Knoll, Florence and Hans, 116

Kojima, Akiko, 25, 234–36

Koken Kogyo Co. Ltd., 221

Kubrick, Stanley: *Dr. Strangelove* (1964), 109

LaBeija, Crystal, 135

Labovsky, Josef, 31

Ladybirds (band), 250

Lady of Chemistry. *See* Mitton, Katherine; Test Tube Girl

Lake, Bunny, 131

Lambert, Susan, 287–88n2

La Roe, Elsa K., 153–54, 183; *Breast Beautiful, The*, 144, 151, 180

latex, 92, 113, 264–65, 306n98

Latinx people, 131, 174, 280

Lavarone, Nicholas, 261–62

Lavender and Lace (periodical), 131, 133, *134*

Lawrie, Lee, 193

lecturers (at DuPont exhibits): hiring of, 64, 298n21; on Miss Chemistry, 65–66; on munitions, 43; on nylon stockings, 38–40, 42, 44, 47–50; reports of, 35–36

Lee, Henry, 170–72

Lee, Linda, 264

Lee, Toni, 131, 265, *266*

Lee's Mardi Gras Boutique, 264

Lenz, Ludwig, 217, 322n14

Leonard, Miss (Shirley May Jones), 130

Letters from Female Impersonators (periodical), 86

Lewis, John E., 135

Life (magazine): advertisements in, 207–8; bombshells in, 14; and *Ebony* (magazine), 125; on silicone, 197, 204; "Tomorrow's Life Today," 105, 107–8, 117; on the topless craze, 240, 248, 251

Linsacrite zelm, 221

Lippincott, Joshua G., 179

Lockheed, 103, 304n43

Loewy, Raymond, 71, 73

lotions, 7, 204

Louise, Tana, 85

Lucite, 36, 46, 48, 66, 71, 130

Lycra, 5, 29, 55, 119, 280, 288n8

MacArthur, Douglas, 219

MacLaine, Shirley, 207

Macy's, 39

Madonna, 20

magazines. See *individual titles*

Maidenform: bust forms by, 177; Chansonette bra by, 123, 307n2; and the Cold War, 109; foundationwear by, 9, 131, 133, 141, 143; Masquerade falsies by, 126–27, 152, 172

Major, Miss (drag performer), 135

makeup, 51–52, 177, 205–11, 235, 275, 300n65. *See also* skincare

mammaplasty. *See* breast implants

Mamou, Judy, 254

Manhattan Project, 50, 195

Mansfield, Jayne, 10, 123, 131, 232

Marion, George, Jr., 50–51

Maryland: Baltimore, *182*, 209, 237

masculinity, 19

Mason, Dian, 54

mastectomies, 138–40, 145, 255, 265

materiality: of Etheron, 181; of foam breast implants, 162–63; of foam foundationwear, 126, 129, 157; of Lucite, 36; of nylon, 23, 29, 31, 39–40, 43, 47–48, 51, 53, 55, 87; of plastics, 3, 76; of polyurethane foam, 24, 92, 97–98, 100, 104, 113, 116, 118–20; of silicone, 25, 190, 202, 217–18, 226, 229, 231, 253, 258, 269, 281

materials (history of), 4

Mattachine Society, 264

Mattel, 267

mattresses, 91–92, 112, 117, 278

Max Factor Corporation, 132, 235

May, Bill, 323n35

May, Elaine Tyler, 289n11

McCoy, John W., *60*

McDonald, Tosha, 240–41, 249, 253

McGregor, Rob, 195, 211–13, 220, 222–23, 225–26, 228

McLeod, Charlotte, 153–54

McLuhan, Marshall, 159–60, 174

Medical Device Regulation Act, 10, 26, 146, 185, 212, 243, 245, 257, 329n26

Meikle, Jeffrey, 62

Mellinger, Frederick, 128–29, 155

Menon, Alka, 18

Merlin, Lee, 11, *12*

metaphor (as method), 19–20, 159, 271

Met Gala, 275

Meyer, Russ: *Faster, Pussycat! Kill! Kill!* (1965), 247, 252; *Mondo Topless* (1966), 153, 247–48, 252

military-industrial complex: and the bombshell, 10, 281; coining of term, 288n11; and foam breast implants, 163, 166, 170, 185; and foam plastics, 9, 92, 94, 102–3, 116, 121; leaders in, 22; and nylon, 5–7, 23; and silicone, 25, 190–91, 195, 204–5, 216, 218, 223; and women's bodies, 2–3, 16, 19–20, 270–71

military technology. *See* bomber planes; nuclear bombs; nylon; parachutes; polyurethane foam; silicone

Miller, Laura, 216

Minor, Audrey, 151, 173–75, 178

Mirage (silicone breast form), 267–68

Miranda, Carmen, 14

Misakura, Marie, 14

misogyny, 44, 216, 237

Miss Atomic Bomb, 11–12

Miss Chemistry. *See* Mitton, Katherine; Test Tube Girl

Miss Universe, 35, 235–36

Mitton, Katherine, 46–47, 64–66. *See also* Test Tube Girl

Mobay Chemical Company: and foam breast implants, 169, 180–82, 185; and foam home furnishings, 107, 112–20; and polyurethane foam, 103–4

models: Black women, 124–25, 275–76; and the bombshell, 11, 14, 160; and breast augmentation surgery, 170; and eugenics, 71–72; and foam plastics, 24, 107, 117, 119; Instagram, 278–79; in Japan, 235–36; and the monokini, 248–49; and nylon stockings, 30, 44–47, 65,

85; Powers, 201–2; and swimwear, 271–73; trans feminine, 135–36. See also *individual models*

Modern Pioneer (performance), 299n32

Modern Plastics (magazine), 100

Moffitt, Peggy, 248

Moltopren, 97

mondo movies. See *Women of the World* (dir. Cavara, Jacopetti, and Prosperi; 1963)

Mondo Topless (dir. Meyer; 1966), 153, 247–48, 252

Monkees: *Head* (dir. Rafelson; 1968), 240–42

monokinis, 240, 248–49, 251

Monroe, Marilyn, 10, 14, 123, 275

Monsanto Company: and foam breast implants, 169, 181; and polyurethane foam, 103, 105–7, 117; and silicone, 221, 237–39

Montiel, Suli, 85–86

Morini, Simona, 203, 212, 223, 243, 255

Murray, Joseph E., 224, 244, 255

Museum of Modern Art (New York), 109, 305n66

Mutou, Yasuo, 233

My Geisha (dir. Cardiff; 1962), 207

National Aniline and Chemical Company, 103

National Association for the Advancement of Colored People (NAACP), 80–81

National Association of Hosiery Manufacturers, 80

National Association of Manufacturers, 62, 298–99n32

National Bureau of Standards, 100

National Geographic (magazine), 56, 57

N-Day, 30, 49, 55

Nearly Me (silicone breast forms), 267

Neff, Hari, 274

Nevada: Las Vegas, 11, 227, 244, 248, 252, 254, 256–57, 274

Newton, Delisa, 154

Newton, Esther, 135–36, 138

New Trends (magazine), 132

New World Through Chemistry, A (DuPont promotional film; 1939), 63

New York City: cosmetic surgery in, 147, 151, 153, 165, 180, 184, 222, 234, 244, 246, 266; fashion in, 74, 85, 271, 276; Greenwich Village, 132; Harlem, 81, 152; modeling in, 201; Rockefeller Center, 193; trans community in, 131, 135, 259, 264, 274, 279

New York Fashion Week, 271, 276

New York Herald Tribune (newspaper), 60–62

New York Hospital–Cornell Medical Center, 163, 221

New York School of Fashion Careers, 78

New York Science Display (1937), 44–45

New York World's Fair (1939), 23, 30–31, 34–36, 38–40, 42–47, 49, 56, 58–61, 64–67, 70, 75, 273

New York World Telegram (newspaper), 38

Niemeyer, Oscar, 165

Nixon, Richard M., 109–11

Noll, Walter, 192–93, 317–18n26

nuclear bombs, 10–13, 109, 159–60, 162, 234–35

"nude" (color): and bust forms, 177; and clothing, 275; and cosmetics, 207, 300n65; and foundationwear, 127, 173; and nylon stockings, 58, 77–80, 87–88, 207

Nude Barre, 88

Nulox, 209

Nye, Gerald P., 33, 42–43

nylon: development of, 3–7, 10, 23–24, 29–31, 34, 56, 291n1; and foam implants, 2, 155, 157, 168, 172–73, 184; and foam plastics, 91, 93, 105, 115, 119, 131; introduction to public, 58–64; legacy of, 26, 271–74, 280; and medical applications (sutures) 67–68; and silicone, 193; as silk replacement, 42–43; wartime uses of, 50–55. *See also* nylon stockings

Nylon Girls, 54

nylon riots, 30, 54

nylon stockings: and the bombshell, 11, 281; colors of, 77–88, 207; development of, 29–31, 34; and eugenics, 70–76; introduction to public, 34–36, 58, 198; materiality of, 42–50; postwar, 108–9, 112; safety of, 40–42, 294n72; and the silk boycott, 36–40; and the Test Tube Girl, 56–58, 64–70; during World War II, 50–55

O-Cedar Dri-Glo (polish), 204

Office of Scientific Research and Development (OSRD), 6

Oldsmobile, 176

On Developing Bosom Beauty (Franklyn), 144, 146–49, 156, 169

On-Hand (lotion), 204

Operation Paperclip, 102

Ordinance Department, 100

Orentreich, Norman, 244, 246–47, 329n35

Organogen, 221, 325n62

Orientalism, 25, 231–34, 237

Ortho Plastic Novelties, *63*
Otálvaro-Hormillosa, Gigi, 251
othering (fluid), 216
Outlaw, The (Hughes; 1943), 123

Pageant (magazine), *17*, 170–71, 173–74
Painter, Nell, 73
Pan-Africanism, 209
Pan Glaze, 7, 198
Pangman, W. John, 151
Papagayo Club, 237
Paper (magazine), 275–76
parachutes, 5–6, 30, 50, 53–54, 93, 296n100
paraffin injections, 217–18, 221
Paris Is Burning (dir. Livingston; 1990), 259
pasties, 130, 240, 252
Pathé newsreels, 70, 74–75, 296n100
pathologization (of women's bodies), 122, 141,
    144, 151, 171, 183, 280, 310n78
patriarchy. *See* cisheteropatriarchy;
    heteropatriarchy
Patterson, Robert P., 196
Pavitt, Jane, 119
Pearl Harbor, 39, 194, 196
Perlon, 93
Pesce, Gaetano: UP 5 chair, 119, *120*
petrocapitalism, 107
Philadelphia Inquirer (newspaper), 199
Pierce, Charles, 241
pin-ups, 10–11, 52, 125, 128, 160
Pittsburgh Courier (newspaper), 51, 54
Planet Pepper, 276–77
Plastic and Reconstructive Surgery (journal), 165
plastics. *See* foam plastics; nylon; silicone; *other
    types of plastic*
plastics race, 103–12
plastic surgery: and butt lifts, 275; and foam plas-
    tics, 1, 4, 7–10, 122, 139, 144–53, 160, 162–66,
    170, 174–75, 180; postwar, 7–10, 185; and race,
    152, 256; and silicone, 215–18, 220–21, 233–37,
    244, 246, 259–60, 268, 279. *See also* Conway,
    Herbert; cosmetic surgery; Edgerton, Milton;
    Franklyn, Robert Alan; La Roe, Elsa K.
Playboy (magazine), 240, *241*, 248, 251
Playtex, 9
Pliny the Elder, 191
polyester, 5, 29, 55, 92, 97, 103, 155, 169, 288n8,
    302n5
Polyfoam, 182–83

polyurethane foam: in bras and falsies, 16, 20,
    122–43, 173, 269, 280–81; development in
    Germany, 6, 91–102, 302n5; in foundation-
    wear, 9, 121–22, 157–58, 178; in furniture,
    112–20, 179; histories of, 4–6, 23–25; implants,
    1–3, 122, 144–57, 162–73, 176, 180–85; legacy
    of, 26, 276–78; and the plastics race, 103–12;
    and silicone, 217, 221, 223, 226, 228–29, 254–56,
    265. *See also* Surgifoam
Ponce, Ruth, 254, 257
Pond's: Angel Face powder and foundation,
    207–8; *208*
Posner, 209
Powers (modeling agency), 201–2
Powers, John Robert, 201
Preciado, Paul B., 111
"Present Status of Silicone Fluid in Soft Tissue
    Augmentation, The" (report), 224
Prince, Virginia, 139
Princeton University, 103
Prosperi, Franco, 231
prosthetics, 4, 9, 16, 18, 26, 131–33, 136, 138–40,
    140, 145–47, 149, 207, 212, 218, 231, 243,
    264–65, 267–68, 279. *See also* Medical Device
    Regulation Act
PRS (journal), 180, 228–29, *230*
publications. *See* African American publications;
    Black publications; trans feminine publica-
    tions; whitestream publications; *individual
    publications*
Putnam, Lou, 198–99, 201–3
*Put the "Soft Sell" of Urethane Foam into Your Fur-
    niture Sales Story* (booklet), 113–14

"Quartermaster Report: Technical Intelligence
    Reports; The German Plastics Industry,"
    94–102
Queen, The (dir. Simon; 1968), 133, 135
Queen Bee (night club), 237
Queens Liberation Front (QLF), 264
queer studies, 3, 19
queering: and bombshells, 10–17, 21, 26, 271, 275;
    and falsies, 131–44; and plastics, 270, 273–74,
    278, 281

racialization: and bodies, 125, 144–45, 151, 157,
    183, 252, 275; and clothing sizing, 83; and foam
    plastics, 9, 104, 113, 116, 122, 162, 178; and
    gender, 19, 22; and Hollywood, 174, 207; and

shapewear. *See* foundationwear

Shimadzu Seisakusho Ltd., 219–20

Shin-Etsu Chemical Company, 215, 220–21, 225, 323n35

Shirakabe, Takeya, 220–21, 225–26, 229

Shirakabe Hospital, 220–22, 225

Silastic silicone, 211, 227–31, 243, 257–58. *See also* silicone

Silico Chemical Company, 262

silicone: in beauty products, 7, 203–11; breast forms, 264–68, 277, 280; development of, 6, 189–94, 217n26; histories of, 2–5, 23–26; medical uses of, 211–13; rubber, 133; safety of, 198, 205, 243; postwar US technology transfer, 197–203; wartime uses of, 194–97, 218n40. *See also* silicone implants; silicone injections

silicone implants, 148, 172, 175, 185, 224, 255–59, 263, 269, 279, 281

silicone injections: development in Japan, 214–19, 231–34; popularity of, 240–47, 252–53, 269; pumping, 279–80; safety of, 253–58, 329n35; technology transfer to US, 219–31, 236–37; in trans feminine communities, 259–64

silicone pumping, 279–80

silk: artificial, 5, 42–43, 47–48, 55; from Europe, 294n61; hosiery, 79; implants, 155, 218; from Japan, 23, 30, 36–40; medical uses, 7; *versus* nylon, 50–51, 87, 115

Simpich, Frederick, 56, 68

Sinatra, Frank, 251

Sinclair, Abby, 154

sizing (of clothing), 77, 83, 85–86, 88

Skidmore, Emily, 16

Skims, 280

skincare, 7, 204–11. *See also* makeup

Smirnoff, 250

Smithsonian National Museum of American History, 31, 127, 130, 177

Snorton, C. Riley, 231

Society of the Second Self, 139

soft power, 24, 92, 104, 108–9, 111, 119–20

Southern General Practitioner of Medicine (journal), 165

Soviet Union: and plastic foams, 94–95, 102, 108–9, 184; and plastic surgery, 165; and silicone, 192, 216n105. *See also* Cold War

spacesuit design, 9

Spano, Vincent, 254

Spanx, 280

spin-offs, 102, 204

Sports Illustrated (magazine), 165

Spun-Lo, 299n32

Standard Hosiery Color Card, 80

STAR (Street Transvestite Action Revolutionaries), 261

Star, Hedy Jo, 131, 151, 153–54

Starr, Vicki, 174, 251

Starrr, Patrick, 277

Stine, Charles M. A., 32

St. James Infirmary, 279

St. Louis, MO, 81, 104

stockings (nylon). *See* nylon stockings

stockings (silk), 36–40, 42, 50, 87

Stone, Sandy, 268–69

Stonewall, 264

strip culture. *See* burlesque performers; topless craze

Stryker, Susan, 14, 16, 19, 260

Sugar & Spice, 83, 85

Sullivan, Eugene, 198

Supreme Commander for the Allied Powers (SCAP), 219

Surgery, Gynecology and Obstetrics (journal): "Augmentation Mammaplasty" (Edgerton and McClary), 182–84

surgical bust forms, 138–39, 141, 145

Surgifoam, 1, 148, 155, 162, 164–70, 172, 180–81, 184. *See also* Breastplasty surgery

Surgitek Perras-Papillon Design Mammary Prosthesis, 265, *267*

Suzuki, Yosuke, 219

Sweater Girl look, 7–8, 20, 123, 129, 147–48, 157, 175. *See also* conical bustline

Syl-Flex (fabric protector), 204

Symmers, William St. Clair, 217, 227–28; "Silicone Mastitis in 'Topless' Waitresses . . . ," 227, 253

Synthetica (*Fortune* magazine map), 62–63

synthetic fibers. *See* nylon

Tana & Mara, 58, 85–86, 88

Tan Confessions (magazine), 82–83, 125–26

Teague, Walter Dorwin, 35, 59, 64, 66, 70–72

Technical Intelligence Industrial Committee, 100

Teflon, 157

Test Tube Girl, 10, 23, 56–58, 66–71, 73–74, 77, 79, 87, 273. *See also* Mitton, Katherine

"Test Tube Girl, The" (story), 69–70

www.ingramcontent.com/pod-product-compliance
Lightning Source LLC
Chambersburg PA
CBHW020452270326
41926CB00008B/575